An Introduction to the
Linear Theories and
Methods of Electrostatic
Waves in Plasmas

An Introduction to the Linear Theories and Methods of Electrostatic Waves in Plasmas

W. D. Jones
Physics Department
University of South Florida
Tampa, Florida

H. J. Doucet

and

J. M. Buzzi
Laboratoire de Physique des Milieux Ionisés
Ecole Polytechnique
Palaiseau, France

PLENUM PRESS • NEW YORK AND LONDON

PHYSICS

Library of Congress Cataloging in Publication Data

Jones, William Denver.

An introduction to the linear theories and methods of electrostatic waves in plasmas.

Bibliography: p.
Includes index.
1. Plasma waves. 2. Ion acoustic waves. I. Doucet, H. J. (Henri J.) II. Buzzi, J. M.
(Jean-Max) III. Title. IV. Title: Linear theories and methods of electrostatic waves in
plasma.
QC718.5.W3J66   1985   530.4′4   85-12424
ISBN 0-306-41961-0

© 1985 Plenum Press, New York
A Division of Plenum Publishing Corporation
233 Spring Street, New York, N.Y. 10013

Printed in the United States of America

FOREWORD

Modern plasma physics, encompassing wave-particle interactions and collective phenomena characteristic of the collision-free nature of hot plasmas, was founded in 1946 when L. D. Landau published his analysis of linear (small-amplitude) waves in such plasmas. It was not until some ten to twenty years later, however, with impetus from the then rapidly developing controlled-fusion field, that sufficient attention was devoted, in both theoretical and experimental research, to elucidate the importance and ramifications of Landau's original work. Since then, with advances in laboratory, fusion, space, and astrophysical plasma research, we have witnessed important developments toward the understanding of a variety of linear as well as nonlinear plasma phenomena, including plasma turbulence. Today, plasma physics stands as a well-developed discipline containing a unified body of powerful theoretical and experimental techniques and including a wide range of applications. As such, it is now frequently introduced in university physics and engineering curricula at the senior and first-year-graduate levels.

A necessary prerequisite for all of modern plasma studies is the understanding of linear waves in a temporally and spatially dispersive medium such as a plasma, including the kinetic (Landau) theory description of such waves. Teaching experience has usually shown that students (seniors and first-year graduates), when first exposed to the kinetic theory of plasma waves, have difficulties in dealing with the required sophistication in multidimensional complex variable (singular) integrals and transforms. In addition, the physics of the rich and subtle phenomena that emerge from the kinetic theory of waves in a collision-free plasma—collisionless (Landau) damping, phase mixing, ballistic (free-streaming) modes, collective modes, etc.—are all new to such students and rather difficult for them to comprehend and assimilate. The usual exposition of this material in plasma physics textbooks to date is much too casual and brief, and rarely, if ever, related in sufficient detail to experimental realities.

Thus, there has been a longstanding need for a text in plasma physics that combines a thorough exposition of the analytical (and computational) tools and basic physical ideas coupled with a critical description of associated experimental results. In focusing on the simplest, yet fundamental, linear electrostatic waves in a stable plasma—electron plasma waves and ion-acoustic waves—the authors of this book have succeeded in fulfilling this need.

Having been themselves important contributors to this area of plasma physics, their exposition of the theory of these waves, and in particular their treatment of the forced excitation of such waves related to experiments, will be much appreciated by students and teachers alike.

ABRAHAM BERS
Massachusetts Institute of Technology
Cambridge, Massachusetts

PREFACE

Many excellent books have been written to give the casual scientific reader a general introduction and overview of basic plasma physics and its varied applications. The present book is intended for a somewhat more selective reader. First, it is intended for the upper-level undergraduate and/or graduate physics or engineering student, for example, who may already have had some introduction to basic plasma physics concepts and theories but who would like to see an in-depth, more operational, treatment, employing some of the current mathematical methods and techniques used, not only in plasma physics but in many other branches of physics and engineering as well. It is also intended for the practicing plasma physicist who, quite likely, has never been exposed to a "working" introduction to the linear theories and methods currently used in plasma physics research. For the benefit of the beginner, we have included detailed discussions of certain experimental work in order to show not only the relevance of the theory to the real world but, also, to introduce the reader to some basic experimental methods and techniques used in plasma physics research. We believe that it is somewhat biased and unsatisfactory to expose the beginning student to only the theoretical side of plasma physics. Too many students never see the description of a real plasma, nor are they introduced to experimental procedures or to the interpretation of experimental data.

In order to present an in-depth and coherent treatment of manageable length, this book considers only a small number of plasma waves, primarily electrostatic waves, and these only for the case where they are marginally stable or damped. Although the first priority is to give the reader a clear insight into the physics of electrostatic waves in plasmas, strong emphasis is given, also, to the methods and procedures themselves. A slight background in plasma physics as well as some knowledge of the theory of complex variables, including the theory of residues, is assumed, although no similar background in plasma waves, mathematical transform techniques, or numerical methods is assumed. Our experience shows that, with only a nominal introduction to the basic properties of plasmas, even a student who has had no formal training in plasma physics can successfully follow a course patterned after this book and that, having done so, he or she can then successfully apply the same general mathematical methods to many other problems in physics and engineering such as electromagnetic waves in plasmas, linear waves in dispersive media, and the general study of linear phenomena in many other fields.

A novel feature of the book is an introductory chapter, called "The Cookbook" and a concluding chapter entitled "Numerical Methods," which together provide a "working" introduction, via nontrivial examples, to some of the main mathematical methods and techniques required for the mathematical treatment of problems of research interest in plasma physics. This includes Fourier, Laplace, and Hilbert transforms in Chapter 1 and numerical methods for the calculation of Hilbert transforms and for "hunting the roots" of dispersion relations in Chapter 11. For the case of Fourier and Laplace transforms, for example, a detailed treatment is given for the generation of electromagnetic waves by an antenna located in space, doing both the forward and inverse Fourier and Laplace transforms in space and time with a comparison being made with the well-known solutions of this problem obtained using more conventional mathematical techniques. Because of the widespread availability of microcomputers, which now make it possible for the student to study many problems that only a few years ago could be studied only with great difficulty, if at all, sample BASIC programs, suitable for microcomputers, are given for the numerical methods discussed in Chapter 11. These programs may be easily translated into FORTRAN for use with larger, faster computers. For convenience, an appendix to Chapter 1 gives a basic summary of the theory of complex variables and of the theory of residues.

In Chapters 2–6 the plasma is treated as being composed of interpenetrating fluids in which wave–particle interactions between waves and plasma electrons and ions can be ignored. Routine use of Fourier and Laplace transforms is made in this part of the book. Chapter 2 considers both electrostatic and electromagnetic waves in infinite, cold, two-component, nonmagnetically supported plasmas; a simple transmission experiment using the extraordinary wave to measure plasma density is described. In Chapter 3, the theoretical study is restricted to the propagation of ion-acoustic waves and electron plasma waves parallel to the applied magnetic field of infinite, warm, two- and three-component, magnetically supported plasmas; detailed descriptions are given of a series of experiments designed to measure the properties of both types of waves. In Chapter 4, the theoretical study considers both the initial-value problem and the boundary-value problem for ion-acoustic wave propagation parallel to the magnetic field of an infinite, warm, two-component, magnetically supported plasma having strong ion-neutral collisions. A detailed description is given of an experimental study of the dispersion and damping of these waves in a collision-dominated discharge plasma for the boundary-value problem. In Chapter 5, a theoretical study is made of the finite-size-geometry effects on ion-acoustic waves and electron plasma waves propagating along both cold and warm magnetically supported plasma columns, infinite in length but finite in radius, for various radial boundary conditions. Detailed descriptions are given of three experimental studies of ion-acoustic waves propagating parallel to the axis of magnetically supported warm-plasma columns. In Chapter 6, a theoretical study is made of the effects of small density gradients on the propagation of ion-acoustic waves in nonuniform, nonmagnetically supported, warm plasmas. Descriptions of two experimental studies of these effects, including observation of wave reflection for the long-wavelength case, are given.

In Chapters 7–11 the plasma is treated from a kinetic-theory point of view, wherein velocity-dependent effects are taken into account by explicitly considering the velocity distributions of the plasma electrons and/or ions. This section of the book is somewhat more demanding, from a mathematical point of view, than the preceding section, with routine use being made of the Fourier, Laplace, and Hilbert transforms as well as of numerical calculations. Before beginning this section the reader may want to make a brief review of the theory of complex variables, paying particular attention to the theory of residues. Chapter 7 gives a detailed mathematical treatment of Landau damping from the boundary-value-problem point of view. The well-known plasma dispersion function is introduced in this chapter as well as a somewhat novel approach allowing the free-streaming and collective contributions to the wave electric field to be shown separately. Both electron-plasma-wave and ion-acoustic-wave propagation are considered in plasmas having various velocity distributions for the ions and electrons. Chapter 8 treats the problem of forced oscillations in one-dimensional, warm plasmas. Chapter 9 examines various computing techniques for electrostatic perturbations of a Maxwellian electron cloud. Chapter 10 considers the properties of ion-acoustic waves in Maxwellian plasmas, from a boundary-value-problem point of view. Detailed numerical calculations are made of the response of the plasma to both dipole and monopole excitation, and the results are used to critically evaluate the landmark experiment of Wong, Motley, and D'Angelo. A detailed description is given of a clear-cut experiment showing Landau damping of ion acoustic waves in a nonisothermal plasma.

It is our experience that the material of this book can be covered in a two-semester course which meets approximately three hours per week. The course can be conveniently divided from a time point of view into two approximately equal parts, the first part consisting of Chapters 1–6. Actually, since Hilbert transforms are not used in the first six chapters, a study of these transforms (which are discussed in Chapter 1) can be conveniently delayed until the beginning of the second part of the course.

WILLIAM DENVER JONES
University of South Florida
Tampa, Florida

CONTENTS

THE COOKBOOK

EVERYTHING YOU ALWAYS WANTED TO KNOW ABOUT FOURIER, LAPLACE, AND HILBERT TRANSFORMS, BUT WERE AFRAID TO ASK . . .

1.1. INTRODUCTION

Electromagnetic waves in plasmas, like electromagnetic waves in space and in all other media, must obey Maxwell's equations. Even for propagation in free space the resulting wave equation is a linear second-order partial-differential equation. Thus, the equations describing wave propagation in a complex medium such as a plasma can be complicated and difficult to solve directly. A technique that is widely used for solving such linear differential equations for the case of *infinite homogeneous* plasmas involves the so-called Fourier and Laplace transforms. Basically, this technique maps the real-space and time variables (r,t) of the wave—which are *differentially* related in the wave equation—to a complex space (ω,k), where the variables ω and k are *algebraically* related. In this way, one arrives relatively easily at an expression for $E(\omega,k)$, for example, where E is the electric field of the wave as expressed in (ω,k) space. Application of Fourier–Laplace transforms to other equations that must be satisfied by the wave–plasma system allows other quantities, such as $f_1(\omega,k)$ and $n_1(\omega,k)$, the first-order perturbations of the particle distribution and particle density, respectively, to be similarly computed in (ω,k) space.

Although the relationship between quantities in the (ω,k) space and the corresponding quantities in the real (r,t) space is not usually obvious, in many cases it is not needed to know the quantities in real space. In the calculation of $E(\omega,k)$, for example, one obtains the so-called dispersion relation, $D(\omega,k) = 0$, which gives many of the basic properties of the wave (phase velocity, growth or damping, resonances, etc.). If this information is all that is needed—and, as we shall see in many nontrivial examples throughout this book, this is indeed often the case—then it is not necessary to calculate $E(r,t)$.

In these cases, we are able essentially to "have our cake and eat it too", i.e., we are able to apply a very powerful mathematical technique without being unduly "penalized". If, however, it *is* needed to know $E(r,t)$, one must make what are called *inverse* Fourier and Laplace transforms on $E(\omega,k)$. In doing this, one immediately encounters equations involving integrals of functions of complex variables. To solve such equations, one must be somewhat adroit in the theory of analytic functions and the theory of residues. It is in these inverse transformations, when they are needed, that one often "pays the price" for the simplicity gained initially by use of the Fourier–Laplace transforms. In some cases, where solutions in real (r,t) space are required, because of the additional complexities that can be introduced by the inverse transformations, it is more appropriate to use other more conventional methods for solving the set of equations describing the wave–plasma system. There are some problems, however, such as the well-known Landau-damping problem (see Chapter 7), where only a careful complete analysis using Fourier–Laplace transform methods gives the proper results.

In keeping with the basic philosophy of "The Cookbook," we will not present a formal description of Laplace–Fourier transform methods; rather we will give a working introduction by briefly describing and applying these techniques to the case of the generation and propagation of electromagnetic waves in free space. In the other chapters of the book, we will then make routine application of these techniques in the solution of numerous nontrivial examples. For in-depth discussions and applications of these techniques in the mathematical analysis of problems arising in several areas of theoretical physcis and engineering, we refer the reader to the excellent books by Whittaker and Watson (1902), Roos (1969), and Rudin (1970). The reader will find a compilation of some of the most useful theorems and properties of the theory of functions of complex variables in Appendix A of this book.

1.2. A BASIC EXAMPLE: ELECTROMAGNETIC WAVE PROPAGATION IN VACUUM

Let us consider that the universe consists only of an antenna in free space, as shown in Fig. 1.1. We imagine a coordinate system as shown, with the antenna lying in the y,z-plane. The $+x$ direction, as indicated by the unit vector $\hat{l}_x$, is perpendicular to the antenna. We will assume that we can move the electrons in our antenna at will. To keep the mathematics as simple as possible, we choose the surface current of our antenna to be an infinitely thin sheet whose direction of flow alternates between the $+y$ and $-y$ directions. Mathematically, we can represent such a current by the equation

$$\mathcal{J} = -en_0 v_0 \delta(x) \sin(\omega_0 t) \, \hat{l}_y, \tag{1.1}$$

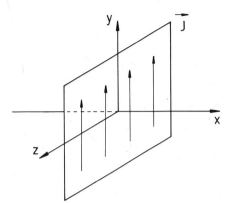

Fig. 1.1. Geometry of the alternating current
sheet used to excite electromagnetic waves.

where $\hat{l}_y$ is a unit vector along the $+y$ axis, $-e$ is the electron charge, n_0 and v_0 are the electron density and speed, respectively, of the electrons carrying the current, ω_0 is the angular frequency of the alternating current, t is the time, and $\delta(x)$ is the usual Dirac delta function.

Given the simple situation described, we may restrict the laws of physics to Maxwell's equations. These equations can be written in MKS units as

$$\nabla \times E = -\frac{\partial B}{\partial t} \tag{1.2}$$

and

$$\nabla \times B = \mu_0 \mathcal{J} + \frac{1}{c^2}\frac{\partial E}{\partial t}. \tag{1.3}$$

Taking the curl of Eq. (1.3) and interchanging the order of the time and space operators in the last term gives,

$$\nabla \times (\nabla \times B) = \mu_0 \nabla \times \mathcal{J} + \frac{1}{c^2}\frac{\partial}{\partial t}\nabla \times E. \tag{1.4}$$

Using Eqs. (1.2) and (1.4), the vector identity

$$\nabla \times (\nabla \times B) = -\overset{\text{2}}{\nabla B} + \nabla(\nabla \cdot B), \tag{1.5}$$

and the fact that $\nabla \cdot B = 0$ gives the wave equation

$$\frac{\partial^2 B}{\partial t^2} - c^2 \underbrace{\Delta B}_{\nabla^2 B} = \mu_0 c^2 \,\nabla \times \mathcal{J}, \tag{1.6}$$

which describes the propagation in vacuum of the electromagnetic waves generated by the alternating sheet of current in our planar antenna. From the geometry of our problem, i.e., from the fact no one place along the z-axis differs physically from any other place along the z-axis, and similarly for the y-axis, we can immediately say that

$$\frac{\partial}{\partial y} = 0 = \frac{\partial}{\partial z}. \tag{1.7}$$

That is, our problem is homogeneous in y and z. Thus, $\nabla = \hat{l}_x \partial/\partial x + \hat{l}_y \partial/\partial y + \hat{l}_z \partial/\partial_z$ becomes simply $\nabla = \hat{l}_x \partial/\partial x$ and, similarly, $\Delta = \partial^2/\partial x^2$. Therefore, Eq. (1.6) can be rewritten as

$$\frac{\partial^2 B}{\partial t^2} - c^2 \frac{\partial^2 B}{\partial x^2} = \mu_0 c^2 \left[\frac{\partial}{\partial x} \hat{l}_x \right] x \vec{J}. \tag{1.8}$$

Using Eq. (1.1) and the fact that $\hat{l}_x \times \hat{l}_y = \hat{l}_z$, Eq. (1.8) becomes

$$\frac{\partial^2 B}{\partial t^2} - c^2 \frac{\partial^2 B}{\partial x^2} = -e\mu_0 c^2 n_0 v_0 \delta'(x) \sin(\omega_0 t) \hat{l}_z, \tag{1.9}$$

where $\delta'(x)$ is the derivative with respect to x of the delta function $\delta(x)$. From the right-hand side of Eq. (1.9) we see that our source generates waves having a magnetic field only in the z-direction. That is, our antenna is a source of polarized electromagnetic waves. Thus, $B_x = 0 = B_y$ and $B = \hat{l}_z B_z$. Omitting the vector notation, we can rewrite Eq. (1.9) as

$$\frac{\partial^2 B_z}{\partial t^2} - c^2 \frac{\partial^2 B_z}{\partial x^2} = -e\mu_0 c^2 n_0 v_0 \delta'(x) \sin(\omega_0 t). \tag{1.10}$$

Equation (1.10) is the equation of propagation for a plane-polarized electromagnetic wave propagating in the x-direction, away from our antenna.

1.3. THE FOURIER–LAPLACE TRANSFORMS

Equation (1.10) is a linear partial-differential equation whose general solution is well known and is given by

$$B_z = B_z^+ \left(t - \frac{x}{c} \right) + B_z^- \left(t + \frac{x}{c} \right), \tag{1.11}$$

where B_z^+ and B_z^- are two arbitrary functions of $(t - x/c)$ and $(t + x/c)$, respectively. For our particular problem, B_z^+ and B_z^- will be determined by requiring that the solution match the boundary conditions of our antenna.

The simple solution given by Eq. (1.11) is obtained because in vacuum there is no damping or distortion of the spatial shape of the wave in one-dimensional propagation. However, if we consider a medium such as a plasma the propagation equation equivalent to Eq. (1.10) will be more complicated, and a simple solution like Eq. (1.11) is no longer possible. In this case, the utilization of Fourier and Laplace transforms is a very useful technique. We will now use these techniques on Eq. (1.10) as an example where the physics and the results are well known. Solving Eq. (1.10) explicitly will allow us to present the main properties of these transforms and the related techniques needed for their manipulation.

1.3.1. Fourier Transform in Space

The Fourier transform of a function $g(x)$ may be defined as

$$g(x) - F \rightarrow G(k) = \int_{-\infty}^{+\infty} g(x)e^{-ikx}\, dx, \tag{1.12}$$

where k is a *real* number. The left-hand side of the above equation is to be read: "The function $g(x)$ is Fourier-transformed into the function $G(k)$." As indicated by the change in symbol for the function (g becomes G), G has a dependence on k completely different from the dependence of g on x. However, in this book, for simplicity, we will often use the same symbol for both functions. That is, we will let the Fourier transform of $g(x)$ be represented by $g(k)$. The right-hand side of Eq. (1.12) gives the "recipe" for transforming $g(x)$ into $G(k)$.

For future reference, we point out that if x and k in Eq. (1.12), which defines a Fourier transform *in space*, are replaced by t and ω, respectively, then Eq. (1.12) becomes the definition of a Fourier transform *in time*. In that case ω, as did k for the spatial case, *must* represent a *real* number.

Since $G(k)$ in Eq. (1.12) is defined by an integral, care must be taken that this integral exists for real values of k. Fortunately, many of the functions that can be used to represent physical processes fulfill the required conditions. Integrable or square-integrable functions in $(-\infty, +\infty)$ have Fourier transforms. For example, the Maxwell distribution function, which is integrable in $(-\infty, +\infty)$, has a Fourier transform given by

$$\exp\left(\frac{-x^2}{a^2}\right) - F \rightarrow a\pi^{1/2} \exp\left(\frac{-k^2 a^2}{4}\right). \tag{1.13}$$

Such functions as the sine and cosine, which are widely used but are not integrable in $(-\infty, +\infty)$, also have Fourier transforms:

$$\cos(k_0 x) - F \rightarrow \pi\delta(k - k_0) + \pi\delta(k + k_0);$$
$$\sin(k_0 x) - F \rightarrow \pi\delta(k - k_0) - \pi\delta(k + k_0). \tag{1.14}$$

Distributions such as the Dirac delta function, $\delta(x)$, the signum or sign function, $\mathrm{sgn}(x)$, and the unit step function (also called the Heaviside function), $\Upsilon(x) = [1 + \mathrm{sgn}(x)]/2$, themselves, have Fourier transforms:

$$\delta(x) - F \rightarrow 1;$$
$$\mathrm{sgn}(x) - F \rightarrow \frac{-2i}{k}; \tag{1.15}$$
$$\Upsilon(x) - F \rightarrow \frac{-i}{k} + \pi\delta(k).$$

For a comprehensive discussion of the theory of distributions and of their extreme importance in mathematical physics, the reader is referred, for example, to Roos (1969).

In Fourier-transforming a differential equation, one often needs not only to Fourier-transform a function itself, but also the derivatives of the function. For example, in Eq. (1.10) the second term is a second-order spatial derivative. Defining $g'(x) = dg/dx$, we can immediately use the basic definition of the Fourier transform given by Eq. (1.12) to find the Fourier transform of $g'(x)$. Using integration by parts, we find that

$$\frac{dg}{dx} - F \rightarrow \int_{-\infty}^{+\infty} g'(x)e^{-ikx}dx = g(x)e^{-ikx}\Big|_{-\infty}^{+\infty} + ik\int_{-\infty}^{+\infty} g(x)e^{-ikx} dx. \tag{1.16}$$

Observation of Eq. (1.16) shows that, for the case where $g(x) \rightarrow 0$ for $|x| \rightarrow \infty$

$$\frac{dg}{dx} - F \rightarrow ikG(k). \tag{1.17}$$

Thus, the Fourier transform of the first derivative of a function is simply related to the Fourier transform of the function itself. Actually, the relationship given by Eq. (1.17) can be true for less stringent conditions on the behavior of $g(x)$ for large x. For example, the relation given by Eq. (1.17) is still valid for the Heaviside step function $\Upsilon(x)$, which has a value of $+1$ at $x \to +\infty$:

$$\Upsilon'(x) - F \to ik\left[\frac{-i}{k + i\pi\delta(k)}\right] = 1 - k\delta(k).$$

But since

$$\int_{-\infty}^{+\infty} f(k)k\delta(k)\,dk = 0,$$

we see that

$$\Upsilon'(x) - F \to 1. \tag{1.18}$$

The Fourier transform of the nth derivative of $g(x)$, when the derivative exists, may be evaluated using the method employed for establishing Eq. (1.17), and is easily given by

$$\frac{d^n g(x)}{dx^n} - F \to (ik)^n G(k). \tag{1.19}$$

Before applying the Fourier transform to Eq. (1.10), we will use the following decomposition, in order to simplify the algebra

$$B_z(x,t) = M(x,t) + M^*(x,t), \tag{1.20}$$

where M^* is the complex conjugate of M. Also, we note that

$$\sin(\omega_0 t) = \frac{i}{2}\exp(-i\omega_0 t) - \frac{i}{2}\exp(i\omega_0 t). \tag{1.21}$$

Using Eqs. (1.20) and (1.21) in Eq. (1.10), we may choose $M(t, x)$ to be the solution of

$$\frac{\partial^2 M}{\partial t^2} - c^2\frac{\partial^2 M}{\partial x^2} = +i\left(\frac{\mu_0 c^2}{2}\right)en_0 v_0 \delta'(x)\exp(-i\omega_0 t). \tag{1.22}$$

Before proceeding with the Fourier transformation of Eq. (1.22), we remind the reader that the Fourier transform is a *linear transform. That is,*

$$ag(x) - F \to aG(k); \tag{1.23}$$

$$f(x) + g(x) - F \to F(k) + G(k), \tag{1.24}$$

where a is a constant with respect to x.

Applying the linear property expressed by Eq. (1.24) allows us to separately Fourier-transform each of the terms in Eq. (1.22). The simplest term to transform is the second term of the left-hand side

$$-c^2 \frac{\partial^2 M(x,t)}{\partial x^2} - F \rightarrow k^2 c^2 M(k,t), \tag{1.25}$$

where we have used the "recipe" given by Eq. (1.19) for Fourier-transforming the derivative of a function and the property of Eq. (1.23). Note, also, that we are using the same symbol, M, to designate the function and its transform. The argument k specifies the transform.

We Fourier-transform the right-hand side of Eq. (1.22) as

$$-i \left(\frac{\mu_0 c^2}{2} \right) e n_0 v_0 \exp(-i\omega_0 t)\, \delta'(x) - F \rightarrow \left(\frac{\mu_0 c^2}{2} \right) e k n_0 v_0 \exp(-i\omega_0 t), \tag{1.26}$$

where the property of Eq. (1.23) has been used, noting that the exponential function is not a function of x and is, thus, an invariant for the transformation. To compute the transformation of $\delta'(x)$ we use Eqs. (1.15) and (1.17) to obtain

$$\delta'(x) - F \rightarrow ik. \tag{1.27}$$

To Fourier transform the first term of the left-hand side of Eq. (1.22), we begin by writing the definition of the partial derivative of $M(x,t)$ with respect to time

$$\frac{\partial M(x,t)}{\partial t} = \lim_{\Delta t \to 0} \left[\frac{M(x, t+\Delta t) - M(x,t)}{\Delta t} \right].$$

Using the properties of Eqs. (1.23) and (1.24) again we see that

$$\frac{\partial M(x,t)}{\partial t} - F \rightarrow \lim_{\Delta t \to 0} \left[\frac{M(k, t+\Delta t) - M(k,t)}{\Delta t} \right].$$

Thus, we can write

$$\frac{\partial M(x,t)}{\partial t} - F \rightarrow \frac{\partial M(k,t)}{\partial t}. \tag{1.28}$$

Note that as before we have used the same symbol M for the transformed function as for the initial function. Repeating the procedure for the second derivative we see that

$$\frac{\partial^2 M(x,t)}{\partial t^2} - F \rightarrow \frac{\partial^2 M(k,t)}{\partial t^2}. \tag{1.29}$$

Finally, using Eqs. (1.25)–(1.27) and (1.29), the Fourier transform in space of Eq. (1.22) may be written as

$$k^2 c^2 M(k,t) + \frac{\partial^2 M(x,t)}{\partial t^2} = \left(\frac{c\mu_0}{2}\right) e n_0 v_0 k c \exp(-i\omega_0 t). \tag{1.30}$$

Comparing Eq. (1.30), our wave-propagation equation in (k,t)-space, with Eq. (1.22), our wave equation in (x,t)-space, we see immediately the advantage of the Fourier transform. Instead of a *partial* differential equation, we now have an *ordinary* differential equation. Moreover, because our original differential equation is *linear* in x, the differential operator becomes a polynomial (i.e., algebraic) operator and, as we shall see later, we will obtain an algebraic equation for the image. If the original equation were not linear, the transform would be integral rather than algebraic because the transform of a product of functions, say $f(x)h(x)$, is a convolution integral

$$g(x)h(x) - F \rightarrow G(k)*H(k) = \int_{-\infty}^{+\infty} G(k - k')H(k')\,dk'. \tag{1.31}$$

Thus, one does not normally apply Fourier transforms to nonlinear differential equations.

Before continuing with the Fourier–Laplace technique of solving our antenna problem [the next step of which will be to Laplace transform in time the already Fourier-transformed-in-space wave equation, Eq. (1.30)], we will digress briefly to: (1) make an additional comment on the existence of the Fourier transform, and (2) give a brief discussion of the *inverse* Fourier transform.

We have already made some general mathematically based comments concerning the existence of the Fourier transform of some function $g(x)$. We now want to make a *plausibility* argument concerning the existence of the Fourier transform of certain such functions. When a function involves the time variable and causality, such as $M(x,t)$ for our antenna problem, as physicists we may argue the following: At any finite time t after being generated, any propagating perturbation is bounded in amplitude and limited in spatial extent because the propagation has started at a given finite time and cannot spread in space faster than the speed of light. Thus, for any realistic representation of an initial perturbation, this argument makes us confident that our use of the Fourier transform is legitimate. In any particular case, however, it is always prudent to check the mathematical justification of this argument!

In the case of an explosive instability, for example, the perturbation becomes infinite after a finite time.

Because of the symmetry between the Fourier transform, which we have just discussed and applied to our antenna problem, and the *inverse* Fourier transform, we pause briefly to introduce the idea of the inverse Fourier transform. Let us begin our discussion of the inverse Fourier transform by reminding the reader that the solution of the relatively simple ordinary differential, Eq. (1.30) (whose relative simplicity resulted from the Fourier transform), will be a function of (k,t), not of (x,t). This solution, $M(k,t)$, which is sometimes called the Fourier "image" of the solution $M(x,t)$ of the original partial-differential equation, is very different from $M(x,t)$. Thus, it is difficult to try to predict the behavior of $M(x,t)$ simply by inspection of $M(k,t)$. The only sure way to know $M(x,t)$ exactly is to compute the so-called inverse Fourier transform of $M(k,t)$.

If $G(k)$ is the Fourier transform of $g(x)$ [see Eq. (1.12)] then the recipe for finding the inverse Fourier transform of $G(k)$ [which is, of course, $g(x)$ itself] is given by the equation

$$G(k) - Fi \rightarrow g(x) = \frac{1}{2\pi} \int_{-\infty}^{+\infty} G(k)e^{ikx}\, dk, \tag{1.32}$$

where, as for the Fourier transform, k is *real*. We again recommend that the reader consult one of the mathematical physics textbooks already cited for a detailed discussion of inverse Fourier transforms.

To explicitly dramatize the extreme difference between a function and its Fourier transform, we consider the following example: Suppose $G(k)$—also sometimes referred to as a *spectrum*—has a so-called Lorentzian shape given by

$$S(k) = \frac{2v}{(k - k_0)^2 + v^2}, \qquad (v > 0). \tag{1.33}$$

We state, without proof, that following the recipe given by Eq. (1.32) gives the inverse Fourier transform of $S(k)$ as

$$s(x) = [\exp(-v|x|)][\exp(-ik_0x)]. \tag{1.34}$$

Figures 1.2(a) and (b) represent $S(k)$ and $s(x)$, respectively. Clearly, the two functions are very different: $s(x)$ is complex, whereas $S(k)$ is real; $s(x)$ is oscillatory, whereas $S(k)$ is not; the narrower is the spectrum $S(k)$, as $v \rightarrow 0$, the broader is $s(x)$ along x; etc. Obviously, unless one has had a lot of experience, one would find it difficult to guess what $s(x)$ would be like, even qualitatively, without doing an inverse Fourier transformation. Certainly a knowledge of

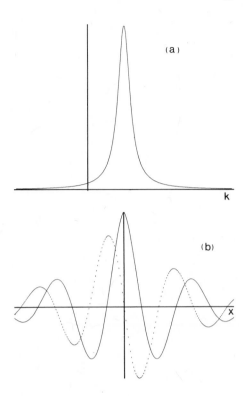

Fig. 1.2. Example of a Fourier transform and its inverse: (a) $S(k)$ given by Eq. (1.33); and (b) its inverse transform $s(x)$ given by Eq. (1.34).

the *quantitative* behavior of the inverse-Fourier-transformed function would in any case require that the actual inversion be made.

In going from Eq. (1.33) to Eq. (1.34), we have expressly not suggested that the reader verify the correctness of Eq. (1.34). We will give a working introduction later in this chapter when we will make both inverse Fourier and inverse Laplace transformations on the solution for our antenna problem, which should enable the reader to fill in the missing steps between Eqs. (1.33) and (1.34).

1.3.2. Laplace Transform in Time

Thus far, by means of a Fourier transform in space, we have reduced our original *partial*-differential equation in real space (r,t) to an *ordinary* differential equation in (k,t)-space. Although this represents a significant reduction in the mathematical complexity of the equation we can do even better. We now want to show that a Laplace transform in time will reduce our ordinary *differential* equation, Eq. (1.30), to an *algebraic* equation in (ω,k)-space. Before doing this, however, it is interesting to note that formally Eq. (1.30) is identical to the equation describing an undamped harmonic oscillator driven at the frequency $\omega_0/2\pi$. Defining $\Omega = kc$ and $\alpha = \mu_0 e n_0 v_0 c$, Eq. (1.30) can be written as

$$\frac{d^2 M}{dt^2} + \Omega^2 M = \frac{\alpha}{2} \Omega \exp(-i\omega_0 t) \qquad (1.35)$$

The solution of Eq. (1.35) is well known from classical mechanics. For $M(t = 0) = 0$ and $M'(t = 0) = 0$, the solution of Eq. (1.35) is given by

$$M(t) = -\frac{\alpha}{2} \left[\frac{\Omega e^{-i\omega_0 t}}{\omega_0^2 - \Omega^2} + \frac{1}{2\Omega} \left(\frac{e^{i\Omega t}}{\omega_0 + \Omega} + \frac{e^{-i\Omega t}}{\omega_0 - \Omega} \right) \right]. \qquad (1.36)$$

Equation (1.36) exhibits the two oscillatory motions that are characteristic of such a driven system: (a) a forced oscillation at the driving frequency $\omega_0/2\pi$, and (b) a transient motion at the natural oscillator frequency $\Omega/2\pi$. Because we have introduced no damping in our antenna problem, the transient motion in Eq. (1.36) goes on forever. At the resonance, i.e., when $\omega_0 = \Omega$, the beating of the driven and of the natural oscillations yields a secular growth (i.e., a growth like t^n, where n is a positive integer), given by

$$M(t) = \frac{\alpha}{4} \left[it \exp(-i\Omega t) - \frac{\cos(\Omega t)}{\Omega} \right]. \qquad (1.37)$$

We will now obtain the solution of Eq. (1.35) by using the Laplace transform. There are several formulations of the Laplace transform; however, in plasma physics, the recipe generally used for making a Laplace transformation in time is given by the equation

$$h(t) - L \rightarrow H(\omega) = \int_0^{+\infty} h(t) e^{i\omega t} \, dt, \qquad (1.38)$$

with $\text{Im}(\omega) > a$, a being *real*. Similar to the defining equation for the Fourier transform, Eq. (1.12), the left-hand side of Eq. (1.38) is to be read: "The function $h(t)$ is Laplace-transformed into the function $H(\omega)$." In general, $H(\omega)$ will not have the same functional dependence on ω as does $h(t)$ on t; however, for simplicity we often use the same symbol for both functions. That is, we let the Laplace transform of $h(t)$ be represented by $h(\omega)$, rather than by $H(\omega)$. The Laplace transform, like the Fourier transform, is a linear transformation and, thus, obeys the properties of linear transformations given by Eqs. (1.23) and (1.24).

Two questions immediately arise when one compares the definitions of the Laplace and Fourier transforms. First, why is the integration in the

Laplace transform only from $t = 0$, and not from $t = -\infty$? Second, why is ω complex, with $\text{Im}(\omega) > a$? The answer to the first question is, that by limiting the integration to $t \geqslant 0$, we will insure that the images in ω-space are *causal* images, i.e., when we compute the inverse of these images, in order to obtain the answer to our problem as a function of time, nothing will appear before time $t = 0$ (the time at which the causal phenomena under investigation is turned on). This avoids such absurdities as the so-called filter paradox: through a noncausal red filter we may see the red light from a flashlight *before* the light is turned on! We will examine this point further when studying the Kramers–Kronig relations in Section 1.4.3.

The answer to the second question—why is ω complex with $\text{Im}(\omega) > a$?— is simply to insure the existence of $H(\omega)$ even when $h(t)$ grows exponentially in time, as is possible for the case of instabilities.

As noted in the discussion of the Fourier transform, one often needs not only to Laplace transform a function itself, but also the derivatives of the function. Using Eq. (1.38) and integration by parts, one can easily show that the Laplace transform of $h'(t) = dh(t)/dt$ is given by

$$\frac{dh(t)}{dt} - L \to \int_0^{+\infty} h'(t)e^{i\omega t}\, dt = -h(t = 0) - i\omega H(\omega), \qquad (1.39)$$

with $\text{Im}(\omega) > a$, where a is *real*. Similarly, doing an integration by parts and using the results given in Eq. (1.39), one can easily show that the Laplace transform of $h'' = d^2h(t)/dt^2$ is given by

$$\frac{d^2h(t)}{dt^2} - L \to -h'(t = 0) + i\omega h(t = 0) - \omega^2 H(\omega), \qquad (1.40)$$

with $\text{Im}(\omega) > a$, a being *real*.

A large number of functions have Laplace transforms. Here we will give the Laplace transforms of only a few functions currently used in plasmas physics research

$$\Upsilon(t) - L \to \frac{i}{\omega},$$

$$\delta(t - t_0) - L \to \exp(i\omega t_0), \qquad t_0 \text{ real}, > 0, \qquad (1.41)$$

$$\exp(i\lambda t) - L \to \frac{i}{\omega + \lambda},$$

where as before $\Upsilon(t)$ is the Heaviside function and $\delta(t - t_0)$ is the Dirac delta function. We refer the reader to Roberts and Kaufman (1966) for a comprehensive listing of Laplace and inverse Laplace transforms.

A final useful transform is the Laplace transform of the so-called convolution product of two causal functions, $h(t)$ and $p(t)$. Such a product is defined by

$$h(t) * p(t) = \int_0^t [h(t')][p(t - t')]\, dt'. \tag{1.42}$$

The recipe for Laplace-transforming this convolution product is given by

$$h(t) * p(t) = \int_0^t h(t')p(t - t')\, dt' - L \rightarrow H(\omega)P(\omega), \tag{1.43}$$

where $H(\omega)$ is the Laplace transform of $h(t)$ [see Eq. (1.38)] and $P(\omega)$ is the Laplace transform of $p(t)$.

Equations (1.38)–(1.43), along with the properties of linear transformations given by Eqs. (1.23) and (1.24), constitute a working introduction to the properties of Laplace transforms. We now want to use some of these properties to Laplace transform our ordinary second-order differential equation, Eq. (1.35), which we will repeat here for convenience

$$\frac{d^2 M}{dt^2} + \Omega^2 M = \frac{\alpha}{2}\Omega \exp(-i\omega_0 t). \tag{1.35}$$

Applying the linear-operator property given by Eq. (1.24) to Eq. (1.35) allows us to Laplace transform each of the terms in Eq. (1.35) separately. The simplest to transform is the second term

$$\Omega^2 M(\Omega, t) - L \rightarrow \Omega^2 M(\Omega, \omega), \tag{1.44}$$

where we have used the linear-operator property given by Eq. (1.23) plus the definition of the Laplace transform given by Eq. (1.38). Note, also, that we have used the same symbol M to designate both $M(\Omega, t)$ and its transform, $M(\Omega, \omega)$.

The first term of Eq. (1.35) is easily transformed by applying Eq. (1.40). Assuming that $M(t = 0) = 0 = M'(t = 0)$ gives

$$\frac{\mathscr{L}\, d^2 M(\Omega, t)}{dt^2} = -\omega^2 M(\Omega, \omega). \tag{1.45}$$

Using the Laplace transform given in Eq. (1.41) for the exponential function allows the Laplace transform of the term on the right-hand side of Eq. (1.35) to be written as

$$\mathscr{L}\left[\frac{\alpha}{2}\Omega \exp(-i\omega_0 t)\right] = \frac{(i\alpha/2)\Omega}{\omega - \omega_0}. \tag{1.46}$$

In arriving at Eq. (1.46), we have also used the linear-operator property given by Eq. (1.23), noting that $(\alpha/2)\Omega$ is an invariant with respect to time. Combining Eqs. (1.44)–(1.46), we can write the Laplace transform of Eq. (1.35) as

$$-\omega^2 M(\Omega,\omega) + \Omega^2 M(\Omega,\omega) = \frac{(i\alpha/2)\Omega}{\omega - \omega_0}. \tag{1.47}$$

Thus, by means of a Laplace transform, we have reduced our *differential* Eq. (1.35) to an *algebraic* equation. Since we have defined $\Omega = kc$, where c is the speed of light in vacuum, we see that k and not Ω is the variable of interest, so that, functionally, we can write $M(\Omega,\omega) = M(k,\omega)$. Thus, we can rewrite Eq. (1.47) as

$$(\omega^2 - \Omega^2)M(\omega,k) = \frac{(-i\alpha/2)\Omega}{\omega - \omega_0}, \tag{1.48}$$

where in Eq. (1.48) we have interchanged the order of appearance of ω and k in the function $M(k,\omega)$. The solution of Eq. (1.48) is easily seen to be given by

$$M(\omega,k) = \frac{-i\alpha}{2}\,\Omega F(\omega), \tag{1.49}$$

where

$$F(\omega) = \frac{1}{(\omega - \omega_0)(\omega + \Omega)(\omega - \Omega)}. \tag{1.50}$$

1.3.3. Inversion of Fourier–Laplace Transforms

As has been pointed out several times, the advantage of Fourier and Laplace transforms is that they allow us to map from an (x,t)-space, where the relationship between the variables is complicated, to an (ω,k)-space where the variables are relatively simply related. If one needs the solution in (x,t)-space, then it is necessary to perform the inverse Fourier and Laplace transforms on the (ω,k)-solution. We will now give the recipe for finding the inverse Laplace transform and apply it to Eq. (1.49).

The inverse Laplace transform of $H(\omega)$ is defined as

$$\mathscr{L}^{-1}[H(\omega)] = h(t) = \frac{1}{2\pi}\int_c H(\omega)e^{-i\omega t}\,d\omega, \tag{1.51}$$

where c is a path lying above all the singularities of $H(\omega)$, as shown in Fig. 1.3. Note that in Fig. 1.3 we use the *complex* plane because ω is a complex variable.

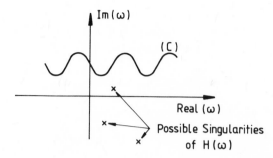

Fig. 1.3. The path of integration c of Eq. (1.51) (the so-called Bromwich path).

Since the singularities of $M(\omega,k)$ [see Eqs. (1.49) and (1.50)] lie on the real axis at $\omega = \omega_0$, $\pm\Omega$, our path of integration c needs to be just slightly above the real axis. Thus, following the recipe given by Eq. (1.51) we can write the inverse Laplace transform of $M(\omega,k)$ [see Eq. (1.49)] as

$$M(k,t) = -\frac{i\alpha}{4\pi}\,\Omega \int_{-\infty+i\varepsilon}^{+\infty+i\varepsilon} F(\omega)e^{-i\omega t}\,d\omega, \qquad (1.52)$$

where ε is an arbitrarily small real positive number. In order to compute $M(k,t)$ we will use the results of the theory of the functions of complex variables, briefly summarized in Appendix A. In the process we will see, as stated earlier, that the Laplace transform has given a causal solution, i.e., nothing happens before the current in the antenna is turned on. From a physics point of view we would be surprised indeed if $M(k,t)$ should have a nonzero value for $t < 0$; however, since t in Eq. (1.52) is allowed to have all values between $+\infty$ and $-\infty$, it is always a prudent check on the general accuracy of the mathematical procedure to demonstrate (rather than assume) that any solution generated by a Laplace transform is, in fact, a causal solution. Thus, we will now demonstrate explicitly that $M(k, t < 0) = 0$.

The integrand of Eq. (1.52) is composed of two parts: the exponential function $\exp(-i\omega t)$ and a simple meromorphic function $F(\omega)$. We begin the evaluation of Eq. (1.52) by defining the integral

$$I(t,R) = I^1(t,R) + I^2(t,R), \qquad (1.53)$$

where

$$I^1(t,R) = \int_{-R+i\varepsilon}^{+R+i\varepsilon} F(\omega)e^{-i\omega t}\,d\omega$$

and   (1.54)

$$I^2(t,R) = \int_c F(\omega)e^{-i\omega t}\,d\omega,$$

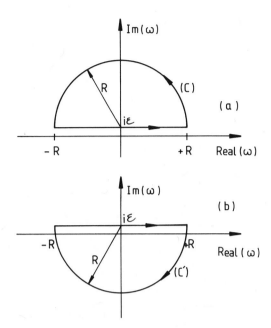

Fig. 1.4. (a) Path of integration c of Eq. (1.54); (b) Path of integration c' of Eq. (1.59).

where C is the semi-circle of radius R shown in Fig. 1.4(a). It will be noted that if we allow $R \to \infty$ then, except for a multiplicative constant, I^1 is the same as $M(k,t)$ [see Eq. (1.52)].

Let us first consider the case where $t < 0$. Since $F(\omega)$ is a meromorphic function, for $R > |\omega_0|, |\Omega|$, we may use the so-called residues theorem (see Appendix A) to conclude that

$$I(t,R) = 0, \qquad (1.55)$$

since all the poles of $F(\omega)$ are outside the contour C [see Fig. 1.4(a)]. Also, since $F(\omega) \to 0$ for $|\omega| \to \infty$ [see Eq. (1.50)] we may apply Jordan's lemma (see Appendix A) to $I^2(t,R)$ to conclude that

$$\lim_{R \to \infty} I^2(t < 0, R) = 0. \qquad (1.56)$$

Thus, if we allow $R \to \infty$, we can write

$$I(t < 0, R \to \infty) = I^1(t < 0, R \to \infty) = \int_{-\infty + i\varepsilon}^{+\infty + i\varepsilon} F(\omega) e^{-i\omega t} \, d\omega = 0. \quad (1.57)$$

Therefore, comparing Eq. (1.57) with Eq. (1.52) we can immediately conclude that $M(k, t < 0) = 0$. Thus, as already anticipated, we see that indeed our

Laplace transform has generated a causal solution: nothing happens before the current in the antenna is turned on (at $t = 0$, by definition).

Let us now consider the case for $t > 0$. In a manner similar to the case for $t < 0$, we define the integral

$$\mathcal{J}(t,R) = \mathcal{J}^1(t,R) + \mathcal{J}^2(t,R), \tag{1.58}$$

where

$$\mathcal{J}^1(t,R) = \int_{-R+i\varepsilon}^{+R+i\varepsilon} F(\omega)e^{-i\omega t}\,d\omega$$

and   (1.59)

$$\mathcal{J}^2(t,R) = \int_{C'} F(\omega)e^{-i\omega t}\,d\omega,$$

where C' is the semi-circle of radius R shown in Fig. 1.4(b). Since $F(\omega)$ is a meromorphic function, for $R > |\omega_0|, |\Omega|$, we may compute $\mathcal{J}^1(t,R)$, using the theory of residues. Doing this, we find that

$$\mathcal{J}^1(t,R) = -2\pi i\left[\frac{e^{-i\omega_0 t}}{(\omega_0 + \Omega)(\omega_0 - \Omega)} + \frac{e^{-i\Omega t}}{(\Omega + \omega_0)2\Omega} + \frac{e^{i\Omega t}}{(\Omega - \omega_0)2\Omega}\right].$$
$$\tag{1.60}$$

Also, as before, since $F(\omega) \to 0$ for $|\omega| \to \infty$, we may apply Jordan's lemma to $\mathcal{J}^2(t,R)$ to show that

$$\lim_{R \to \infty} \mathcal{J}^2(t > 0, R) = 0. \tag{1.61}$$

Therefore, for $R \to \infty$

$$\mathcal{J}(t > 0, R \to \infty) = \mathcal{J}^1(t, R \to \infty) = \int_{-\infty+i\varepsilon}^{+\infty+i\varepsilon} F(\omega)e^{-i\omega t}\,d\omega. \tag{1.62}$$

Thus, comparing Eq. (1.62) with Eq. (1.52) and using the result of Eq. (1.60), we can write for $t > 0$ that

$$M(k, t > 0) = -\frac{\alpha}{2}\left[\frac{\Omega e^{-i\omega_0 t}}{\omega_0^2 - \Omega^2} + \frac{1}{2\Omega}\left(\frac{e^{i\Omega t}}{\omega_0 + \Omega} + \frac{e^{-i\Omega t}}{\omega_0 - \Omega}\right)\right]. \tag{1.63}$$

The solution obtained is identical to the solution $M(t)$ given by Eq. (1.36). It is left as an exercise for the reader to show that when $\omega_0 = \Omega$ [giving rise to a *second*-order pole for $F(\omega)$] application of the theory of residues in the same manner as used to obtain Eq. (1.63) gives the result shown earlier by Eq. (1.37) for the case of resonant excitation of the antenna.

In order to complete the transformation of the solution back to real (r,t)-space, we must now do the inverse *Fourier* transform on Eq. (1.63). For that purpose, let us rewrite Eq. (1.63) in the form

$$M(k, t > 0) = \frac{\alpha}{4} e^{-i\omega_0 t} \left[\frac{1 - \exp[-i(\Omega - \omega_0)t]}{\Omega - \omega_0} + \frac{1 - \exp[i(\Omega + \omega_0)t]}{\Omega + \omega_0} \right],$$

(1.64)

where we have used the mathematical identity,

$$\frac{1}{\omega_0^2 - \Omega^2} = -\frac{1}{2\Omega} \left(\frac{1}{\Omega - \omega_0} + \frac{1}{\Omega + \omega_0} \right).$$

(1.65)

We leave it as an exercise for the reader to show, using the theory of residues and the recipe given by Eq. (1.32) for inverse Fourier transforms, that

$$\frac{1 - \exp[-i(\Omega - \omega_0)t]}{\Omega - \omega_0} - Fi \rightarrow \frac{-i}{c} \exp\left(\frac{i\omega_0 x}{c} \right) u(x - ct);$$

$$\frac{1 - \exp[i(\Omega + \omega_0)t]}{\Omega + \omega_0} - Fi \rightarrow \frac{i}{c} \exp\left(\frac{-i\omega_0 x}{c} \right) u(x + ct),$$

(1.66)

where $u(x - x_0) = 1$ for x between 0 and x_0 and $= 0$ elsewhere. Thus, using Eq. (1.66) and noting that the exponential function $\exp(-i\omega_0 t)$ is an invariant with respect to the spatial inverse Fourier transformation, we can write M in real-space and time coordinates as

$$M(x, t > 0) = \frac{i\alpha}{4c} \left[\exp[-i(\omega_0 t + k_0 x)] u(x + ct) - \exp[-i(w_0 t - k_0 x)] u(x - ct) \right],$$

(1.67)

where $k_0 = \omega_0/c$.

Using our original definition of B_z in terms of M and its complex conjugate [see Eq. (1.20)] we now write the equation for the time and spatial variation of the magnetic field component of the electromagnetic waves generated by our antenna in vacuum as

$$B_z(x, t > 0) = \frac{\alpha}{2c} \left[\sin(\omega_0 t - k_0 x) u(x - ct) - \sin(\omega_0 t + k_0 x) u(x + ct) \right].$$

(1.68)

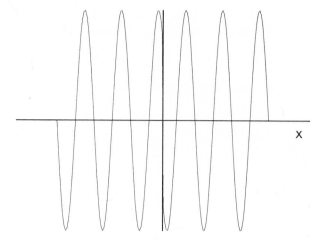

Fig. 1.5. Wave generation described by Eq. (1.68) showing B_z as a function of x at a given time.

Thus, as shown in Fig. 1.5 which gives the shape of the magnetic field of the wave as a function of x, at a given time, electromagnetic waves are generated and propagate away from the antenna in both the backward and forward directions.

1.4. LAPLACE TRANSFORMS AND CAUSALITY

In Section 1.3.3., after Laplace inversion of the Fourier–Laplace transform function $M(\omega,k)$, we found that $M(k, t < 0) = 0$. At that time it was stated without proof that this intuitively reasonable result, which we referred to as causality, was a direct result of having made a *Laplace* transform in time (in which t varies from 0 to ∞) rather than a *Fourier* transform in time (in which t varies from $-\infty$ to $+\infty$). The solution found predicted one wave propagating in the $+x$ direction in the positive $(x > 0)$ half-space of our universe and another wave propagating in the $-x$ direction in the negative $(x < 0)$ half-space. That is, both waves were *outgoing* waves, *leaving* the antenna as expected. We now want to show that if a *Fourier* transform in time is used we will have *both incoming and outgoing* waves predicted to occur in both half-spaces, i.e., we will have standing waves, so that causality can no longer be ascertained. Thus, in a sense, causality manifests itself mathematically in our example in a somewhat subtle manner as the distinction between incoming and outgoing waves. In fact, the example that we now consider is an indication of a more general rule: problems involving causality may be *Laplace-*

transformed in time, but *never Fourier*-transformed in time. Although it might seem intuitively obvious that one should not make a transformation over all of time ($-\infty$ to $+\infty$) for a causal problem that physically involves only part of time (0 to $+\infty$), more than one famous scientist has suffered the embarrassment of having to be pulled out of such a trap by a more knowledgable colleague!

1.4.1. Comparison of Fourier and Laplace Transforms

The recipe for finding the Fourier transform *in time* of a function $f(t)$ has already been briefly discussed in Section 1.3.1 [see Eq. (1.12) and the immediately following discussion]. Thus, the Fourier transform in time of the function $g(t)$ can be written as

$$g(t) - F \rightarrow G(\omega) = \int_{-\infty}^{+\infty} g(t)e^{+i\omega t}\,dt, \tag{1.69}$$

where ω is a *real* number. Application of Eq. (1.69) to our already Fourier-transformed-in-space wave equation, Eq. (1.35), can be done in a straightforward way, and it is left as an exercise for the reader to show that this gives

$$(-\omega^2 + \Omega^2)M = \frac{\alpha}{2}\Omega(2\pi)\delta(\omega - \omega_0), \tag{1.70}$$

from which

$$M(\omega,k) = \frac{\alpha\pi\Omega\delta(\omega - \omega_0)}{\Omega^2 - \omega^2}. \tag{1.71}$$

For comparison with the previous case, in which a *Laplace* transform in time was made on Eq. (1.35), see Eqs. (1.48) and (1.49), respectively.

In analogy with the inverse Fourier transform in space, we can write the inverse Fourier transform *in time* of $G(\omega)$ as

$$G(\omega) - Fi \rightarrow g(t) = \frac{1}{2\pi}\int_{-\infty}^{+\infty} G(\omega)e^{-i\omega t}\,dt. \tag{1.72}$$

Since the mathematics are straightforward, we leave it as an exercise for the reader to show that application of Eq. (1.72) to Eq. (1.71) gives the inverse Fourier transform in time of $M(\omega,k)$ as

$$M(k,t) = \frac{(\alpha\Omega/2)\exp(-i\omega_0 t)}{\Omega^2 - \omega_0^2}. \tag{1.73}$$

Observation of Eq. (1.73) shows that, in contrast with what was found when *Laplace* transforms were made, $M(k,t)$ is not equal to zero for $t < 0$. In addition, in contrast with what is known from the classical solution of this problem, see Eq. (1.36) in Section 1.3.2., Eq. (1.73) gives no transient behavior. Thus, our use of a *Fourier* transform in time has led to a complete loss of causality as well as the disappearance of the transient motion of our *dissipation-free* oscillator.

Let us now perform the inverse Fourier transform in k of Eq. (1.73). To do this we must evaluate

$$M(x,t) = \frac{\alpha}{4c} (K_1 + K_2) \exp(-i\omega_0 t),$$ (1.74)

where

$$K_1 = \frac{1}{2\pi} \int_{-\infty}^{+\infty} \frac{e^{ikx}}{k - \omega_0/c} \, dk,$$ (1.75)

and

$$K_2 = \frac{1}{2\pi} \int_{-\infty}^{+\infty} \frac{e^{ikx}}{k + \omega_0/c} \, dk.$$ (1.76)

Making the change of variable, $k' = k - \omega_0/c$, Eq. (1.75) becomes

$$K_1 = e^{i\omega_0 x/c} \frac{1}{2\pi} \int_{-\infty}^{+\infty} \frac{e^{ik'x}}{k'} \, dk'.$$ (1.77)

Then, using Eq. (1.15) giving the Fourier transform of $\text{sgn}(x)$, we may evaluate Eq. (1.77) as

$$K_1 = \exp\left(\frac{i\omega_0 x}{c}\right) \frac{i}{2} \, \text{sgn}(x).$$ (1.78)

Similarly, for K_2 we obtain

$$K_2 = \exp\left(\frac{-i\omega_0 x}{c}\right) \frac{i}{2} \, \text{sgn}(x).$$ (1.79)

Substituting from Eqs. (1.78) and (1.79) into Eq. (1.74) gives

$$M(x,t) = \frac{i\alpha}{8c} \, \text{sgn}(x) \, \{\exp[-i(\omega_0 t - k_0 x)] + \exp[i(\omega_0 t + k_0 x)]\},$$ (1.80)

where $k_0 = \omega_0/c$. Since, $B_z(x,t) = M + M^* = 2\text{Re}(M)$, we obtain, finally, for the perturbed magnetic field

$$B_z(x,t) = \frac{\alpha}{4c} \text{sgn}(x)[\sin(\omega_0 t - k_0 x) + \sin(\omega_0 t + k_0 x)]. \tag{1.81}$$

Observation of this equation shows that we have a standing-wave pattern for both $x < 0$ and $x > 0$. That is, in both half-spaces we have waves propagating in both directions. This nonphysical result is the consequence of the loss of causality of Eq. (1.73), and as we will show in the next section is due to the loss of the transient motion in Eq. (1.73). Our conclusion is that, in contrast to a Fourier transform in time, a Laplace transform in time yields to causal solutions of linear differential equations describing nondissipative systems.

1.4.2. Causality and Transient Motion

In this section, we will show more precisely how causality is related to the transient motion in a nondissipative system. Let us consider Eq. (1.64) again, which is the causal solution of the differential equation, Eq. (1.35). We are looking, now, for an expression of $M(x,t)$ for $t \to +\infty$, i.e., for the asymptotic behavior of M. If we use the relation

$$\lim_{t \to \infty} \frac{\exp(i\lambda t)}{\lambda} = i\pi\delta(\lambda), \tag{1.82}$$

then Eq. (1.64) becomes

$$\lim_{t \to \infty} M(k,t) = \frac{\alpha}{4} e^{-i\omega_0 t} \left[\frac{1}{\Omega - \omega_0} + i\pi\delta(\Omega - \omega_0) + \frac{1}{\Omega + \omega_0} - i\pi\delta(\Omega + \omega_0) \right]. \tag{1.83}$$

Thus, the inverse Fourier transform of Eq. (1.83) is given by

$$\lim_{t \to \infty} M(k,t) = \frac{\alpha}{4c} e^{-i\omega_0 t}(L_1 + L_2), \tag{1.84}$$

where

$$L_1 = \frac{1}{2\pi} \int_{-\infty}^{+\infty} \left[\frac{1}{k - k_0} + i\pi\delta(k - k_0) \right] e^{ikx}\, dk, \tag{1.85}$$

and

$$L_2 = \frac{1}{2\pi} \int_{-\infty}^{+\infty} \left[\frac{1}{k + k_0} - i\pi\delta(k + k_0) \right] e^{ikx}\, dk. \tag{1.86}$$

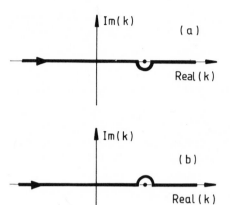

Fig. 1.6. (a) Integration path for L_1 of Eq. (1.85); (b) Integration path for L_2 of Eq. (1.86).

Therefore, the integration path is defined, as shown in Fig. 1.6, and is different for L_1 and L_2. It is interesting, also, to compare Eqs. (1.74) and (1.83). Keeping the transient motion of the oscillator leads to integrals L_1 and L_2 which are different from K_1 and K_2. As we will show, these integrals give causal solutions. It is interesting, also, to note that the distributions appearing in Eqs. (1.85) and (1.86) may be replaced by the distributions

$$P\frac{1}{k-k_0} + i\pi\delta(k-k_0) = \lim_{v\to+0}\frac{1}{k-(k_0+iv)}$$
$$P\frac{1}{k+k_0} - i\pi\delta(k+k_0) = \lim_{v\to+0}\frac{1}{k+(k_0+iv)}.$$

(1.87)

Then, using these expressions in the definitions of L_1 and L_2, the method of residues and Jordan's lemma quickly yield the result

$$M(x, t\to\infty) = \frac{i\alpha}{4c}\left[\exp[-i(\omega_0 t + k_0 x)]\Upsilon(-x) - \exp[i(\omega_0 t - k_0 x)]\Upsilon(x)\right].$$

(1.88)

Comparison of Eq. (1.88) with Eq. (1.67) shows that we have obtained the correct solution for very large values of t. As stated at the beginning of this section, this causal solution has been obtained by keeping the transient solution in our equations. The Laplace transform is an easy mathematical tool by which one can guarantee the retention of causality in a causality problem; this is the main reason for which it should be well understood as being distinct from the Fourier transform. This is essential when considering non-dissipative systems, because they involve integrals having singularities right on the real axis, and the meaning of such integrals is generally not obvious. The Laplace transform in these cases allows one to remove the ambiguity of such integrals. As shown here, the physical meaning of these mathematical subtleties is related to the fact that the general solution of a system of linear equations driven by an external force is the sum of a particular solution, coming from the source term, and of the general solution (the transient response of the system). In the case of nondissipative systems the transient response never disappears and should be kept in the equations since it contains the information necessary for obtaining the causal solutions.

In dissipative systems the situation is different because the damping causes the transient solution to disappear for $t \to \infty$. However, in this situation, the integrals that one must evaluate are no longer singular and can be interpreted in a straightforward manner. In one sense one can say that the damping keeps the information on causality because one knows that the observed signal is damping away in time. It will be particularly interesting for the reader to solve again the propagation problem we have been studying in this chapter but with some arbitrary damping being introduced in Maxwell's equations. The reader will then verify that for $t \to \infty$ the solutions are simpler and the integrals have a simpler meaning; as the damping is made to decrease, one regains expressions such as Eq. (1.87) for the nondissipative case.

1.4.3. Kramers–Kronig Relations

As shown in the previous sections of this chapter, causal solutions are very important to obtain in nondissipative systems. It would be very convenient if by some test we could determine if a function of ω will lead to a causal solution without having to perform the inverse transform. Fortunately, this criteria does exist and is known as the Kramers–Kronig relations.

Let us consider a function $f(t)$ bounded in time such that its Laplace transform $F(\omega)$ exists for $\text{Im}(\omega) \geq 0$:

$$F(\omega) = \int_0^{+\infty} f(t) e^{i\omega t} \, dt. \tag{1.89}$$

Because $F(\omega)$ exists for $\text{Im}(\omega) = 0$, $f(t)$ is called a stable function. Since $F(\omega)$ is a Laplace transform it is analytic in the upper half-plane, $\text{Im}(\omega) \geq 0$. Furthermore, $F(\omega) \to 0$ for $|\omega| \to \infty$, because $f(t < 0) = 0$. Therefore, we may apply Cauchy's formula (see Appendix A.3.)

$$F(\omega) = \frac{1}{2\pi i} \oint_c \frac{F(\omega')}{\omega' - \omega} \, d\omega, \tag{1.90}$$

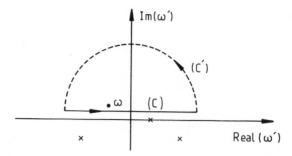

Fig. 1.7. Path of integration c of Eq. (1.90); the dashed half circle corresponds to integration path c'.

where c is the path defined in Fig. 1.7, and $\mathrm{Im}(\omega) \geq 0$. Since $|F(\omega)| \to 0$ for $|\omega| \to \infty$ in the upper half-plane, the integration along the path c' is zero (Theorem 1, Appendix A). Therefore,

$$F(\omega) = \frac{1}{2\pi i} \int_{-\infty}^{+\infty} \frac{F(\omega')}{\omega' - \omega}\, d\omega. \qquad (1.91)$$

Moreover, for $\mathrm{Im}(\omega) = 0$, Eq. (1.91) may be explicitly defined as

$$F(\omega) = \frac{1}{2\pi i}\left[P \int_{-\infty}^{+\infty} \frac{F(\omega')}{\omega' - \omega}\, d\omega + i\pi F(\omega)\right], \qquad (1.92)$$

where P indicates that the integral is taken in the principal-value Cauchy sense (see Appendix A.10). Let us define $R(\omega)$ and $I(\omega)$, the real and imaginary parts of $F(\omega)$, for real ω. Then Eq. (1.92) becomes

$$R(\omega) + iI(\omega) = \frac{1}{2\pi i} P \int_{-\infty}^{+\infty} \frac{R(\omega')}{\omega' - \omega}\, d\omega' + \frac{1}{2\pi} P \int_{-\infty}^{+\infty} \frac{I(\omega')}{\omega' - \omega}\, d\omega'$$

$$+ \frac{1}{2} R(\omega) + \frac{i}{2} I(\omega). \qquad (1.93)$$

Splitting Eq. (1.93) into real and imaginary parts, we obtain

$$R(\omega) = \frac{1}{\pi} P \int_{-\infty}^{+\infty} \frac{I(\omega')}{\omega' - \omega}\, d\omega; \qquad I(\omega) = -\frac{1}{\pi} P \int_{-\infty}^{+\infty} \frac{R(\omega')}{\omega' - \omega}\, d\omega. \qquad (1.94)$$

The relations given by Eq. (1.94) are known as the Kramers–Kronig relations. They show that for the Laplace transform of a stable system the real and imaginary parts are related in a well-defined manner. The Kramers–

Kronig relations may be used for testing the causal nature of a transform, or for the establishment of particular properties of integrals in causal transforms.

1.4.4. The Red-Filter Paradox

Let us suppose that at time $t = 0$ we produce an intense short pulse of light whose intensity can be represented by the equation

$$S(t) = S_0 \delta(t), \tag{1.95}$$

where $\delta(t)$ is the Dirac delta function. The Fourier transform in time of Eq. (1.95) is given by

$$S(\omega) = \int_{-\infty}^{+\infty} S_0 \delta(t) \exp(i\omega t) \, d\omega = S_0. \tag{1.96}$$

Equation (1.96) shows that our light source emits equally at all frequencies, i.e., we have a source of white light.

We now suppose that we want to regard this pulse of light through a filter whose spectral response can be represented by the function $F(\omega)$. To find the light signal transmitted by the filter we first multiply the spectral intensity $S(\omega)$ of the source by the spectral response $F(\omega)$ of the filter to obtain $R(\omega) = S(\omega)F(\omega)$. $R(\omega)$ is the Fourier transform in time of the signal predicted to be transmitted by the filter. We then make the inverse Fourier transform of $R(\omega)$ to find $R(t)$, the real-time signal that the mathematics predicts will be transmitted by the filter. We now want to show that if $F(\omega)$ is chosen in an arbitrary way, even though the choice may seem a priori reasonable, the mathematical result can be without physical significance.

Let us choose a so-called Lorentz spectrum to approximate the transmission characteristics of our filter. Then $F(\omega)$ is given by

$$F(\omega) = \frac{2v}{(\omega - \omega_0)^2 + v^2}. \tag{1.97}$$

Figure 1.2(a) shows a graph of this function. Observation of this figure might very well lead one to believe that Eq. (1.97) represents an acceptable approximation to the actual transmission characteristics of the filter. Using Eq. (1.97), we can immediately write the Fourier transform of the signal which theoretically would be transmitted by a filter having a spectral response given by Eq. (1.97). Doing this gives

$$R(\omega) = S(\omega)F(\omega) = 2vS_0 \left[\frac{1}{(\omega - \omega_0)^2 + v^2} \right]. \tag{1.98}$$

Making an inverse Fourier transform of Eq. (1.98) gives

$$R(t) = \frac{vS_0}{\pi} \int_{-\infty}^{+\infty} \frac{\exp(i\omega t)}{[\omega - (\omega_0 + iv)][\omega - (\omega_0 - iv)]} d\omega, \qquad (1.99)$$

which gives the intensity, as a function of time, of the light signal which is theoretically predicted to be seen by an observer using the filter to observe the light pulse produced at $t = 0$.

Using the method of residues and Jordan's lemma (see Appendix A), the integration on the right-hand side of Eq. (1.99) can be done without difficulty. For $t < 0$ we apply Jordan's lemma in the upper half-plane, $\text{Im}(\omega) > 0$. We leave it as an exercise for the reader to show that this gives

$$R(t < 0) = \frac{vS_0}{\pi} 2\pi i \frac{\exp[-i(\omega_0 - iv)t]}{2iv} = S_0 e^{-i\omega t} e^{vt}. \qquad (1.100)$$

For $t > 0$, we apply Jordan's lemma in the lower half-plane $\text{Im}(\omega) < 0$. Again, we leave it as an exercise for the reader to show that the result is given by

$$R(t > 0) = S_0 e^{-i\omega_0 t} e^{-vt}. \qquad (1.101)$$

Finally, observation of Eqs. (1.100) and (1.101) shows that they can be compactly combined into a single equation

$$R(t) = S_0 e^{-i\omega_0 t} e^{-v|t|}, \qquad (1.102)$$

which is valid for all values of time. Both Eqs. (1.100) and (1.102) show that our mathematical exercise has given us a theoretical result that is not physically reasonable: if we use the filter to look at the light source at negative times $(t < 0)$, i.e., before the light pulse is produced at $t = 0$, we will see light! That is, the result violates causality. If ω_0 (the frequency of maximum transmissivity of the filter) is chosen to correspond to red light, the theoretical result predicts that red light should be seen coming through the filter before the white-light pulse is produced. This absurdity is sometimes referred to as the red-filter paradox.

The purpose of the preceding exercise was to show the importance of the Kramers–Kronig relations in connection with causality. We leave it as an exercise for the reader to show that $F(\omega)$ does not satisfy the Kramers–Kronig relations. Thus, $F(\omega)$ is not a causal function and it is, therefore, not surprising that the result violates causality.

We now want to consider a spectral response function $F'(\omega)$ which *is* a causal function. This function is given by

$$F'(\omega) = \frac{i}{\omega - (\omega_0 - iv)}.\qquad(1.103)$$

The Fourier transform of the signal predicted to be seen [see Eq. (1.98)], using a filter having the spectral response function given by Eq. (1.103) is given by

$$R(\omega) = \frac{iS_0}{\omega - (\omega_0 - iv)},\qquad(1.104)$$

the inverse Fourier transform of which is given by

$$R(t) = \frac{iS_0}{2\pi} \int_{-\infty}^{+\infty} \frac{\exp(-i\omega t)}{\omega - (\omega_0 - iv)}\, d\omega.\qquad(1.105)$$

We leave it for the reader to show that application of Jordan's lemma and the method of residues to the integral on the right-hand side of Eq. (1.105) gives the results

$$R(t < 0) = 0,\qquad(1.106)$$

$$R(t > 0) = S_0\exp(-i\omega t)\exp(-vt).$$

Thus, using a causal function, we have obtained a causal result—no light is predicted to be seen through the filter before the white-light pulse is produced.

The reader may wonder why red light is seen through the red filter after the source of white light is turned off. The explanation is that the filter has been excited by the flash of white light and that this excitation fades away only in a finite time. This property is, in fact, the result of the modelization of the real physical properties of the filter by $F'(\omega)$. These properties must be carefully studied in order to be able to make the correct choice of $F'(\omega)$.

It is of interest to note that $F'(\omega)$ is related to $F(\omega)$ in a simple way. To show this, we first multiply both the numerator and denominator of the right-hand side of Eq. (1.103) by $[\omega - (\omega_0 + iv)]$. It is not difficult to show then that $F'(\omega)$ can be written as

$$F'(\omega) = \frac{v}{(\omega - \omega_0)^2 + v^2} + i\frac{\omega - \omega_0}{(\omega - \omega_0)^2 + v^2}.\qquad(1.107)$$

Except for a factor of two, the first term on the right-hand side of Eq. (1.107) is just $F(\omega)$. Thus, we see that, by adding an imaginary part to $F(\omega)$, we have rendered it causal. We leave it for the reader to show that the imaginary part of $F'(\omega)$ can be expressed as

$$\mathrm{Im}[F'(\omega)] = \frac{\omega - \omega_0}{(\omega - \omega_0)^2 + v^2} = -\frac{v}{\pi} \int_{-\infty}^{+\infty} \frac{d\omega'}{[(\omega' - \omega_0)^2 + v^2](\omega' - \omega_0)}.$$

(1.108)

Equation (1.108) shows that the imaginary part of $F'(\omega)$ is related to its real part by the recipe given by the Kramers–Kronig relations, thus proving that $F'(\omega)$ is a causal function, as we have already claimed.

1.5. HILBERT TRANSFORMS

Hilbert transforms are defined as integrals of the type

$$Q(z) = \int_{-\infty}^{+\infty} \frac{f(u)}{u - z} \, du.$$

(1.109)

These integrals are frequently encountered in physics, particularly in plasma physics in the linear theory of waves in hot plasmas. In Section 1.5.1, we will show that this kind of transform appears when one calculates the average of a large ensemble of oscillators simulating the response of a plasma to an external perturbation. In the following sections, we will then discuss from a physicist's point of view the most important mathematical properties of Hilbert transforms.

1.5.1. Averages

Let us again consider Eq. (1.64). We may consider $M(k,t)$ to be the solution of the differential equation for an undamped harmonic oscillator which is driven at a frequency $\omega_0/2\pi$ and which has a natural frequency Ω. Very often in physics one must consider not a single oscillator but rather a large assembly of oscillators, each having a different frequency. Let $f(\Omega)$ be the distribution function of these oscillators in frequency. Then

$$N = \int_{-\infty}^{+\infty} f(\Omega) \, d\Omega$$

(1.110)

is the total number of oscillators in the assembly and

$$dN = f(\Omega) \, d\Omega,$$

(1.111)

is the number of oscillators having a natural frequency between Ω and $\Omega + d\Omega$.

We are now interested in computing the average value of $M(\Omega,t)$. This is given by

$$\langle M \rangle = \frac{1}{N} \int_{-\infty}^{+\infty} M(\Omega,t) f(\Omega) \, d\Omega. \tag{1.112}$$

It is often interesting to know this average for times much greater than the average period of the oscillator assembly. In order to find such an asymptotic average for M, we again use the relation

$$\lim_{t \to \infty} \frac{\exp(i\lambda t)}{\lambda} = i\pi\delta(\lambda). \tag{1.113}$$

Doing this, we find, for $t \to \infty$ that

$$\langle M(\Omega, t \to \infty) \rangle = \frac{\alpha}{4N} e^{-i\omega_0 t} [H'(\omega_0) + H''(\omega_0)], \tag{1.114}$$

where

$$H'(\omega_0) = \int_{-\infty}^{+\infty} f(\Omega) \left[P \frac{1}{\Omega - \omega_0} + i\pi\delta(\Omega - \omega_0) \right] d\Omega, \tag{1.115}$$

and

$$H''(\omega_0) = \int_{-\infty}^{+\infty} f(\Omega) \left[P \frac{1}{\Omega + \omega_0} - i\pi\delta(\Omega + \omega_0) \right] d\Omega. \tag{1.116}$$

As shown previously

$$P \frac{1}{\Omega \mp \omega_0} \pm i\pi\delta(\Omega - \omega_0) = \lim_{v \to 0} \frac{1}{\Omega \mp (\omega_0 + iv)}. \tag{1.117}$$

Therefore

$$H'(\omega_0) = \lim_{v \to 0} \int_{-\infty}^{+\infty} \frac{f(\Omega)}{\Omega - (\omega_0 + iv)} \, d\Omega \tag{1.118}$$

and

$$H''(\omega_0) = H'(-\omega_0)^*. \tag{1.119}$$

It is important to note that H' is a perfectly defined integral for any real value of ω_0 and that this integral is different from the one we would have obtained if we had ignored the transient motion. In that case it is easy to show that the result is given by

$$\langle M(\mathbf{\Omega}, t \to \infty) = \frac{\alpha}{4\mathcal{N}} e^{-i\omega_0 t} [G'(\omega_0) + G'(-\omega_0)], \qquad (1.120)$$

where

$$G'(\omega_0) = P \int_{-\infty}^{+\infty} \frac{f(\mathbf{\Omega})}{\mathbf{\Omega} - \omega_0} d\mathbf{\Omega}. \qquad (1.121)$$

The two integrals are different since

$$H'(\omega_0) = G'(\omega_0) + i\pi f(\omega_0). \qquad (1.122)$$

The last remark that needs to be made is that H' and G' are both Hilbert transforms and, as we have shown here, although their formal definition is the same, being given by

$$\int_{-\infty}^{+\infty} \frac{f(\mathbf{\Omega})}{\mathbf{\Omega} - \omega_0} d\mathbf{\Omega}, \qquad (1.123)$$

their explicit definition is different. This shows that in any particular physical problem where this kind of integral appears care must be taken in order to preserve causality.

1.5.2. Basic Properties of Hilbert Transforms

We now want to look at some of the basic properties of Hilbert transforms [see Eq. (1.09)]. First, we will show that this definition is purely formal, that is, Eq. (1.09) is not the definition of a single function. In order to simplify we will require that $f(u)$ in Eq. (1.09) be an entire function that is integrable in the range $-\infty$ to $+\infty$.

1.5.2.1. The Functions, Q^+, Q^- and Q^0. Let us define the function $Q^0(z)$, of real z as

$$Q^0(z) = P \int_{-\infty}^{+\infty} \frac{\varphi(u)}{u - z} du, \qquad \text{Im}(z) = 0. \qquad (1.124)$$

It is interesting to note that the integrand is perfectly regular at $u = z$. To show that, we may write $Q^0(z)$ as

$$Q^0(z) = P \int_{-\infty}^{+\infty} \frac{\varphi(u) - \varphi(z)}{u - z} du + \varphi(z) P \int_{-\infty}^{+\infty} \frac{1}{u - z} du. \qquad (1.125)$$

Since for $\mathrm{Im}(z) = 0$

$$P \int_{-\infty}^{+\infty} \frac{1}{u - z} \, du = 0, \tag{1.126}$$

$Q^0(z)$ is also defined by

$$Q^0(z) = P \int_{-\infty}^{+\infty} \frac{\varphi(u) - \varphi(z)}{u - z} \, du. \tag{1.127}$$

Since $\varphi(u)$ is an entire function

$$\varphi(u) - \varphi(z) = \varphi'(z)(u - z) + \tfrac{1}{2}\varphi''(z)(u - z)^2 + \cdots, \tag{1.128}$$

so that the integrand is perfectly regular for $u = z$. $Q^0(z)$ is therefore defined by the proper integral

$$Q^0(z) = \int_{-\infty}^{+\infty} \frac{\varphi(u) - \varphi(z)}{u - z} \, du. \tag{1.129}$$

Let us define a second integral $Q^+(z)$ as

$$Q^+(z) = \int_{-\infty}^{+\infty} \frac{\varphi(u)}{u - z} \, du, \qquad \mathrm{Im}(z) > 0. \tag{1.130}$$

Using the same procedure as for Q^0, we may write

$$Q^+(z) = \int_{-\infty}^{+\infty} \frac{\varphi(u) - \varphi(z)}{u - z} \, du + \varphi(z) \int_{-\infty}^{+\infty} \frac{1}{u - z} \, du. \tag{1.131}$$

However, since $\mathrm{Im}(z) > 0$

$$\int_{-\infty}^{+\infty} \frac{1}{u - z} \, du = i\pi, \tag{1.132}$$

so that

$$Q^+(z) = \int_{-\infty}^{+\infty} \frac{\varphi(u) - \varphi(z)}{u - z} \, du + i\pi\varphi(z). \tag{1.133}$$

Similarly, we may define the function Q^- as

$$Q^-(z) = \int_{-\infty}^{+\infty} \frac{\varphi(u)}{u - z} \, du, \qquad \mathrm{Im}(z) < 0, \tag{1.134}$$

and show that

$$Q^-(z) = \int_{-\infty}^{+\infty} \frac{\varphi(u) - \varphi(z)}{u - z} \, du - i\pi\varphi(z). \tag{1.135}$$

Now we will show that the three functions, Q^0, Q^+, and Q^-, although all derived from the same formal Hilbert transform, are not the same function. First, it is easy to show that $Q^+(z)$ and $Q^-(z)$ are analytic in the upper and lower complex-z planes, respectively. One way to do that is as follows: dQ^+/dz may be calculated as

$$\frac{dQ^+}{dz} = \frac{d}{dz} \int_{-\infty}^{+\infty} \frac{\varphi(u)}{u - z} \, du = -\int_{-\infty}^{+\infty} \frac{\varphi(u)}{(u - z)^2} \, du. \tag{1.136}$$

Integration of Eq. (1.136) by parts gives

$$\frac{dQ^+}{dz} = \int_{-\infty}^{+\infty} \frac{\varphi'(u)}{(u - z)} \, du. \tag{1.137}$$

Now we may examine the behavior of $Q^\pm(z)$ when $\mathrm{Im}(z) \to 0$. From the expressions we have obtained for $Q^\pm$ we may write

$$\lim_{z \to 0^+} Q^+(z) = Q^0(z) + i\pi\varphi(z),$$

$$\lim_{z \to 0^-} Q^-(z) = Q^0(z) - i\pi\varphi(z). \tag{1.138}$$

Therefore, the functions Q^+, Q^-, and Q^0 are three distinct functions.

1.5.2.2. Analytic Continuation of Hilbert Transforms. As shown in the previous section, $Q(z)$ represents the three distinct functions

$$Q^+(z), \qquad \text{defined for } \mathrm{Im}(z) > 0,$$

$$Q^0(z), \qquad \text{defined for } \mathrm{Im}(z) = 0, \tag{1.139}$$

$$Q^-(z), \qquad \text{defined for } \mathrm{Im}(z) < 0.$$

Moreover, $Q^\pm$ are analytic in their respective half-planes of definition. It would be interesting to perform an analytic continuation of $Q^\pm$ to the *full* complex plane. As the reader is probably aware, it is often necessary when one moves an integration path in the complex plane that the functions involved be defined in the full complex plane. Such an analytic continuation is easy to do using the theorems given in Appendix A. For example, let us define

$$P(z) = Q^-(z) + 2\pi i \varphi(z), \qquad \text{for } \text{Im}(z) < 0. \tag{1.140}$$

Then

$$\lim_{\text{Im}(z) \to 0^-} P(z) = \lim_{\text{Im}(z) \to 0^+} Q^+(z). \tag{1.141}$$

From Theorem 3 of Appendix A, it follows that $P(z)$ is the analytic continuation of $Q^+(z)$ in the lower half-plane. Therefore, the definition of Q^+ as an entire function in the complex-z plane can be expressed as

$$Q^+(z) = \int_{-\infty}^{+\infty} \frac{\varphi(u)}{u - z} \, du, \qquad \text{Im}(z) > 0,$$

$$Q^+(z) = P \int_{-\infty}^{+\infty} \frac{\varphi(u)}{u - z} \, du + i\pi\varphi(z), \qquad \text{Im}(z) = 0, \tag{1.142}$$

$$Q^+(z) = \int_{-\infty}^{+\infty} \frac{\varphi(u)}{u - z} \, du + 2i\pi\varphi(z), \qquad \text{Im}(z) < 0,$$

and similarly for $Q^-(z)$

$$Q^-(z) = \int_{-\infty}^{+\infty} \frac{\varphi(u)}{u - z} \, du - 2i\pi\varphi(z), \qquad \text{Im}(z) > 0,$$

$$Q^-(z) = P \int_{-\infty}^{+\infty} \frac{\varphi(u)}{u - z} \, du - i\pi\varphi(z), \qquad \text{Im}(z) = 0, \tag{1.143}$$

$$Q^-(z) = \int_{-\infty}^{+\infty} \frac{\varphi(u)}{u - z} \, du, \qquad \text{Im}(z) < 0.$$

A more compact definition of these functions may be given by using the Landau paths L^+ and L^- as defined in Fig. 1.8. Doing this gives

$$Q^\pm(z) = \int_{L^\pm} \frac{\varphi(u)}{u - z} \, du. \tag{1.144}$$

Finally, using the definitions of $Q^\pm$ and noticing that Q^0 is also an entire function of z, we establish the so-called Plemelj formulas

$$Q^+(z) - Q^-(z) = 2i\pi\varphi(z),$$
$$Q^+(z) + Q^-(z) = 2Q^0(z), \tag{1.145}$$

defined everywhere in the complex z-plane.

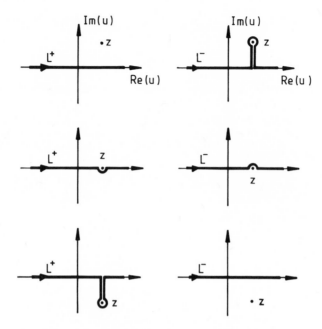

Fig. 1.8. Landau paths $L^\pm$ as a function of $\mathrm{Im}(z)$.

1.5.2.3. Hilbert Transform as a Fourier–Laplace Transform. The Hilbert transforms, $Q^\pm(z)$, can be seen to be, also, Fourier–Laplace transforms. Noting that

$$\frac{1}{u-z} = i \int_0^{+\infty} e^{i(z-u)p}\,dp, \qquad \mathrm{Im}(z) > 0, \tag{1.146}$$

and using this integral representation of $1/(u-z)$ in the definition of $Q^+(z)$,
one obtains

$$Q^+(z) = i \int_0^{+\infty} \left[\int_{-\infty}^{+\infty} \varphi(u) e^{-ipu}\,du \right] e^{ipz}\,dp. \tag{1.147}$$

Thus, $Q^+(z)$ is i times the Laplace transform in p of the Fourier transform
in u of the function $\varphi(u)$. This property again illustrates the relation between
Hilbert transforms and causality. It is useful, also, for computing some Hilbert
transforms, in particular the plasma dispersion function $Z(z)$. In fact, the
plasma dispersion function $Z(z)$, as tabulated by Fried and Conte (1960), is
defined as

$$Z^+(z) = \pi^{-1/2} \int_{-\infty}^{+\infty} \frac{e^{-u^2}}{u-z}\,du, \qquad \mathrm{Im}(z) > 0. \tag{1.148}$$

As a Fourier–Laplace transform, Eq. (1.148) may be written for $\text{Im}(z) > 0$ as

$$Z^+(z) = i\pi^{-1/2} \int_0^{+\infty} \left[\int_{-\infty}^{+\infty} e^{-u^2} e^{-ipu} \, du \right] e^{ipz} \, dp. \qquad (1.149)$$

Using Eq. (1.15) the Fourier transform is easily evaluated to give

$$Z^+(z) = i \int_0^{+\infty} e^{-p^2/4} e^{ipz} \, dp. \qquad (1.150)$$

Introducing the new variable, $v = p/2 - iz$, Eq. (1.150) becomes

$$Z^+(z) = 2ie^{-z^2} \int_{-iz}^{+\infty - iz} e^{-v^2} \, dv. \qquad (1.151)$$

Considering the closed path defined by Fig. 1.9 and applying Cauchy's theorem we have

$$\int_{-iz}^{R-iz} e^{-v^2} \, dv + \int_{R-iz}^{R} e^{-v^2} \, dv + \int_{R}^{0} e^{-v^2} \, dv + \int_0^{-iz} e^{-v^2} \, dv = 0. \qquad (1.152)$$

Moreover, since

$$\lim_{R \to \infty} \int_{R-iz}^{R} e^{-v^2} \, dv = 0, \qquad (1.153)$$

using the fact that $\exp(-v^2)$ is an even function of v, we obtain

$$\int_{-iz}^{\infty - iz} e^{-v^2} \, dv = \int_0^{\infty} e^{-v^2} \, dv + \int_0^{iz} e^{-v^2} \, dv. \qquad (1.154)$$

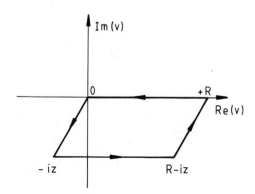

Fig. 1.9. The path of integration for Eq.(1.152).

Finally,

$$\mathcal{Z}^+(z) = i\pi^{1/2}e^{-z^2}[1 + \text{erf}(iz)], \tag{1.155}$$

where

$$\text{erf}(z) = \frac{\pi^{1/2}}{2} \int_0^z e^{-u^2} \, du. \tag{1.156}$$

It should be noted that the result we have obtained is an entire function of z. Although we started the calculation in the half-plane of definition of $\mathcal{Z}^+$, we have obtained the analytic continuation which is valid everywhere in the complex-z plane.

1.5.2.4. Asymptotic Expansion of Hilbert Transforms. One frequently needs approximate expressions of Hilbert transforms for quick calculations of dispersion relations in hot plasmas. Asymptotic expansions of $Q^\pm(z)$, valid for large values of z and for $|\text{Im}(z)| \ll |\text{Re}(z)|$, are useful for that. For small values of $\text{Im}(z)$, we may define $Q^\pm$ as

$$Q^\pm(z) = P \int_{-\infty}^{+\infty} \frac{\varphi(u)}{u - z} \, du \pm i\pi\varphi(z). \tag{1.157}$$

For $z > u$ we may write

$$\frac{1}{u - z} = -\frac{1}{z} \frac{1}{1 - u/z} = -\frac{1}{z}\left(1 + \frac{u}{z} + \frac{u^2}{z^2} + \cdots\right), \tag{1.158}$$

and assuming that $\varphi(u) \simeq 0$ for $u \geqslant z$, $Q^\pm$ can be approximated as

$$Q^\pm(z) \simeq -\frac{1}{z} \int_{-\infty}^{+\infty} \left[\varphi(u) + \frac{u\varphi(u)}{z} + \frac{u^2\varphi(u)}{z^2} + \cdots\right] du \pm i\pi\varphi(z),$$

or alternatively as

$$Q^\pm(z) \simeq -\frac{1}{z}\left(a^{(0)} + \frac{a^{(1)}}{z} + \frac{a^{(2)}}{z^2} + \cdots\right) \pm i\pi\varphi(z), \tag{1.159}$$

where the $a^{(n)}$ are the moments of the function $\varphi(u)$

$$a^{(n)} = \int_{-\infty}^{+\infty} \varphi(u)u^n \, du. \tag{1.160}$$

Since the expansion of $1/(u - z)$ in powers of u/z is not uniformly convergent on the integration path, the expansion we have obtained is not convergent. In fact, the reader may show that the expansion obtained for a particular $\varphi(u)$ is divergent. However, if z is large enough, the sum

$$S_n(z) = \sum_{m=0}^{n} \frac{a^{(m)}}{z^m}, \tag{1.161}$$

may be very close to $\text{Re}[Q^+(z)]$. But one should keep in mind that for a given value of z, S_n diverges for $n > \mathcal{N}(z)$. Therefore, increasing the number of terms in the series does not always improve the approximation. This is true, in particular, when $|z|$ is not very large.

In the case of the plasma dispersion function, the asymptotic expansion is given by

$$Z^+(z) \simeq -\frac{1}{z} - \frac{1}{2z^3} - \frac{3}{4z^5} - \cdots + i\pi^{1/2}e^{-z^2}, \tag{1.162}$$

valid for $|z| \gg 1$ and $|\text{Im}(z)| \ll |\text{Re}(z)|$

1.5.3. Other Properties of Hilbert Transforms

In this section, we give the reader some properties of Hilbert transforms that are useful in particular applications. The illustrations that we will use for establishing these properties of Hilbert transforms are based on the properties already stated in the preceding sections.

1.5.3.1. Properties for Complex z. First, we recall the Plemelj equations

$$\begin{align}
Q^+(z) - Q^-(z) &= 2i\pi\varphi(z), \\
Q^+(z) + Q^-(z) &= 2Q^0(z).
\end{align} \tag{1.163}$$

Then, using the fact that $Q^\pm$ are Fourier–Laplace transforms, we have

$$Q^\pm(z) = \pm i \int_0^{+\infty} e^{\pm zp} \left[\int_{-\infty}^{+\infty} \varphi(u)e^{\mp iup} \, du \right] dp, \tag{1.164}$$

these definitions being valid for Q^+ for $\text{Im}(z) > 0$, and for Q^- for $\text{Im}(z) < 0$.

1.5.3.2. Properties for Real z. For real values of z, one has

$$Q^\pm(z) = [Q^\mp(z)]^*, \tag{1.165}$$

and using the Kramers–Kronig relations one has

$$\mathrm{Re}[Q^{\pm}(z)] = P \int_{-\infty}^{+\infty} \frac{\varphi(z')}{z'-z} \, dz' = \pm\left(\frac{1}{\pi}\right) P \int_{-\infty}^{+\infty} \frac{\mathrm{Im}[Q^{\pm}(z')]}{z'-z} \, dz',$$

$$\mathrm{Im}[Q^{\pm}(z)] = \pm\pi\varphi(z) = \mp\left(\frac{1}{\pi}\right) P \int_{-\infty}^{+\infty} \frac{\mathrm{Re}[Q^{\pm}(z')]}{z'-z} \, dz'.$$

(1.166)

Moreover, if $\varphi(u)$ is an even function, then

$$Q^{\pm}(-z) = -Q^{\mp}(z); \tag{1.167}$$

$$Q'^{\pm}(-z) = +Q^{\mp}(z); \tag{1.168}$$

$$Q^{\pm}(-z^*) = -[Q^{\pm}(z)]^*; \tag{1.169}$$

$$Q'^{\pm}(-z^*) = [Q'^{\pm}(z)]^*. \tag{1.170}$$

1.6. APPENDIX A: FUNCTIONS OF COMPLEX VARIABLES

The theory of functions of complex variables is used extensively in plasma physics. In this appendix we give basically a *reference list* of definitions and theorems, relative to the theory of functions of complex variables, which are presently in common use in plasma physics. For more general and detailed treatments of the theory of functions of complex variables, we refer the reader to such books as Whittaker and Watson (1902), Roos (1969), and Rudin (1970).

A.1. Definitions

Let C be the set of the complex numbers in the entire complex plane defined by the x and iy coordinates, and let $z = x + iy$ represent any point in C. Let $f(z)$ be an application from C to C, i.e., let $f(z)$ be a function of the complex variable z. We then say that $f(z)$ is *analytic at point* z_0 in C if

$$f'(z_0) = \frac{df(z)}{dz}\bigg|_{z=z_0} = \lim_{z \to z_0} \left[\frac{f(z) - f(z_0)}{z - z_0}\right] \tag{1.171}$$

exists *and* if this limit is independent of $\arg(z - z_0)$.

Let us define a *region* to be a nonempty, connected subset of the complex plane. Then, if $f(z)$ is *analytic and a single-valued function* of z for any z belonging to a region D of C, $f(z)$ is said to be *holomorphic* or *analytic in* D. We will denote the class of all holomorphic functions in D as $A(D)$. If $D = C$, then $f(z)$ is said to be an *entire function*. For n being any positive integer, $f(z) = z^n$ is single-valued and analytic in the whole complex plane and is thus said to be an entire function. On the other hand, $f(z) = z^n$ where n is any negative integer, is single-valued, and is analytic in the whole complex plane, except at $z = 0$ and is, thus, said to be holomorphic in D where D is a region consisting of all the points in the entire complex plane except $z = 0$.

A.2. Cauchy–Riemann Conditions

Let $f(z) = g(x, y) + ih(x, y)$, where $z = x + iy$ and g and h are real functions. If $f(z)$ belongs to $A(D)$, then the real and imaginary parts of $f(z)$ satisfy the so-called Cauchy–Riemann conditions

$$\frac{\partial g}{\partial x} = \frac{\partial h}{\partial y}, \qquad \frac{\partial g}{\partial y} = -\frac{\partial h}{\partial x}. \tag{1.172}$$

The Cauchy–Riemann conditions represent the necessary and sufficient conditions for analyticity. It is interesting to note that the Cauchy–Riemann equations imply that

$$\frac{\partial^2 g}{\partial x^2} + \frac{\partial^2 g}{\partial y^2} = 0, \qquad \frac{\partial^2 h}{\partial x^2} + \frac{\partial^2 h}{\partial y^2} = 0. \tag{1.173}$$

Therefore, any analytical function is a solution of the so-called two-dimensional Laplace equation and is thus a *harmonic* function. However, the reciprocal is not true, i.e., not all harmonic functions are analytic.

Application of the Cauchy–Riemann conditions to the function $f(z) = x + iy$ shows immediately that this function is analytic. However, application of the Cauchy–Riemann conditions to the complex conjugate of this function, i.e., to $f(z) = z^* = x - iy$ shows that this function is not analytic since $\partial g/\partial x = 1$, whereas $\partial h/\partial y = -1$. The Cauchy–Riemann conditions can be applied to show that, in general, if $f(z)$ is analytic, then $f(z^*)$ is not analytic.

A.3. Cauchy's Theorem, Morera's Theorem, and Cauchy's Formula

Cauchy's Theorem: If $f(z)$ belongs to $A(D)$ and if $f'(z)$ is continuous in D, then for any closed path Γ in D we have

$$\oint_{\Gamma} f(z) \, dz = 0. \tag{1.174}$$

Morera's Theorem: If $f(z)$ is a continuous, complex function in D, and if $f(z)$ is such that Eq. (1.174) is satisfied for every closed triangle in D, then $f(z)$ is analytic in D.

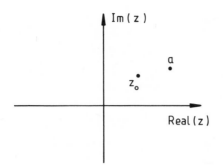

Fig. 1.10. Geometry of a Taylor expansion.

Cauchy's Formula: If $f(z)$ belongs to $A(D)$ and if $f'(z)$ is continuous in D, then

$$\frac{1}{2\pi i} \oint_\Gamma \frac{f(z)}{z - z_0}\, dz = f(z_0),\qquad(1.175)$$

for any closed path Γ in D around the point z_0. Note that Eq. (1.175) is useful not only for evaluating closed-path integrals in the complex plane but also for making integral representation of analytic functions.

Corollary 1: If $f(z)$ belongs to $A(D)$, then all derivatives of $f(z)$ exist and are analytic in D.

A.4. Taylor Expansion of an Analytic Function

Let us consider Fig. 1.10, where we have an analytic function in the region D, and let a and z_0 be two points in D. Taylor's expansion allows us to represent $f(z_0)$ in terms of the derivatives of $f(z)$ at point a as the infinite series

$$f(z_0) = \sum_{n=0}^{n=\infty} f^{(n)}a(z_0 - a)^n,\qquad(1.176)$$

where $f^{(n)} = d^n f/dz^n$.

A.5. Laurent Expansion

Let us consider Fig. 1.11 where we have a function $f(z)$ that has an isolated singularity at the point a but which is analytic at all other points of the region D in which a is located. Then we may express $f(z_0)$ by an expansion around a called the Laurent expansion, which is defined by the equation

$$f(z_0) = \sum_{n=0}^{\infty} A_n(z_0 - a)^n + \sum_{m=1}^{\infty} \frac{B_m}{(z_0 - a)^m},\qquad(1.177)$$

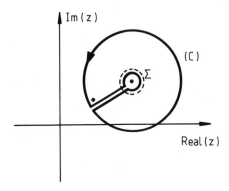

Fig. 1.11. Geometry of a Laurent expansion.

where

$$A_n = \frac{1}{2\pi i} \oint_c \frac{f(z)}{(z - z_0)^n}\, dz;$$

$$B_m = \frac{1}{2\pi i} \oint_\Sigma f(z)(z - z_0)^{m-1}\, dz,$$

(1.178)

for the paths of integration C and Σ shown in Fig. 1.11.

A.6. Classification of Isolated Singularities

The Laurent expansion offers an easy method for the classification of isolated singularities. This expansion consists of two parts: the sum over the A_n terms and the sum over the B_m terms. When all the B_m terms are zero, then the function is analytic with no singularities, and Laurent's expansion is the same as Taylor's expansion. *The singularities of the function are therefore represented by the B_m terms.* If the sum over the B_m terms is finite, i.e., limited to $m = M$, then the singular point a is said to be a *pole of order M.* If the sum over the B_m terms is infinite, then the point a is said to be an *essential singularity.*

A.7. Meromorphic Functions

A function $f(z)$ which has poles only in a region D of the complex plane is said to be *meromorphic in D.* Rational fractions are examples of functions that are meromorphic in the full complex plane.

A.8. Residues

The coefficient B_1 of the first term of the singular part of the Laurent expansion [see Eqs. (1.177) and (1.178)] is called the residue and is defined by

$$B_1 = \frac{1}{2\pi i} \oint_\Sigma f(z)\, dz. \tag{1.179}$$

One notes that

$$\text{Residue}[f(z),a] = B_1. \tag{1.180}$$

Residues Theorem: If $f(z)$ is a meromorphic function in the region D, then for any closed path Σ in D

$$\oint_\Sigma f(z)\, dz = 2\pi i \sum_m \text{Residues}[f(z),a_m], \tag{1.181}$$

where the a_m are the poles inside the closed path Σ.

Because $f(z)$ is a meromorphic function the residues $B_1(a_m)$ are easily computed using the equation

$$B_1 a_m = \frac{1}{(n-1)!} \left[\frac{d^{n-1}}{dz^{n-1}} (z-a_m)^n f(z) \right]_{z=a_m} \tag{1.182}$$

where n is the order of the pole a_m.

Equation (1.182) can be easily verified by using the definition of B_m given by Eq. (1.178) and the fact that around a pole a_m of order n, $f(z)$ may be expressed by

$$f(z) = \frac{g(z)}{(z-a_m)^n}, \tag{1.183}$$

where $g(z)$ is an analytic function.

A.9. Transformation of the Path of Integration

The use of either Cauchy's equation or of the residues theorem requires that the paths of integration be closed. Very often in plasma physics, however, the paths of integration required in nontrivial problems are not closed paths but are, rather, open paths such as $(-\infty + ia, +\infty + ia)$, which is a path parallel to the real axis. Therefore, because of the extreme usefulness of Cauchy's equation and of the residues theorem in problems involving complex variables, we give two very useful theorems for transforming open-path integrals into closed-path integrals.

Let $f(z)$ be a function that is meromorphic in the entire complex plane. We define $\mathcal{J}$ as

$$\mathcal{J} = \oint_c f(z)\, dz, \tag{1.184}$$

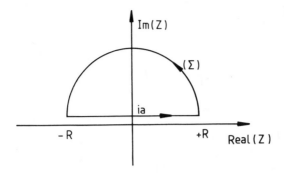

Fig. 1.12. Path of integration of Eq. (1.184).

where c is the closed path shown in Fig. 1.12. This path is made up of two parts: the path, $(-R + ia, +R + ia)$, and the half-circle Σ of radius R centered at $z = ia$. Since this is a *closed*-path integration, we may use the residues theorem to say that

$$\mathcal{J} = 2\pi i \sum_m \text{Residues}[f(z), a_m]. \tag{1.185}$$

We now define two integrals, the *sum* of which is equal to the closed-path integral given by Eq. (1.184) but which, *separately*, are *open*-path integrals:

$$\mathcal{J}_1 = \int_{-R+ia}^{+R+ia} f(z)\, dz;$$
$$\mathcal{J}_2 = \int_\Sigma f(z)\, dz. \tag{1.186}$$

There are two commonly used theorems, each of which, under certain rather general conditions, causes $\mathcal{J}_2$ to be zero. When $\mathcal{J}_2 = 0$, since $\mathcal{J}_1 + \mathcal{J}_2 = \mathcal{J}$, we see that $\mathcal{J}_1 = \mathcal{J}$. That is, we are effectively able to transform an open-path integral into a closed-path integral, so that the very powerful Cauchy theorem, Cauchy formula, and residues theorem can be used to evaluate $\mathcal{J}_1$, a type of open-path integral frequently encountered in plasma physics problems. The two theorems that allow us to do that are:

Theorem 1: If in the half-plane defined by $\text{Im}(z) \geq a$, $zf(z) \to 0$ uniformly as $|z| \to \infty$, then

$$\lim_{R \to \infty} \int_\Sigma f(z)\, dz = 0. \tag{1.187}$$

Theorem 2 (Jordan Lemma): If in the half-plane defined by $I(z) \geq a$, for *any real* $p > 0$, $f(z) \to 0$ uniformly as $|z| \to \infty$, then

$$\lim_{R \to \infty} \int_\Sigma f(z) e^{ipz}\, dz = 0. \tag{1.188}$$

A.10. Principal Part (in the Cauchy Sense)

An integral such as

$$I = \int_{-\infty}^{+\infty} \frac{1}{x}\, dx, \tag{1.189}$$

is not defined as a *proper* integral in the *Riemann* sense. However, this integral can be defined as a *principal-part* integral in the *Cauchy* sense if

$$I = \lim_{\substack{\varepsilon \to 0 \\ R \to \infty}} \left[\int_{-R}^{-\varepsilon} \frac{1}{x}\, dx + \int_{+\varepsilon}^{+\infty} \frac{1}{x}\, dx \right] \tag{1.190}$$

exists. In the case of the function $1/x$, this limit exists and is 0. Therefore

$$I = P \int_{-\infty}^{+\infty} \frac{1}{x}\, dx = 0, \tag{1.191}$$

where P indicates that the integral is taken in the principal-value Cauchy sense.

More generally, if the integrand $f(x)$ has a singularity at $x = c$, the principal part of the integral in the Cauchy sense is defined by

$$I = P \int_a^b f(x)\, dx = \lim_{\varepsilon \to 0} \left[\int_a^{c-\varepsilon} f(x)\, dx + \int_{c+\varepsilon}^b f(x)\, dx \right], \tag{1.192}$$

if the limit exists.

A.11. Analytic Continuation

Very often in physics the original definition of a particular function is limited to a smaller region of the complex plane than the region where this function may be defined. In order to simplify (or, in some cases to make possible) some calculations, we need to extend the region of definition of the function. This process, for analytic functions, is called *analytic continuation*. As an example let us consider the function $F(z)$ defined by

$$F(z) = 1 + z + z^2 + z^3 + \cdots, \tag{1.193}$$

where z is defined to be in the circular region $|z| < 1$. Now the function $G(z)$ defined by

$$G(z) = \int_0^{+\infty} e^{(z-1)x}\, dx, \tag{1.194}$$

for $\mathrm{Re}(z) < 1$ defines the same function in a larger region of the complex plane. Finally, $H(z) = 1/(1-z)$ defines the same function everywhere in the complex plane, except at $z = 1$. $G(z)$ is said to be the analytic continuation of $F(z)$ outside the unit circle centered at the origin, and $H(z)$ is said to be the analytic continuation of $G(z)$ outside the half-plane $\mathrm{Re}(z) < 1$.

Following are *some basic properties of analytic continuation*:

Theorem 3: Let E be a region inside the intersection of two regions F and G. If f belongs to $A(F)$ and g to $A(G)$, and if $f = g$ in the region E, then $f = g$ in the whole intersection of F and G. Moreover, f is the analytic continuation of g in F and g is the analytic continuation of f and G.

Corollary 2: If $f(z) = 0$ in a subregion of F, then $f = 0$ in the whole complex plane.

Corollary 3: If $g(z)$ is the analytic continuation of $f(z)$ in G, then this analytic continuation is unique in G.

Riemann Principle: Let f and g be two analytic functions defined, respectively, in the regions F and G of the complex plane. Let C be a path common to F and G, where f and g are continuous. If $f = g$ on C, then f is the analytic continuation of g in F, and g is the analytic continuation of f in G.

Schwartz Reflection Principle: Let S be a segment of the real axis in the complex plane, let F be a region in the upper-half complex plane bounded on the real axis by S and let G be a region in the lower-half complex plane bounded on the real axis by S. If f belongs to $A(F)$ and if, on the real axis, f is real, then $f^*(z^*)$ is the analytic continuation of f in G.

WAVES IN A CONDUCTIVITY-TENSOR-DEFINED MEDIUM
A COLD-PLASMA EXAMPLE

2.1. INTRODUCTION

The aim of this chapter is to introduce and briefly describe the propagation of electromagnetic waves in a medium where anisotropic conductivity tensor σ can be defined. This approach provides a valid description of the wave phenomena possible in many plasmas of interest. Since such a description can be found in many places [see, for example, Denisse and Delcroix (1961), Stix (1962), Allis, Buchsbaum, and Bers (1963), Quemada (1968), Chen (1974), and Nicholson (1983)], we present here only the cold-plasma theory as an example of the utility of this approach. Special emphasis will be given to the concept of dispersion relations and their importance to wave propagation in plasmas. Also, as will be true throughout the rest of the book, we will routinely use the mathematical techniques of Fourier and Laplace transforms.

2.2. WAVES IN IDEALIZED MEDIA

Within the context of our interest in wave propagation in plasmas, we will now briefly review wave propagation in an idealized medium. The properties of wave propagation in a plasma will be described using the "charges in vacuum" model. In this description, one uses Maxwell's equations, where the particle dynamics are represented by the charge and current densities of all the charged particles. This model is useful when the particle motion can be solved self-consistently with the fields. This is possible in the general case of a linear homogeneous medium, as we will show in this section.

Explicitly including the ideas of external charges ρ_{ex} and external currents j_{ex}, which can represent, for example, wave excitation, Maxwell's equations can be written in MKS units as

$$\nabla \times E = -\frac{\partial B}{\partial t}, \tag{2.1}$$

$$\nabla \times B = \frac{1}{c^2}\frac{\partial E}{\partial t} + \mu_0 j + \mu_0(j_{ex}), \tag{2.2}$$

$$\nabla \cdot B = 0, \tag{2.3}$$

$$\nabla \cdot E = \frac{\rho + (\rho_{ex})}{\varepsilon_0}, \tag{2.4}$$

where E is the electric-field-intensity vector, and B is the magnetic induction vector. We have put in parentheses those charges and currents which can be applied from outside the plasma (in conductors surrounding the plasma, on grids inside the plasma, etc.). Since this description applies only to charges in vacuum, the vectors D and E, and B and H, are connected, respectively, through the vacuum dielectric constant ε_0 and the vacuum magnetic permeability μ_0 by

$$D = \varepsilon_0 E; \tag{2.5}$$

$$B = \mu_0 H, \tag{2.6}$$

where D and H are the electric displacement and magnetic-field-intensity vectors, respectively. (In MKSA units, $\varepsilon_0 = 10^{-9}/(36\pi) \simeq 8.86 \times 10^{-12}$ and $\mu_0 = 4\pi \times 10^{-7}$.)

We will now make some assumptions concerning the properties of the medium in order to obtain a description of our charged-particle dynamics. The first assumption is that, without any external perturbation, the medium is in an equilibrium state. Moreover, we will assume that this equilibrium is space invariant, i.e., that we have an infinite homogeneous equilibrium in space.

The second basic assumption is that any perturbations to this equilibrium, i.e., E, B, j, ρ, etc., are small, so that the response of the medium will be linear. This assumption may allow us to establish a general relationship between the fields and the particle motion. For example, we can write that

$$j(r,t) = \sigma(r,t)E(t), \tag{2.7}$$

where the conductivity σ is not a function of E or B. Moreover, we may assume that σ is a tensor, so that the vectors j and E are not generally colinear.

The relation given by Eq. (2.7), however, is not the most general one for a linear homogeneous medium. As a matter of fact, Eq. (2.7) states that the current at a given position r and at a given time t, is a function of the electric field at this position and time. One may expect, however, because of the charged-particle motion, that the current is not a function of only the local electric field. Since $j(r,t)$ also depends on the electric field $E(r',t' < t)$, at any other place r' and time $t' < t$, a more general relation than that given by Eq. (2.7) may be used:

$$j(r,t) = \int d^3r' \; dt' \; \sigma(r,r',t,t') E(r',t'). \tag{2.8}$$

Moreover, since we are assuming a time-independent homogeneous equilibrium, the conductivity should be invariant to translation in space and time, so that it should be a function only of $r - r'$ and $t - t'$. Thus, for this quite general case, Eq. (2.8) should be written as

$$j(r,t) = \int_V d^3r' \int_0^t dt' \; \sigma(r - r', t - t') E(r',t'). \tag{2.9}$$

The complexity of Eq. (2.8) appears only in (r,t) space. Since Eq. (2.8) is a convolution product, a Fourier–Laplace transform of this relation yields the familiar equation

$$j(\omega,k) = \sigma(\omega,k) E(\omega,k), \tag{2.10}$$

so that we may define the relation between the current and the electric field in (r,t) space as the convolution product

$$j(r,t) = \sigma(r,t) * E(r,t), \tag{2.11}$$

where the symbol $*$ stands for a convolution product.

It is interesting to compare wave propagation in a plasma with the usual description of wave propagation in dielectrics. In this latter case, one introduces the electric displacement vector D which is related to the local instantaneous electric field intensity E by

$$D = \varepsilon_0 \varepsilon * E, \tag{2.12}$$

where ε is the relative dielectric tensor of the material. In Eq. (2.12) we have also assumed that similar to the conductivity, the relative dielectric tensor ε is a nonlocal function, i.e., the product $\varepsilon * E$ is a convolution product. Maxwell's equations for wave propagation in a dielectric can be written as

$$\nabla \times E = -\frac{\partial B}{\partial t}, \tag{2.13}$$

$$\nabla \times H = \frac{\partial D}{\partial t} + (j_{ex}), \tag{2.14}$$

$$\nabla \cdot B = 0, \tag{2.15}$$

$$\nabla \cdot D = (\rho_{ex}), \tag{2.16}$$

with B and D defined by Eqs. (2.6) and (2.12), respectively. It is possible, using this set of equations, to describe wave propagation in plasmas in a way similar to what is done for wave propagation in dielectrics. Such a description can be found in many books. However, it appears to us that this dielectric model does not provide any more information than does the "charges in vacuum" model presented in this chapter and can, in some cases, introduce misleading ideas if the analogy between a plasma and a dielectric is pushed too far, since the microscopic properties of plasmas and of classical dielectrics are quite different.

2.3. WAVES IN PLASMAS

We are now interested to know under what circumstances the above set of equations for wave motion in a classical dielectric can be valid for wave motion in a plasma (a kind of dielectric) characterized by the conductivity tensor defined by Eq. (2.7). Comparing the set of Eqs. (2.1)–(2.4) for a conducting medium with the set of Eqs. (2.13)–(2.16) for a dielectric, shows that the two sets of equations will give identical descriptions if D in the second set of equations is chosen in such a way that Eqs. (2.2) and (2.14), and Eqs. (2.4) and (2.16), respectively, are identical. We will now show that the equality of these two pairs of equations will require that there be no net plasma production or loss in time, and that the plasma dielectric tensor ε be related to the plasma conductivity tensor σ in a particular way. Requiring that Eqs. (2.2) and (2.14) be identical requires that

$$\frac{\partial D}{\partial t} = \frac{\varepsilon_0 \partial E}{\partial t} + j. \tag{2.17}$$

Taking the divergence of both sides of Eq. (2.17) and interchanging the order of the time and space operators gives

$$\frac{\partial (\mathbf{\nabla} \cdot D)}{\partial t} = \frac{\varepsilon_0 \partial (\mathbf{\nabla} \cdot E)}{\partial t} + \mathbf{\nabla} \cdot j. \tag{2.18}$$

Now, substituting in Eq. (2.18) from Eqs. (2.4) and (2.16) for $(\mathbf{\nabla} \cdot E)$ and $(\mathbf{\nabla} \cdot D)$, respectively, gives

$$\frac{\partial \rho}{\partial t} + \mathbf{\nabla} \cdot j = 0. \tag{2.19}$$

But Eq. (2.19) is just the conservation equation for charged particles. Thus,

Eqs. (2.2) and (2.14) will be identical if there is no net plasma production or loss in time.

A comparison of Eqs. (2.4) and (2.16) shows that, in order for these two equations to be equal, ε must be chosen such that

$$\frac{\varepsilon_0 \partial E}{\partial t} + \sigma * E = \frac{\varepsilon_0 \partial(\varepsilon * E)}{\partial t}. \tag{2.20}$$

Taking the Fourier–Laplace transform of Eq. (2.20), we obtain a general relation between the conductivity and the relative dielectric constant given by

$$\varepsilon(\omega,k) = 1 + \chi(\omega,k),$$

with $\qquad\qquad\qquad\qquad\qquad\qquad\qquad\qquad\qquad\qquad\qquad$ (2.21)

$$\chi(\omega,k) = \frac{\sigma(\omega,k)}{i\omega\varepsilon_0},$$

where χ is the susceptibility.

2.3.1. Some General Remarks on Solving Maxwell's Equation

We must now solve Maxwell's equations, which are partial differential equations. We see that these equations are linear with respect to the various fields, and that neither the space coordinate $r(x,y,z)$ nor time t appears explicitly in these equations. Thus, we can use Fourier and Laplace transforms in space and time, respectively, to transform the above set of linear partial differential equations into a set of linear algebraic equations. After verification that the quantity under consideration is "well-behaved" at infinity, and that its initial value is zero (see Chapter 1), we can, via Fourier and Laplace transforms, effectively replace the space differential operator ∇ by ik and the time differential operator $\partial/\partial t$ by $-i\omega$. Thus, making Fourier transforms in space and Laplace transforms in time allows us to write Maxwell's equation in (ω,k) space as the following set of algebraic equations

$$k \times E = \omega B, \tag{2.22}$$

$$k \times H = -\omega D + (j_{\text{ex}}), \tag{2.23}$$

$$k \cdot B = 0, \tag{2.24}$$

$$k \cdot D = -i(\rho_{\text{ex}}), \tag{2.25}$$

where $D = \varepsilon_0\varepsilon \cdot E$ and $B = \mu_0 H$ [see Eqs. (2.12) and (2.6), respectively].

2.3.2. *Dispersion Relations*

Equations $(2.22)-(2.25)$ constitute a set of linear equations with respect to the fields for any pair of (ω,k) values. When there are no external charges and currents, this set of linear equations is homogeneous and of first degree with respect to the four fields (E,H,B,D), and we have the same number of equations as variables. In general, for any given pair of (ω,k) values, there will exist nontrivial solutions for the fields only if the determinant of the coefficients is zero. This introduces a functional relationship between ω and k. Such a relationship is a condition of compatability between ω and k for having nonzero fields in the medium, i.e., for having waves. This relationship

$$\mathscr{D}(k,\omega) = 0 \tag{2.26}$$

is called the *dispersion relation*. (In general, dispersion refers to the dependence of wave velocity v_φ on wave frequency. Since

$$v_\varphi = \frac{\omega}{k}, \tag{2.27}$$

Eq. (2.26) can be thought of as representing a relationship between the phase velocity v_φ and the angular frequency ω, thus explaining the term *dispersion relation*.) The dispersion relation is very important because it contains the linear properties of the medium with respect to waves.

2.3.3. *Purely Electrostatic and Purely Electromagnetic Waves*

Before considering general wave propagation in which the waves can have both electrostatic and electromagnetic character, we will first make a few remarks and observations about the special cases of purely electrostatic and purely electromagnetic modes. For simplicity, except for any external currents producing dc magnetic fields, we will set $\rho_{ex} = 0 = j_{ex}$. Doing this the set of Fourier-and-Laplace-transformed Maxwell equations describing general wave propagation in a plasma becomes

$$k \times E = \omega B, \tag{2.28}$$

$$k \times B = -\omega\mu_0 D, \tag{2.29}$$

$$k \cdot B = 0, \tag{2.30}$$

$$k \cdot D = 0, \tag{2.31}$$

where

$$D = \varepsilon_0\varepsilon \cdot E. \tag{2.32}$$

By definition, purely electro*static* waves have no electric field component perpendicular to the direction of propagation. Thus, since k is in the direction of propagation, E and k are parallel. Therefore, from Eq. (2.22), $B = 0$. That is, the magnetic field of the wave is zero (however, we can still have an externally applied dc magnetic field B_0). Thus, a purely electrostatic mode has no magnetic effect associated with it. Such a mode does not always exist. For example, there are no such modes in vacuum. (In vacuum $\varepsilon = 1$ is a scalar quantity, so that by Eq. (2.32) D is parallel to E, which is parallel to k. Thus, by Eq. (2.31), D and therefore E must be zero.) But in a plasma, where ε is not a scalar quantity, solutions can exist with E parallel to k. In this case, Maxwell's Eqs. (2.1)–(2.4) can be shown to reduce to Poisson's equation (Eq. 2.4)

$$\nabla \cdot E = \frac{\rho}{\varepsilon_0}, \tag{2.33}$$

since Eq. (2.1) shows that E is parallel to k, and Eq. (2.2) is equivalent to the conservation of electric charge.

Purely electro*magnetic* modes, by definition, have purely transverse electric fields. That is, E is perpendicular to k so that $k \cdot E = 0$, and Eqs. (2.31) and (2.32) give $\rho = 0$. Thus, there is no space charge induced in such a mode. This is the classical behavior of a wave in vacuum.

2.3.4. *General Dispersion Relation with* $\rho_{ex} = 0 = j_{ex}$

Equations (2.28)–(2.32) describe the general propagation of electromagnetic waves in plasmas. For the cold-plasma case to be discussed here, we will assume that there are no external charges or currents, i.e., $\rho_{ex} = 0 = j_{ex}$. We will denote $E(k, \omega)$ and $B(k, \omega)$ simply as E and B, and we will introduce the so-called relative dielectric tensor ε defined by

$$\varepsilon = 1 + \frac{\sigma}{i\omega\varepsilon_0}, \tag{2.34}$$

where σ has been determined from Eq. (2.11), after making a Laplace transform in time. Then, cross-multiplying Eq. (2.22) on the left by the vector k, and using Eqs. (2.29) and (2.32), we get

$$k \times k \times E + \frac{\omega^2 \varepsilon \cdot E}{c^2} = 0, \tag{2.35}$$

where $[c = (\mu_0\varepsilon_0)^{-1/2}]$ is the speed of light in a vacuum. Introducing the so-called optical index vector $\mathcal{N} = ck/\omega$, Eq. (2.35) becomes

$$(\mathcal{N}\mathcal{N} - \mathcal{N}^2 I + \varepsilon) \cdot E = 0, \tag{2.36}$$

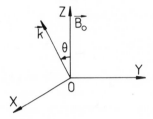

Fig. 2.1. For wave propagation in a magnetically supported plasma, the angle of propagation Θ with respect to the magnetic field is important. The general dispersion relation given by Eq. (2.37) assumes that B_0 is along the z-axis of a rectangular coordinate system and that k is in the x-z plane, as shown.

which represents three linear homogeneous equations. The condition of compatibility of these three equations gives the dispersion relation

$$\mathscr{D}(k,\omega) = \left| \mathcal{N}_m \mathcal{N}_n - \mathcal{N}^2 \delta_{mn} + \varepsilon_{mn} \right| = 0, \qquad m,n = 1,2,3, \qquad (2.37)$$

which is a 3×3 determinant. Equation (2.37) is usually presented in another form by introducing the angle Θ which the vector k (or $\mathcal{N}$) makes with the applied external magnetic field B_0. If, as shown in Fig. 2.1, we put the external magnetic field B_0 along the z axis of a rectangular-coordinate system and the wave vector k in the xz-plane, then Eq. (2.37) can be written in determinant form as

$$\begin{vmatrix} \varepsilon_{xx} - \mathcal{N}^2 \cos^2 \Theta & \varepsilon_{xy} & \varepsilon_{xz} + \mathcal{N}^2 \sin \Theta \cos \Theta \\ \varepsilon_{yx} & \varepsilon_{yy} - \mathcal{N}^2 & \varepsilon_{yz} \\ \varepsilon_{zx} + \mathcal{N}^2 \sin \Theta \cos \Theta & \varepsilon_{zy} & \varepsilon_{zz} - \mathcal{N}^2 \sin^2 \Theta \end{vmatrix} = 0. \quad (2.38)$$

Equation (2.38) is a very general equation for electromagnetic wave propagation in a multicomponent, infinite plasma, including thermal effects, relativistic effects, external magnetic fields, etc. The next step in solving this general equation is to find the relative dielectric tensor ε for the specific plasma of interest.

2.4. WAVES IN A COLD PLASMA

As an example of the application of the general theory, we will now study the propagation of electromagnetic waves in a so-called cold plasma consisting of electrons and one specie of ions. The procedure will be, first, to determine the dielectric tensor of the cold plasma. This will then be used to find, from the general dispersion relation given by Eq. (2.38), the dispersion relation for the cold plasma. Finally, this dispersion relation will be examined to determine the various kinds of wave motion possible in such a plasma.

For general wave propagation the modes will be partially electrostatic and partially electromagnetic; that is to say, the electric field will be neither

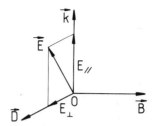

Fig. 2.2. For general wave propagation the modes will be partially electrostatic and partially electromagnetic. Equations (2.22)–(2.25) show that $E \perp B$ and that k, B, and D are mutually perpendicular as shown.

parallel nor perpendicular to the wave number k. In this general situation, we can choose a rectangular-coordinate system in the following way: from Eqs. (2.24)–(2.25) it can be seen that the three vectors k, B, and D, are along three mutually perpendicular directions. Also, Eq. (2.22) shows that E is perpendicular to B; that is, E is in the D-k plane. These directional relationships cna be represented in the manner shown in Fig. 2.2.

2.4.1. Cold-Plasma Dielectric Tensor

A cold-plasma model is used when describing wave propagation in plasmas in which thermal effects are not important. In this case the particle behavior can be described by the equation of conservation of particles and the equation of motion, respectively, as

$$\frac{\partial n_n}{\partial t} + \nabla \cdot (n_n v_n) = 0, \tag{2.39}$$

$$\frac{\partial v_n}{\partial t} = \frac{q_n(E + v_n \times B)}{m_n}, \tag{2.40}$$

where $n = e, i$ for electrons and ions, respectively, E is the electric field of the wave, and B is the total magnetic field. In this formulation the cold-plasma approximation is made by not including a thermal-pressure term in Eq. (2.40). We will assume further that we do not have relativistic effects so that the magnetic force on particles due to the wave is negligible with respect to the electric force. Thus, we will replace the total magnetic field B by the external applied magnetic B_0, which will be assumed to be constant.

We will now determine the plasma relative dielectric tensor arising from the charge and current densities. These are given, respectively, by

$$\rho = \sum_n n_n q_n, \tag{2.41}$$

$$j = \sum_n q_n(n_n v_n). \tag{2.42}$$

First, we note that multiplying Eq. (2.39) by q_n and summing over all particle species gives the charge-conservation equation

$$\frac{\partial \rho}{\partial t} + \nabla \cdot j = 0. \tag{2.43}$$

But this relationship is already included in Maxwell's equations, as we have seen earlier [see Eq. (2.19) and the accompanying discussion]. Thus, Eq. (2.39) is already contained implicitly in our formulation so that in order to compute the relative dielectric tensor we need consider only the equation of motion, i.e., Eq. (2.40), which we rewrite as

$$\frac{\partial v}{\partial t} = \left(\frac{q}{m}\right)(E + v \times B_0), \tag{2.40}$$

where we have omitted the species subscript n. Making a Laplace transform in time, this equation can be written as

$$v = a + v \times b, \tag{2.44}$$

where $a = i(qE/m\omega)$, $b = i(qB/m\omega) = i(\Omega/\omega)$, and $\Omega = (qB/m)$ is the angular cyclotron frequency of a given specie of particle. Solving Eq. (2.44) gives the particle velocity for each specie in matrix form as

$$v = \left[\frac{1}{1 + b^2}\right] \begin{bmatrix} 1 & b & 0 \\ -b & 1 & 0 \\ 0 & 0 & 1 + b^2 \end{bmatrix} a. \tag{2.45}$$

Or, since a is proportional to E, we can write that $v = \mu \cdot E$, where μ is defined as the particle's mobility tensor. Thus, the total current density given by Eq. (2.42) can be written as

$$j = \sum_n n_n q_n v_n = \left(\sum_n n_n q_n \mu_n\right) \cdot E. \tag{2.46}$$

Comparing Eqs. (2.46) and (2.11), we see that the conductivity tensor σ can be written in terms of the particle mobility tensor as

$$\sigma = \sum_n n_n q_n \mu_n, \tag{2.47}$$

where μ_n for each particle can be obtained directly from Eq. (2.45). Using Eq. (2.34), which relates ε and σ, we can write the following matrix expression for the relative dielectric tensor

$$\varepsilon = \begin{bmatrix} \pi_1 & -i\pi_2 & 0 \\ i\pi_2 & \pi_1 & 0 \\ 0 & 0 & \pi_3 \end{bmatrix}, \tag{2.48}$$

where

$$\pi_1 = 1 + \frac{\sum_n \omega_{pn}^2}{\Omega_n^2 - \omega^2},$$

$$\pi_2 = \frac{\sum_n \Omega_n \omega_{pn}^2}{\omega(\Omega_n^2 - \omega^2)}, \tag{2.49}$$

$$\pi_3 = 1 - \frac{\sum_n \omega_{pn}^2}{\omega^2},$$

with $\omega_{pn} = (n_n q_n^2 / \varepsilon_0 m_n)^{1/2}$, the angular plasma frequency of particle n.

2.4.2. Cold-Plasma Dispersion Relation

Using Eq. (2.48) to replace the relative dielectric matric element in Eq. (2.38) permits the dispersion relation for a cold two-component plasma to be written as

$$A\mathcal{N}^4 - B\mathcal{N}^2 + C = 0, \tag{2.50}$$

where

$$A = \pi_1 \sin^2 \Theta + \pi_3 \cos^2 \Theta,$$

$$B = (\pi_1^2 - \pi_2^2) \sin^2 \Theta + \pi_1 \pi_3 (1 + \cos^2 \Theta), \tag{2.51}$$

$$C = \pi_3 (\pi_1^2 - \pi_2^2),$$

and Θ, as defined earlier, is the wave propagation angle with respect to the applied magnetic field B_0.

The discriminant of Eq. (2.50) is given by

$$D = (\pi_1^2 - \pi_2^2 - \pi_1 \pi_3)^2 \sin^4 \Theta + 4\pi_2^2 \pi_3^2 \cos^2 \Theta. \tag{2.52}$$

Since D is always > 0, there are two solutions for $\mathcal{N}^2$

$$\mathcal{N}^2 = \frac{B \pm (D)^{1/2}}{2A}. \tag{2.53}$$

Thus, in principle, for any angle Θ there are always two different modes of propagation possible.

It can be shown that an interesting alternative way to write Eq. (2.50) is in terms of the propagation angle Θ as

$$\tan^2 \Theta = \frac{-\pi_3(\mathcal{N}^2 - \pi_1^2)^2 - \pi_2^2}{(\pi_1\mathcal{N}^2 - \pi_1^2 - \pi_2^2)(\mathcal{N}^2 - \pi_3)}. \tag{2.54}$$

This equation, in some cases, allows the various modes to be discussed somewhat more easily than does Eq. (2.50). We will now present a short outline of the modes possible in our two-component cold plasma of ions and electrons. Detailed discussions of these modes are available in many other books [see, for example, Stix (1962), Allis, Buchsbaum, and Bers (1963), and Quemada (1968)].

2.4.3. Cutoffs and Resonances

We will say that there is a *resonance* when the optical index $\mathcal{N} = ck/\omega$ goes to infinity. This occurs when the phase velocity $v_\varphi = \omega/k$ goes to zero. From Eq. (2.54) we can see that this condition for resonance occurs when

$$\tan^2 \Theta = \frac{-\pi_3}{\pi_1}. \tag{2.55}$$

Thus, resonances depend on the angle Θ, in contrast to what will now be found for cutoffs.

We will say that there is a *cutoff* when the optical index $\mathcal{N}$ goes to zero (as does k). This occurs when the phase velocity goes to infinity. From Eq. (2.50) it can be seen that when the cutoff occurs, $C = 0$. Using Eq. (2.51) then gives the cutoff conditions as

$$\pi_3(\pi_1^2 - \pi_2^2) = 0. \tag{2.56}$$

In contrast to what was found for resonances, the conditions for cutoffs have no angle dependence.

It is of interest to note that a plasma behaves as a very strange dielectric under wave propagation conditions near resonance. Under these conditions, the index of refraction $\mathcal{N}$ can be much less than unity. Thus, light, for example, can have a phase velocity inside the plasma that is much higher than the speed of light in vacuum, can experience total external reflection when incident upon a plasma *from vacuum*, and so on.

As noted already, for any given angle of propagation with respect to the externally applied magnetic field there are always two possible modes. Also noted was the fact that these modes can exhibit cutoffs and resonances. For purposes of discussion and classification, it is convenient to choose two extreme directions of propagation: parallel and perpendicular to the applied magnetic field. In the following we will briefly examine the main characteristics of these two categories of wave propagation. The properties of wave propagation at any other angle with respect to the magnetic field can always be determined from an examination of the cold-plasma dispersion relation given either by Eq. (2.50) or (2.54).

2.4.4. Propagation Parallel to the Applied Magnetic Field

For $\Theta = 0$, Eq. (2.54) gives

$$\pi_3[(\mathcal{N}^2 - \pi_1^2)^2 - \pi_2^2] = 0. \tag{2.57}$$

This equation has two solutions. Letting $\pi_3 = 0$ gives a standing-wave mode whose dispersion relation [see Eq. (2.49)] is given by

$$1 - \frac{\omega_0^2}{\omega^2} = 0, \tag{2.58}$$

where $\omega_0 = (\omega_{pi}^2 + \omega_{pe}^2)^{1/2}$ is called the *total (angular) plasma frequency*. Thus, this mode is nothing more than a (nonpropagating) normal mode of oscillation of the entire plasma parallel to the magnetic field, and is purely electrostatic in nature.

Choosing the second solution of Eq. (2.57), i.e., $[(\mathcal{N}^2 - \pi_1^2)^2 - \pi_2^2] = 0$, gives

$$\mathcal{N}_\pm^2 = \pi_1 \pm \pi_2. \tag{2.59}$$

Choosing the $+$ sign gives $\mathcal{N}_+^2 = \pi_1 + \pi_2$. Again, using Eq. (2.49) we find the dispersion relation for this mode, which is called the right-hand mode, to be given by

$$\mathcal{N}_+^2 = 1 - \frac{\omega_0^2}{(\omega - \Omega_{ce})(\omega + \Omega_{ci})}. \tag{2.60}$$

Closer examination of this mode reveals that it is a propagating wave whose electric vector E rotates in the same direction as the cyclotron motion of electrons in the same magnetic field. Examination of Eq. (2.60) shows that this mode exhibits a resonance (i.e., $\mathcal{N} \to \infty$, $v_\varphi \to 0$) when ω approaches the electron frequency Ω_{ce}. It exhibits a cutoff ($\mathcal{N} \to 0$, $v_\varphi \to \infty$) when ω approaches ω_R where

$$\omega_R = \tfrac{1}{2}[(\Omega_{ce}^2 + 4\omega_{pe}^2)^{1/2} + \Omega_{ce}]. \tag{2.61}$$

Equation (2.60) shows that $\mathcal{N}_+^2 < 0$ in the frequency range $\Omega_{ce} < \omega < \omega_R$, indicating that the right-hand mode does not propagate in this frequency range since the index of refraction $\mathcal{N} = ck/\omega$ must be a real quantity.

Choosing the $-$ sign in Eq. (2.59) gives the dispersion relation for the left-hand mode

$$\mathcal{N}_-^2 = \boxed{1} - \frac{\omega_0^2}{(\omega + \Omega_{ce})(\omega - \Omega_{ci})}. \tag{2.62}$$

As this wave propagates in the plasma parallel to the applied magnetic field, its electric-field vector E rotates in the same direction as the cyclotron motion of positive ions in the same magnetic field. Examination of Eq. (2.62) shows that the left-hand mode exhibits a resonance $(\mathcal{N} \to \infty)$ for $\omega \to \Omega_{ci}$ and a cutoff for $\omega \to \omega_L$, where ω_L is defined by

$$\omega_L = \tfrac{1}{2}\left[(\Omega_{ce}^2 + 4\omega_{pe}^2)^{1/2} - \Omega_{ce}\right]. \tag{2.63}$$

Again, as for the right-hand mode, there is a range of frequencies, $\Omega_{ci} < \omega < \omega_L$, for which the left-hand mode does not propagate. Because of the rotating behavior of their electric fields, both the right-hand and left-hand modes are said to be circularly polarized.

It is interesting to note that at the low-frequency limit $(\omega \to 0)$, $\mathcal{N}_+$ and $\mathcal{N}_-$ approach the same limit, called the *Alfven limit*, given by

$$\mathcal{N}_{\mathrm{Alfven}}^2 = 1 + \frac{\omega_0^2}{\Omega_{ce}\Omega_{ci}}. \tag{2.64}$$

This represents a limit of $\mathcal{N}$ and, thus, of the phase velocity. This limiting phase velocity is called the *Alfven velocity*. When $\omega_0^2 \gg \Omega_{ce}\Omega_{ci}$ (i.e., high plasma density, weak magnetic field), for example, the Alfven velocity can be written as

$$v_\varphi = v_{\mathrm{Alfven}} = \left[\frac{B^2}{\mu_0 n m_i}\right]^{1/2}. \tag{2.65}$$

The characteristics of wave propagation parallel to the magnetic field are summarized in the $\omega - k$ diagram shown in Fig. 2.3.

2.4.5. Propagation Perpendicular to the Applied Magnetic Field

For $\Theta = \pi/2$, Eq. (2.54) gives

$$(\pi_1\mathcal{N}^2 - \pi_1^2 + \pi_2^2)(\mathcal{N}^2 - \pi_3) = 0. \tag{2.66}$$

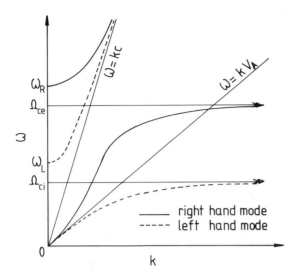

Fig. 2.3. Wave propagation parallel to B_0 in a magnetically supported plasma. Both the right-hand and left-hand modes are circularly polarized, have frequency ranges of nonpropagation, and exhibit cutoff and resonance behavior as shown.

This equation, as for the parallel propagation case, has two solutions. Choosing the second factor in Eq. (2.66) to be zero gives $N^2 = \pi_3$. Using Eq. (2.49) gives the dispersion relation of this so-called *ordinary wave* as

$$N_0^2 = 1 - \frac{\omega_0^2}{\omega^2}, \qquad (2.67)$$

where ω_0 is again the total plasma frequency. Equation (2.67) shows that the ordinary mode "sees" an index of refraction N_0 that is independent of the value of the magnetic field, depending only on the properties of the plasma itself. No propagation occurs for frequencies less than $\omega = \omega_0$. At $\omega = \omega_0$, the ordinary wave exhibits a cutoff ($N = 0$). No resonance is observed for this wave. Notice that $N < 1$ for all frequencies of interest. This wave is purely electromagnetic, in contrast to the extraordinary wave that we will discuss now.

The second solution for perpendicular propagation corresponds to the so-called *extraordinary mode* and is given by choosing the first factor of Eq. (2.66) to be zero. This gives the dispersion relation

$$N^2 = \frac{\pi_1^2 - \pi_2^2}{\pi_1}. \qquad (2.68)$$

This mode exhibits resonances $(\mathcal{N} \to \infty)$ at two hybrid frequencies near the cyclotron frequencies. The first occurs at the so-called upper-hybrid frequency

$$\Omega_{UH} = (\omega_{pe}^2 + \Omega_{ce}^2)^{1/2}. \tag{2.69}$$

The second resonance occurs at the so-called lower-hybrid frequency

$$\Omega_{LH} = \left[\frac{(\omega_{pe}^2 + \Omega_{ce}\Omega_{ci})(\Omega_{ce}\Omega_{ci})}{\omega_{pe}^2 + \Omega_{ce}^2} \right]^{1/2}. \tag{2.70}$$

There are two cutoff frequencies, ω_R and ω_L, which are the same as the cutoff frequencies seen already for the propagation of the right-hand and left-hand modes, respectively, parallel to the applied magnetic field. Also, as observed for the right-hand and left-hand modes, the low-frequency limit of the phase velocity is the Alfven velocity. In contrast to the right-hand and left-hand modes, however, the extraordinary mode has both electrostatic and electromagnetic properties. That is, both magnetic and space-charge effects are involved. The electric field E of the extraordinary wave rotates in a plane perpendicular to the direction of the magnetic field, as shown in Fig. 2.4. However, as E rotates, its magnitude varies in a regular manner in such a way that the extraordinary wave is said to be elliptically polarized. It can be shown that the polarization is given by $E_x/E_y = i(\pi_2/\pi_1)$, where $i = (-1)^{1/2}$ indicates that the x and y components of the field rotate always 90° out of phase with respect to each other.

The characteristics for wave propagation perpendicular to the external magnetic field are summarized in the $\omega - k$ diagram shown in Fig. 2.5.

2.5. APPLICATIONS OF THE COLD-PLASMA-THEORY RESULTS

We will now describe how the ordinary and extraordinary waves can be used to determine or estimate the density of both steady-state and time-varying plasmas immersed in a magnetic field. Since such techniques are

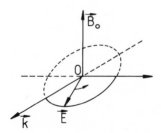

Fig. 2.4. The electric field E of the extraordinary wave rotates in such a way that the tip of E traces out an ellipse in a plane perpendicular to B_0.

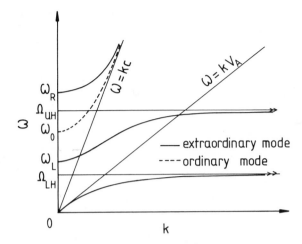

Fig. 2.5. Wave propagation perpendicular to B_0 in a magnetically supported plasma. The dispersion of the ordinary wave, which is purely electromagnetic, is independent of the strength of the magnetic field, depending only on the plasma density. In contrast, the extraordinary wave is both electromagnetic and electrostatic and has a dispersion depending on both magnetic field strength and plasma density.

described in great detail elsewhere [see, for example, Heald and Wharton (1965)], the descriptions given here are brief and are given only to demonstrate how the properties of the waves, as predicted by the cold-plasma-theory dispersion relation, are important from a practical point of view. One experimental technique which is used for both types of waves is called *microwave interferometry*. This is the technique used in the next two sections.

2.5.1. Using the Ordinary Wave to Measure Plasma Density

The cutoff and dispersive characteristics of the ordinary wave are routinely used in the determination of the electron density of a plasma immersed in a magnetic field. One of the main features of this method is that it, like the propagation characteristics of the wave itself, is independent of the value B_0 of the magnetic field.

Figure 2.6 shows in part how microwave interferometry is used with the ordinary wave to determine or estimate the average density of a plasma "slab." With the interferometer, which can be made using microwave devices operating in the wavelength range of a few millimeters to a few centimeters, the optical path length of the ordinary wave propagating through the plasma is compared with the path length of a portion of the same microwave propagating in a waveguide of variable length (not shown) external to the plasma. That the optical path length of an ordinary wave of a given frequency in the

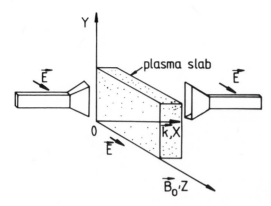

Fig. 2.6. In using the ordinary wave to measure plasma density the electric field E of the wave is polarized to be *parallel* to B_0. An interference between this wave and one propagating outside of the plasma can be used to determine the average density of the plasma.

plasma is a function of the plasma density (but not of the magnetic field) can be easily seen from the ordinary wave dispersion relation given by Eq. (2.67) which shows that the index of refraction presented to the ordinary wave by the plasma depends only on the plasma frequency, i.e., on the plasma density. The average electron density can be obtained for the plasma slab by using the equation

$$\psi = 2\pi \int\limits_{\text{plasma path}} k(x)\,dx \;=\; 2\pi \int\limits_{\text{plasma path}} (\omega/c) \left[1 - \frac{n_0^2}{n(x)^2} \right]^{1/2} dx, \qquad (2.71)$$

where ψ is the phase shift observed for the wave propagating in the plasma relative to the wave propagating external to the plasma, and n_0 is the so-called *critical plasma density*

$$n_0 \simeq \frac{m_e \varepsilon_0 \omega^2}{e^2}, \qquad (2.72)$$

for the microwave frequency ω used. This critical density corresponds to the frequency for which cutoff $(\mathcal{N} \to 0)$ occurs for the ordinary wave [see Eq. (2.67)]. If the local plasma density $n(x)$ exceeds the critical density at any point in the plasma for a given frequency ω, then $\mathcal{N}^2 < 0$ and the wave is reflected rather than transmitted. In this case only a lower estimate of the plasma density, given by Eq. (2.72), can be made.

Density variation with time can also be measured using this technique. When the denisty changes, interference "fringes" appear in the detected signal that results from a summation of the transmitted signal and the external microwave signal, as shown in Fig. 2.7. Such interference fringes can be "unfolded" using Eq. (2.71) to determine the variation of the average plasma density with time during the time that $n(x)$ is everywhere $< n_0$, and to deter-

Fig. 2.7. Using the ordinary wave to learn information about the density of a plasma whose density is changing in time. The flat central portion of the curve corresponds to the time during which the plasma density is higher than the critical density given by Eq. (2.72).

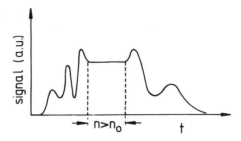

mine a lower limit to the plasma density during any time when $n(x)$ at any point in the plasma is $> n_0$.

2.5.2. Using the Extraordinary Wave to Measure Plasma Density

The extraordinary wave can be used, also, for density measurements. In this case the electric field E of the linearly polarized microwaves is made to be perpendicular to the direction of the external magnetic field, as indicated in Fig. 2.8. Both this measurement and the preceding one using the ordinary wave can be made by the same microwave apparatus, which is constructed in such a way that the waveguide and, thus, the direction of the E field, can easily be rotated by 90°. By using a somewhat more sophisticated microwave system, both waves (usually at different frequencies) can be transmitted and detected simultaneously (Heald and Wharton, 1965).

This method can be used for lower-denisty measurements since limited propagation, even for $\omega < \omega_R$, is possible: because of the resonance, the k

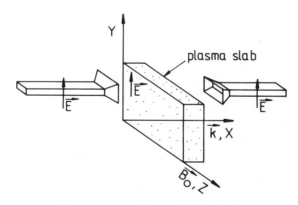

Fig. 2.8. In using the extraordinary wave to measure plasma density the electric field E of the wave is polarized to be *perpendicular* to B_0. This wave can be used for lower-density measurements than the ordinary wave, since some limited propagation is possible, even for $\omega < \omega_R$.

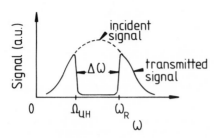

Fig. 2.9. A simple transmission experiment using the extraordinary wave to measure plasma density. In this technique the transmitter frequency is swept from below Ω_{UH} to above ω_R. The nonpropagating frequency band $\Delta\omega = \omega_R - \Omega_{UH}$ (see Fig. 2.5) is simply related to the plasma density by Eq. (2.73).

value corresponding to a given frequency ω is larger than ω/c below the resonance frequency (which is never true for the ordinary mode). But resonance also introduces absorption and some reflection, so that application with $\omega < \omega_R$ is somewhat limited.

This method can be used, also, for frequencies higher than ω_R, and results similar to those obtained with the ordinary mode can be expected, except that they now depend upon the value B_0 of the magnetic field.

Another interesting way to measure plasma density using the extraordinary wave is just to make a simple transmission experiment and sweep the transmitter frequency ω from values below Ω_{UH} to values above ω_R. For the case where $\Omega_{ci} < \omega_{pe} \ll \Omega_{ce}$, for example, an easily observed cutoff band of frequencies appears, with the cutoff bandwidth $\Delta\omega$ being proportional to the electron density (see Section 2.6), as given by the equation

$$\Delta\omega = \omega_R - \Omega_{UH} \simeq \frac{\omega_{pe}^2}{2\Omega_{ce}} = \frac{e}{2B_0\varepsilon_0}\, n_0, \qquad (2.73)$$

where n_0 is, as before [see Eq. (2.72)], the critical plasma density for the wave frequency ω used. Figure 2.9 shows qualitatively the kind of experimental results that are predicted for this method of measuring plasma denisty.

2.5.3. Some General Comments on the Use of Microwaves to Measure Plasma Parameters

The literature abounds with descriptions of experimental methods, many of them very sophisticated, for measuring plasma parameters using microwaves. The very excellent book by Heald and Wharton (1965), for example, contains several hundred references to such measurements. [This book is of particular interest for novices in this field (although it also has interest for experts), because it not only summarizes the basic theory of the interaction of electromagnetic waves with plasmas, it also contains extensive descriptions and photographs of the microwave hardware and special circuits useful in many practical experimental applications that exploit this interaction for plasma diagnostic purposes.]

Because of the subtleties of behavior of microwaves and the myriads of ways in which real plasmas deviate from even sophisticated plasma models, the implementation and correct interpretation of microwave diagnostics, although widely used, is fraught with difficulties and pitfalls.

2.6. SELECTED EXPERIMENT: A SIMPLE TRANSMISSION EXPERIMENT USING THE EXTRAORDINARY WAVE TO MEASURE PLASMA DENSITY

Equation (2.73) of Section 2.5.2 predicts that the electron density of a plasma can be obtained by making a determination of $\Delta\omega$, the frequency range over which the extraordinary wave will not propagate in a plasma. Mills and Schmitt (1969) used this technique to measure the density of a magnetically supported cylindrical column of plasma generated by the thermal ionization of a cesium vapor (a so-called Q machine). Figure 2.10 shows a diagram of their experiment. By means of a properly designed microwave horn, the electric field of the extraordinary wave was linearly polarized to be perpendicular to the direction of the uniform magnetic field supporting the plasma column (in the figure, the axes of the plasma column and of the magnetic field are perpendicular to the page). The frequency of the extraordinary wave was "swept" by placing a time-varying ("sawtooth") voltage on the reflector plate of the microwave generator (klystron). Figure 2.11 shows that over a certain range of frequencies the transmitted power is strongly reduced in a manner qualitatively similar to what is theoretically predicted (see Fig. 2.9).

Using the known actual density profile of their plasma column rather than assuming a uniform density, as is done for Eq. (2.73), Mills and Schmitt

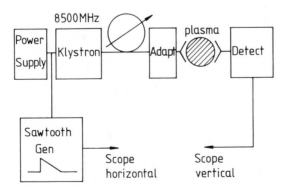

Fig. 2.10. Diagram of an experiment to measure the power transmission of the extraordinary wave through a cesium plasma as a function of wave frequency. [After Mills annd Schmitt (1969).]

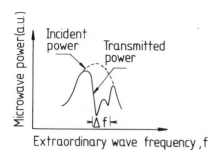

Fig. 2.11. The oscilloscope trace shows strong attenuation of the extraordinary wave over a frequency range indicated by Δf. The dashed curve shows the microwave power incident on the plasma column. [After Mills and Schmitt (1969).]

(1969) showed theoretically that $\Delta\omega = \omega_{pe}^2/\Omega_{ce}$. This result differs from Eq. (2.73) by a factor of 2, and gives the electron density in terms of Δf as,

$$n_e = 10^{-8} f_{ce} \Delta f, \qquad (2.74)$$

where f_{ce} in the cyclotron frequency of the plasma electrons in the magnetic field. By making Δf measurements for a wide variety of plasma conditions, these workers found that over a density range of $10^8 - 10^{10}$ cm^{-3} the values found by the microwave transmission experiment were in good agreement with values found using other independent measurement techniques.

Note that the transmission pattern seen in Fig. 2.11 lacks the symmetry of the "ideal" pattern seen in Fig. 2.9. Part of the asymmetry is because the frequency corresponding to maximum power output of the klystron was not the same as the frequency in the center of the Δf region. The reason for the observed regrowth in transmission near the center of Δf was not explained. The close agreement between the results of these measurements with the results of other independent measurements, however, suggests that the experiment was working basically according to the theory.

ELECTROSTATIC WAVES IN A WARM PLASMA
A FLUID-THEORY EXAMPLE

3.1. INTRODUCTION

In Chapter 2 we studied general wave propagation in an infinite, cold, two-component plasma which was characterized by charged particles having no energy except the energy alternately gained and lost as a result of their participation in the wave motion. In this chapter, we restrict our study to the behavior of electrostatic waves propagating parallel to the applied magnetic field in so-called warm plasmas, which are characterized by charged particles having *thermal* energy. Whereas, for the cold plasma, we found only one purely electrostatic mode, which was simply a (nonpropagating) normal mode of oscillation of the plasma, we find that in a warm plasma the thermal pressure of the particles makes it possible for more than one purely electrostatic mode to exist and causes these modes to be propagating modes.

The two modes of interest here are the so-called *electron plasma waves*, which occur at frequencies higher than the *electron* plasma frequency, and *ion-acoustic waves*, which occur at frequencies lower than the *ion* plasma frequency. Subject to the restrictions set forth in the model to be described in the following section, we will first derive the general dispersion relation for purely electrostatic waves in an *n*-component warm plasma. We then describe the propagation of electron plasma waves in an *n*-component plasma, one component of which must be a group of electrons. Next, we look at the propagation of ion-acoustic waves in a two-component electron–ion plasma with $T_e \gg T_i$. Finally, we look briefly at the propagation of ion-acoustic waves in a three-component plasma consisting of electrons and two species of ions (the ions of one specie being negatively charged, those of the other being positively charged) with $T_e \gg T_i$.

3.2. DISPERSION RELATION FOR PURELY ELECTROSTATIC WAVES IN A WARM PLASMA

We begin by defining the waves and the plasma of interest, i.e., we define a model. At this point we are free to make any assumptions, consistent with nature, that we want. A good rule-of-thumb is to keep the assumptions as simple as possible without changing the physics of interest. This policy generally greatly simplifies the mathematics involved and prevents the results from being cluttered with phenomena that are not of interest or relevance to the immediate problem.

We will assume, first, that we are interested only in undamped modes, i.e., modes that are undamped both in time and in space. This will allow us to use only real values for the wave frequency and wave number. We then assume that we can Fourier-analyze any physical quantity A (magnetic field, particle velocity, particle density, etc.) of interest. This is equivalent to assuming that A has the form

$$A(r,t) = A_0 + A_1 \exp[i(k \cdot r - \omega t)], \tag{3.1}$$

where A_0 describes the physical quantity when no wave is present (whose value we will assume we know), and A_1 is the perturbation or change in the physical quantity produced by the wave. We will also assume that the plasma is homogeneous and infinite in extent, so that A_0 does not depend on r. We will assume small perturbations so that we may use linear theory. Thus, when A_0 is nonzero, we assume that $A_1 \ll A_0$. An n-component plasma is assumed with each charge specie m being characterized by its charge q_m, mass m_m, mean velocity v_m, and temperature T_m. The applied external dc (i.e., steady-state) magnetic field B_0 is assumed to be uniform. There is assumed to be no dc electric field, i.e., $E_0 = 0$. The waves are assumed to be purely electrostatic, so that $B_1 = 0$, and only propagation parallel to the magnetic field is assumed to be of interest, so that $k//B_0$. The particles are assumed to be nonrelativistic, so that the force on each particle in the plasma can be written as

$$F = q(E_1 + vxB_0), \tag{3.2}$$

where E_1 is the wave electric field. The same assumptions made in Chapter 2 in arriving at the particle-conservation equation and the equation of motion for each specie are made here. That is, there are no particle-source terms, no correlation among particles, and the only interactions between particles are limited to space-charge effects and collisions between particles of the same specie m, the latter interactions leading to the pressure terms P_m.

Having verbally described the system of interest, we must now make a corresponding mathematical description. From the verbal description and from Chapter 2 we can write the particle-conservation equation for each charge specie m as

$$\frac{\partial n_m}{\partial t} + \nabla \cdot (n_m v_m) = 0. \tag{3.3}$$

Similarly, the equation of motion for each specie can be written as

$$n_m m_m \left(\frac{\partial v_m}{\partial t} + v_m \cdot \nabla v_m \right) = n_m q_m (E_1 + v_m \times B_0) - \nabla P_m, \qquad (3.4)$$

where ∇P_m is the pressure gradient.

As done in Chapter 2 these equations will be linearized, keeping only the first- order terms. The pressure gradient is then given, assuming an adiabatic compression, by

$$\nabla P_m = \gamma_m K T_m \nabla n_{1m}. \qquad (3.5)$$

In Eq. (3.5), γ_m is called the compression coefficient, K is Boltzmann's constant, T_m is the temperature, and n_{1m} is the density perturbation produced by the wave.

Since we have assumed that we know the value of all the A_0 physical quantities of interest, Eqs. (3.3) and (3.4) constitute a set of two equations in three unknowns for each charge specie: n_{1m}, v_{1m}, and E_1. Thus, we need a third independent equation involving these unknowns. At this point, we have not explicitly used Maxwell's equations although, as was demonstrated in Chapter 2, Eq. (2.19) (the particle-conservation equation) is implicitly contained in Maxwell's equations when particle-source terms are not present. We now use Maxwell's $\nabla \times H$ equation to obtain a third independent equation. Starting with this equation

$$\nabla \times H = \frac{\partial(\varepsilon_0 E)}{\partial t + j}, \qquad (3.6)$$

we first recall that in our model, $E_0 = 0$ and $B_1 = 0 = H_1$. Therefore, Eq. (3.6) gives

$$j = i\omega \varepsilon_0 E, \qquad (3.7)$$

where we have made a Fourier transform in time (or a Laplace transform in time, assuming a zero perturbation at time $t = 0$). Vector-multiplying both sides of Eq. (3.7) by (ik) and using the relationship

$$j = \sum_{m=1}^{n} q_m n_m v_m,$$

gives

$$ik \cdot E_1 = \frac{1}{\varepsilon_0 \omega} \sum_{m=1}^{n} q_m k \cdot (n_m v_m). \qquad (3.8)$$

But from Eq. (3.3), after making a Fourier transform in time (or a Laplace transform in time, assuming a zero perturbation at time $t = 0$), and a Fourier transform in space, we see that

$$k \cdot (n_m v_m) = \omega n_m, \tag{3.9}$$

so that Eq. (3.8) becomes

$$ik \cdot E = \frac{1}{\varepsilon_0} \sum_{m=1}^{n} n_m q_m = \frac{\rho}{\varepsilon_0}, \tag{3.10}$$

where ρ is the net charge density. It is not difficult to convince oneself that Eq. (3.10), which is known as Poisson's equation, cannot be obtained from Eqs. (3.3) and (3.4) and, thus, is the third independent equation needed.

Keeping only terms of first-order (a process called linearization) in Eq. (3.7), we see that $j_1 // E_1$. But from our assumptions, $E_1 // k // B_0$. Thus, $j_1 // B_0$, so that in Eq. (3.4) $v_{1m} \times B_0 = 0$ for all particle species. Replacing each of the variables in linearized Eqs. (3.3), (3.4), and (3.10) by an expression of the form given by Eq. (3.1), we obtain for *each* specie of particle the set of so-called *linearized* equations of interest

$$(\omega - k v_{0m}) n_{1m} - k n_{0m} v_{1m} = 0, \tag{3.11}$$

$$n_{0m}(\omega - k v_{0m}) v_{1m} - k n_{1m} V_m = \frac{i q_m}{m_m} n_{0m} E_1, \tag{3.12}$$

$$ikE_1 = \frac{1}{\varepsilon_0} \sum_{m=1}^{n} n_{1m} q_m, \tag{3.13}$$

where in Eq. (3.12)

$$V_m = \left(\frac{\gamma_m K T_m}{m_m} \right)^{1/2} \tag{3.14}$$

is called loosely the *thermal speed* of the particles of specie m. Note that since all vector quantities are parallel to B_0 there is no need for vector notation in these equations.

Solving Eq. (3.11) for v_1, substituting into Eq. (3.12), and then solving the resulting equation for the perturbed density n_{1m} for each particle specie, gives

$$n_{1m} = \frac{q_m n_{0m}}{m_m} \frac{ikE_1}{(\omega - k v_{0m})^2 - k^2 V_m^2}. \tag{3.15}$$

Substituting n_{1m} of Eq. (3.15) into the linearized Poisson equation, i.e., Eq. (3.13), gives

$$\mathscr{D}(k,\omega)E_1 = 0, \tag{3.16}$$

where

$$\mathscr{D}(k,\omega) = 1 - \sum_{m=1}^{n} \frac{\omega_{pm}^2}{(\omega - kv_{0m})^2 - k^2 V_m^2}, \tag{3.17}$$

and $\omega_{pm} = (n_m q_m^2/\varepsilon_0 m_m)^{1/2}$ is the plasma frequency of the particles of specie m.

Observation of Eq. (3.16) shows that in order to find a nontrivial value for E_1 the coefficient $\mathscr{D}(k,\omega)$ of E_1 must be zero. That is, using Eq. (3.17)

$$1 - \sum_{m=1}^{n} \frac{\omega_{pm}^2}{(\omega - kv_{0m})^2 - k^2 V_m^2} = 0. \tag{3.18}$$

Equation (3.18) is the sought after dispersion relation for purely electrostatic wave propagation along the direction of the magnetic field of a magnetically supported n-component warm plasma. As noted in Chapter 2, the dispersion relation contains all of the linear properties of the defined plasma with respect to the defined waves.

3.3. ELECTROSTATIC MODES IN A WARM PLASMA

Equation (3.18) contains all the information that can be known concerning not only the number of purely electrostatic modes possible in the defined plasma but, also, the linear properties of these waves. We will be interested first of all to know how many modes are predicted to be possible. Observation of Eq. (3.18) shows that it is an ordinary algebraic equation in ω and k. We look first at the case where none of the particle species has a mean velocity v_{0m}, i.e., where there is no net "drift" or "streaming" of any of the particle species with respect to the laboratory reference frame. For this case, for a plasma composed of n species of charged particles, Eq. (3.18) can be viewed as an algebraic equation in ω^2/k^2, with the highest-order term being $(\omega^2/k^2)^n$. Such an equation then has n solutions for ω^2/k^2. Each of these solutions corresponds to two modes characterized by two different values of ω/k. The corresponding pairs of ω,k values can be chosen as (ω,k) and $(\omega,-k)$. These two modes are identical except that they are propagating in opposite directions. From the laboratory frame these would be viewed as identical modes. Thus, for this case, we have n different modes. For the more general case where there *is* a *nonzero* mean plasma velocity, Eq. (3.18) can be viewed as an algebraic equation in k whose highest-order term is k^{2n}. Thus, for this general case, there will be $2n$ solutions, (ω,k), corresponding to $2n$ modes. The modes for these two cases, as the reader might suspect, are closely related. The connection is that, when there is a nonzero plasma velocity, the two oppositely propagating modes (which were identical in frequency and speed for the *zero-drift-velocity* case) now have their frequencies and speeds, as viewed from the laboratory frame, Doppler-shifted by $\pm k v_{0m}$ and $\pm v_{0m}$, respectively, assuming that all particle species have the same mean velocity $v_{0m} = v_0$. When these velocities become different, one mode can be unstable.

In the following three sections we will describe some purely electrostatic modes that are well known and which have $\omega - k$ regimes where relatively little damping occurs, so that their description by the present theory is valid in those regimes. Other low-frequencey modes exist but are highly damped and cannot be correctly described by the present theory. A comparison of the results of the present theory with the results of a more precise kinetic theory, which is able to handle wave-particle damping interactions, will be given in Chapter 7.

3.3.1. Electron Plasma Waves

We look now at the propagation of electron plasma waves. Since by definition this mode involves the motion of electrons, one component of the plasma must be electrons. The composition of most plasmas of interest will, in fact, be electrons and one specie of positive ions; however, electron plasma waves are possible in n-component plasmas where the value of n is arbitrary. To see this we look at the general dispersion relation given by Eq. (3.18). The dispersion relation for electron plasma waves is obtained by considering the high-frequency behavior of Eq. (3.18). Because $\omega_{pe}^2 \gg \omega_{pi}^2$ and $V_e^2 \gg V_i^2$, we see that for high frequencies the electron term in the sum part of Eq. (3.18) is much greater than any other term. Thus, for high frequencies, Eq. (3.18) can be approximated by

$$1 - \frac{\omega_{pe}^2}{\omega^2 - k^2 V_e^2} = 0. \tag{3.19}$$

For planar waves of high frequency the compression of the electrons due to the wave is, to a close approximation, one-dimensional and adiabatic. For this case the (adiabatic) compression coefficient γ_m in Eq. (3.14) is equal to 3, as we will find in the microscopic theory, and Eq. (3.19) can be written as

$$\omega^2 = \omega_{pe}^2 + 3k^2 \frac{KT_e}{m_e}, \tag{3.20}$$

where k is the wave number, and K is Boltzmann's constant. Equation (3.20) is the well-known Bohm–Gross dispersion relation for electron plasma waves

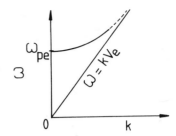

Fig. 3.1. The Bohm–Gross dispersion relation for electron plasma waves.

(Bohm and Gross, 1949). An ω-k plot of Eq. (3.20) is given in Fig. 3.1. The electron plasma wave can propagate only for frequencies $\omega > \omega_{pe}$. This represents a cutoff ($\mathcal{N} \to 0, v_0 \to \infty$) for this mode. Thus, for low frequencies the phase velocity of the electron plasma wave is greater than c, the speed of light. However, as the wave frequency increases to higher and higher values, the phase velocity decreases toward the thermal speed V_e of the electrons and, as we shall see in Chapter 7, a wave–particle interaction between the wave and the electrons occurs that causes a strong damping known as Landau damping. Since we have assumed no damping in our model the present description is no longer valid at such high frequencies, as indicated by the dashed part of the curve in Fig. 3.1.

It is of interest to compare the dispersion relation for a warm plasma given by Eq. (3.19) with the corresponding dispersion relation for a cold plasma given by Eq. (2.58). Basically, the two are seen to be the same, except for the V_e^2 term in the denominator of Eq. (3.19). It is this term, which is related to the thermal energy of the electrons, that supplies the thermal pressure that makes it possible for this purely electrostatic mode to *propagate* in a warm plasma.

3.3.2. Ion-Acoustic Waves in a Two-Component Electron–Ion Plasma with $T_e \gg T_i$

As a second example of purely electrostatic waves in a warm plasma we look at ion-acoustic waves in a nondrifting, two-component, electron–ion plasma. For such a plasma, Eq. (3.18) can be written as

$$1 - \frac{\omega_{pe}^2}{\omega^2 - k^2 V_e^2} - \frac{\omega_{pi}^2}{\omega^2 - k^2 V_i^2} = 0. \tag{3.21}$$

We will now assume that $\omega \ll \omega_{pe}$ and that $T_e \gg T_i$. With these assumptions, $V_i \ll \omega/k \ll V_e$, so that Eq. (3.21) becomes

$$1 + \frac{\omega_{pe}^2}{k^2 V_e^2} - \frac{\omega_{pi}^2}{\omega^2 - k^2 V_i^2} = 0, \tag{3.22}$$

which is the classical dispersion relation for ion-acoustic waves. For frequencies $\omega \ll \omega_{pi}$, the phase velocity ω/k of ion-acoustic waves is given by

$$\frac{\omega}{k} = \left(\frac{\gamma_e n_e K T_e + \gamma_i n_i K T_i}{n_i m_i + n_e m_e}\right)^{1/2},$$

where $\gamma_i = 3$, corresponding to a one-dimensional adiabatic compression of the ions, and $\gamma_e = 1$, corresponding to an isothermal compression of the electrons. (Because of the large thermal speed of the electrons and the relatively low frequencies of the waves, no localized heating or cooling of the electrons occurs as a result of the wave motion.) For most plasmas charge neutrality requires that $n_e \sim n_i = n_0$; also, $m_i \gg m_e$. Then

$$\frac{\omega}{k} = Cs_i = \left(\frac{KT_e + \gamma_i KT_i}{m_i}\right)^{1/2}, \tag{3.23}$$

In many plasmas of interest, such as the discharge type, $\gamma_i K T_i \ll K T_e$. For such plasmas Eq. (3.23) can be simplified and the speed of ion-acoustic waves is given with good accuracy by the equation

$$Cs = \left(\frac{KT_e}{m_i}\right)^{1/2}. \tag{3.24}$$

This compressional-type wave in a plasma is analogous to an ordinary sound wave in air, thus explaining why this wave is sometimes called an ion sound wave. The ions of the plasma provide most of the inertia of the wave, while the electrons provide most of the thermal pressure to drive the wave.

When the frequency of an ion-acoustic wave approaches the ion plasma frequency ω_{pi}, a decoupling occurs between the electrons and ions. As a result the thermal pressure of the electrons is no longer effective in driving the wave and the phase velocity of the wave decreases toward the ion thermal speed [see Eqs. (3.23) and (3.14)]. Again, as for the electron plasma wave under similar circumstances (i.e., when the wave speed is of the same order as the particle speed), Landau damping of the wave occurs (in this case, due to a wave—particle interaction between the wave and the *ions*), and the present description of the wave is no longer valid. In fact, our primary motivation here in assuming that $T_e \gg T_i$ was to guarantee that Landau damping would not be an important effect. Experimentally, ion-acoustic waves are difficult to observe in plasmas where the electron temperature is of the same order as the ion temperature. Wave motion with Landau damping present will be considered in Chapter 7. An $\omega - k$ plot of Eq. (3.22) is given in Fig. 3.2 where, as for the electron plasma wave, the dashed part of the curve corresponds to the frequency range where Landau damping occurs.

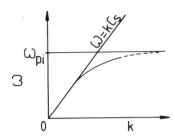

Fig. 3.2. Ion-acoustic wave dispersion relation for a two-component, ion–electron warm plasma.

3.3.3. Ion-Acoustic Waves in a Three-Component Electron–Positive-Ion– Negative-Ion Plasma with $T_e \gg T_i$

As a final example of purely electrostatic waves in warm plasmas we consider a three-component plasma consisting of electrons, positive ions, and negative ions. For this case, Eq. (3.18) can be written as

$$1 + \frac{\omega_{pe}^2}{k^2 V_e^2} - \frac{\omega_{p-}^2}{\omega^2} - \frac{\omega_{p+}^2}{\omega^2} = 0, \qquad (3.25)$$

where the $+$ and $-$ subscripts refer to the positive and negative ions, respectively, and, as in the previous example, we have assumed that the electrons are much hotter than the ions (i.e., $T_e \gg T_+, T_-$), so that $V_+, V_- \ll \omega/k \ll V_e$. For the low-frequency limit, i.e., for $\omega \ll \omega_{p+}, \omega_{p-}$, the phase velocity of the ion-acoustic wave can be written as

$$\frac{\omega}{k} = V_e \left(\frac{\omega_{p+}^2 + \omega_{p-}^2}{\omega_{pe}^2} \right)^{1/2}. \qquad (3.26)$$

Again, as in Eq. (3.23), both the masses and the relative densities of the components are involved in the wave speed. Of particular interest is the case where the electron density is very *small* compared to the ion densities. Since charge-neutrality requires that $n_e = n_+ - n_-$, such a plasma would correspond to a situation where $n_- \sim n_+$. Experimentally, plasmas with a large fraction of negative ions have been produced in several gases including H_2, CsCl, O_2, and I_2. Ion-acoustic waves have been studied in iodine plasmas having β values as low as 3×10^{-3}, where $\beta = n_e/n$, with $n = n_e + n_+ + n_- \sim n_+ + n_-$ (Doucet, 1970). For small β, the low-frequency limit of the ion-acoustic-wave phase velocity in such a plasma can be written as (D'Angelo, Von Goeler et al., 1966)

$$\frac{\omega}{k} \sim \left(\frac{KT_e/M}{\beta} \right)^{1/2} = \frac{Cs}{\beta^{1/2}}, \qquad (3.27)$$

where the "average" mass M is defined as $1/M = 1/m_+ + 1/m_-$, and Cs can be thought of as the classical ion sound speed in a normal $(T_e \gg T_i)$, two-component, electron–ion plasma having ions of mass M. Equation (3.27) gives an ion sound speed that is higher by a factor of $(\beta)^{-1/2}$ than the classical value. If $m_+ \sim m_-$, this speed is easily much higher than the thermal speeds of the ions, so that the wave is not Landau damped by the *ions*. If, however, β is too small, the phase velocity given by Eq. (3.27) can approach the thermal speed of the *electrons* and Landau damping due to wave–electron interactions can occur.

3.4. SELECTED EXPERIMENTS

Jones and Alexeff (1966), Doucet *et al.* (1968), Alexeff *et al.* (1968), and Joyce *et al.* (1969) have made a series of studies of the propagation of ion-acoustic waves in simple, discharge-type, two-component, ion–electron, collisionless plasmas having $T_e \gg T_i$. These studies, some of which will be described here, have not only delineated the macroscopic and microscopic properties of these waves but, also, have led to a variety of diagnostic techniques that use these waves to measure the basic properties of the plasmas in which they propagate. Electron plasma waves, being characterized by much higher speeds and frequencies than ion-acoustic waves, are more difficult to study in finite-size plasmas, such as most laboratory plasmas. Also, over most of their available frequency range, they experience electron–wave interactions that lead to strong Landau damping (see Chapter 7) making their observation difficult. Nevertheless, some interesting experiments have bee made, and we will briefly describe some work which has been done by Derfler and Simonen (1966).

3.4.1. Ion-Acoustic Waves

Figure 3.3 shows the type of experimental apparatus used for the experiments to be described here. The measurements were made in a spherical glass discharge tube some 20 cm in diameter. The background impurity pressure was maintained at approximately 10^{-6} Torr by continuous pumping and liqiud-nitrogen trapping, while the rare-gas operating pressures were approximately 10^{-3} Torr (1 Torr = 1 mmHg). The anode–cathode assembly, seen on the right-hand side of the tube, provides a source of energetic electrons that, by way of ionizing collisions with the background gas atoms, generates the plasma. By operating in the emission-limited regime of the hot cathode, plasma generated in this manner can be very quiescent and can operate stably for long periods of time. Plasma densities of up to a few times 10^9 cm^{-3} and electron temperatures of a few eV are typical. Figure 3.3 shows that propaga-

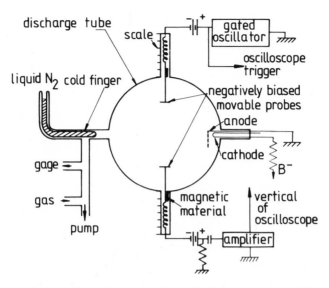

Fig. 3.3. Schematic of experimental apparatus for studying ion-acoustic waves. (After Jones and Alexeff, 1966.)

tion occurs between negatively biased transmitting and receiving flat-disk, movable probes. With such an arrangement, so-called time-of-flight measurements (in which the position, amplitude, and shape of the wave as a function of propagation time are measured) can be made. The waves are generated at the upper probe by voltage pulses, voltage steps, or sinusoidal voltage bursts. The arrival of an ion wave at the bottom probe produces a perturbation of the ion current to this probe. Figure 3.4 shows some typical results. Figure 3.4(a) shows the detector probe response (upper trace) to step-function voltages

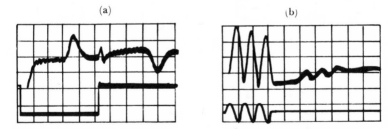

Fig. 3.4. Production of ionic sound waves: (a) The lower trace shows two voltage steps of opposite polarity as the driving voltages on the emitter probe. The upper trace shows the two corresponding ion waves, also of opposite polarity, arriving at the detector. Note the time separation of the ion waves and the initial transient signals due to direct coupling between emitter and detector probes. (b) The lower trace shows a sine wave burst as a driving voltage. The upper trace shows the direct-coupled signal and the later-arriving ionic sound wave signal. (After Jones and Alexeff, 1966.)

(lower trace) placed on the emitter probe. Figure 3.4(b) shows, similarly, the response to a sinusoidal driving voltage. Note that in each case there is a *directly coupled* signal on the detector probe. Thus, a time-of-flight technique is needed to separate the directly coupled signal from the later-arriving ion wave signal.

For plasmas having $T_e \gg T_i$, and in which the ion-neutral collision rate is negligible, the theoretical phase velocity of ionic sound waves in a plasma is given by Eq. (3.24) in Section 3.3.2 as

$$v_\varphi = \left(\frac{KT_e}{m_i}\right)^{1/2}. \tag{3.28}$$

Figure 3.5 shows some time-of-flight data taken for ion-acoustic wave propagation in plasmas made from the rare gases. In this figure the experimental velocities are given by the slopes of the curves. The velocity dependence on ion mass is immediately obvious.

Figure 3.5 shows, also, that the linear extrapolation of the experimental curves gives a *finite probe separation* at $t = 0$. This is evidence for *sheaths* on the negatively biased probes. The total thickness of each sheath, assuming that the emitter and receiver sheaths are equal, is about 7–10 Debye lengths.

Figure 3.6 shows time-of-flight data that confirm the square-root dependence of the phase velocity on T_e given by Eq. (3.28). Actually, for the lowest values of T_e shown in Fig. 3.6, the inequality $T_e \gg T_i$ is no longer valid. For these values the following equation [see Eq. (3.23), where we have assumed $n_i \sim n_e$, $\gamma_e = 1$, $\gamma_i = 3$]

$$v_\varphi = \left(\frac{KT_e + 3KT_i}{m_i}\right)^{1/2} \tag{3.29}$$

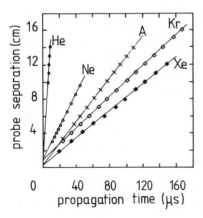

Fig. 3.5. Time-of-flight data for ionic sound wave propagation in rare-gas plasmas using pulses. Time is measured from the driving pulse to the leading edge of the response signal. For the four heaviest gases the electron temperature was about 1 eV, whereas for He it was about 9 eV. (After Jones and Alexeff, 1966.)

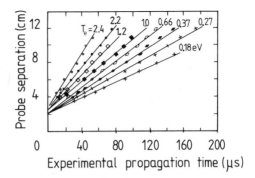

Fig. 3.6. Typical time versus distance plots showing the ion wave velocity dependence on electron temperature. The gas used was xenon. (After Alexeff et al., 1968.)

must be used in order to obtain agreement between experiment and theory. Again, as in Fig. 3.5, the slopes of the curves in Fig. 3.6 give the experimental velocities of the ion-acoustic waves. Figure 3.7 shows a plot of these experimental velocities versus the experimental electron temperatures. In agreement with Eq. (3.29), extrapolation of the line-of-best-fit in Fig. 3.7 to $T_e = 0$ gives a finite intercept on the vertical axis. Setting this intercept equal to $(3KT_i/m_i)^{1/2}$—see Eq. (3.29)—gives $T_i \simeq 0.05$ eV, a value in good agreement with other independent determinations of this quantity.

Using a transmitter consisting of only a single *thin wire*, Joyce et al. (1969) were able to observe the dispersion that is theoretically predicted to occur as $\omega \to \omega_{pi}$ (see Fig. 3.2). A comparison of their measurements and the behavior that is theoretically predicted for the conditions of their experiment is shown in Fig. 3.8.

Using the theoretical model described in Section 3.2 Doucet et al. (1968) showed that a simple relationship exists between the electrostatic potential V_1 of an ion-acoustic wave and the ion density perturbation n_1 produced by the wave given by

$$\frac{eV_1}{KT_e} = \gamma_e \frac{n_1}{n_0}. \tag{3.30}$$

Fig. 3.7. Determination of plasma ion temperature by measuring the ion-acoustic wave velocity dependence on electron temperature. The theoretical curve assumes the ion temperature value calculated from the vertical axis intercept of the least-squares line of best fit. The gas used was xenon. (After Alexeff et al., 1968.)

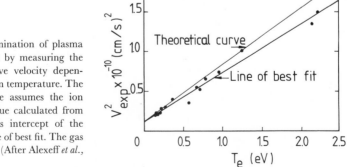

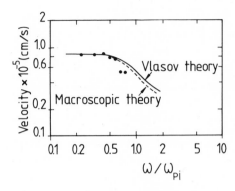

Fig. 3.8. A comparison of experimental and theoretical dispersion for ion-acoustic waves. The set of points are the ion wave velocities measured experimentally. The two curves show the theoretically expected dispersion for the conditions of the experiment. (After Joyce et al., 1969.)

The relationship expressed by Eq. (3.30) was investigated experimentally by Wong et al. (1964) and then by Doucet et al. (1968). In this latter work simultaneous measurements were made of the absolute magnitudes of V_1 and n_1/n_0. T_e was measured by means of a standard Langmuir probe. eV_1/KT_e was then compared to n_1/n_0 to yield an experimental value for γ_e. Figure 3.9 shows the basic features of the experiment. The ion-acoustic waves were generated by applying 30-kHz, ~50-V amplitude sine wave bursts to the movable, negatively biased transmitter probe shown in the figure. The small, positively biased receiver probe was used to measure n_1/n_0 of Eq. (3.30). The probe was biased just above the plasma potential, a region of the characteristic where the current to the probe is relatively insensitive to small fluctuations in the probe–plasma potential. By using both the dc and ac inputs to the oscilloscope, relative values could be immediately determined for the unperturbed and perturbed electron currents, respectively, to the probe. Since

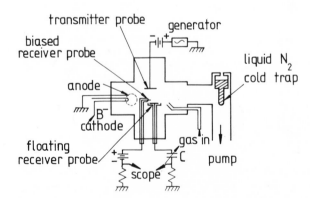

Fig. 3.9. Schematic of the apparatus used for studying ion-acoustic waves. (After Doucet et al., 1968.)

the electron current to the probe is directly proportional to the local electron density, accurate relative values were obtained for the undisturbed plasma density n_0 and for the density perturbation n_1 produced at the probe by the propagating ion-acoustic waves. This small probe was also used in the conventional Langmuir manner to measure the plasma electron temperature.

The *floating* receiver probe shown in Fig. 3.9 was used to measure the *potential* of the propagating ion-acoustic wave. This probe, like the transmitter probe, is a flat copper disk a few centimeters in diameter. Unlike the transmitter probe, however, it was found necessary to insulate entirely the floating receiver probe, except for the surface facing the transmitter, in order to prevent capacitive loss of the detected voltage signal to ground. It was necessary, also, to use a large series resistor to reduce transient currents to the probe to small enough values to prevent appreciable excursions of the probe from floating potential.

Figure 3.10 shows plots of eV_1/KT_e and n_1/n_0, as a function of transmitter–receiver probe separation. It is noted that the two sets of points seem to lie on the same curve indicating that $\gamma_e \sim 1$, in agreement with our discussion of γ_e in Section 3.3.2.

Extrapolation of the straight line in Fig. 3.10 to zero probe separation, if valid, indicates that the wave is a very *low-amplitude* one. Even at the point of generation, eV_1/KT_e is seen to be less than 0.1. Since $T_e \sim 4.3$ eV, and since the driving voltage at the transmitter is on the order of 100 V, this indicates that the amplitude of the generated wave is of the order of only one-half of one percent of the applied voltage. Thus, the coupling to the plasma is extremely weak, and the propagating wave is quite *linear* throughout its entire range of propagation.

Fig. 3.10. Experimental ion-acoustic wave potential and plasma density perturbation as a function of distance. The clustering of both sets of points about the same visual line of best fit seems to indicate an experimental value of $\Upsilon_e \sim 1$. The lengths of the bars labeled $\Upsilon_e = \frac{5}{3}$ and $\Upsilon_e = 3$ show the average vertical separation of the two sets of data points for Υ_e having those values. Note that extrapolation to zero probe separation leads to a ratio of <0.1, indicating linear waves throughout the region of propagation. (After Doucet *et al.*, 1968.)

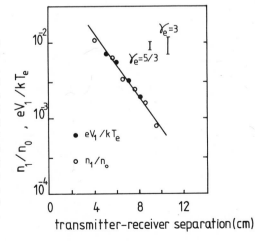

3.4.2. Electron Plasma Waves

In Section 3.3.1 we found that under certain conditions the behavior of electron plasma waves is described by the Bohm–Gross dispersion relation given by Eq. (3.20). This equation can be rewritten as

$$v_\varphi^2 = \frac{\omega^2}{k^2} = \frac{3KT_e/m_e}{1 - \omega_{pe}^2/\omega^2}. \tag{3.31}$$

Equation (3.31) predicts a low-frequency cutoff $(v_\varphi \to \infty)$ at $\omega = \omega_{pe}$ and, for ω a few times larger than ω_{pe}, predicts that $v_\varphi \to (3KT_e/m_e)^{1/2}$, the thermal speed of the electrons. As we have noted already this latter condition leads to an important wave–particle interaction that produces a strong wave damping called Landau damping (see Chapter 7). Thus, there is only a small frequency "window" between ω_{pe} and a few times ω_{pe} where these waves can be expected to be experimentally observable. Figure 3.11 shows an $\omega - k$ plot of Eq. (3.31).

Figure 3.11 shows the diagram of an experimental device used by Derfler and Simonen (1966) to study electron plasma wave propagation in a sodium plasma produced by the resonance ionization of a sodium vapor created between two hot tantalum plates. The resulting plasma had a density of 2×10^7 cm^{-3} and an electron temperature of 0.2 eV, giving, respectively, an electron plasma frequency of 34 MHz and an electron thermal speed of 3×10^6 cm/s. Due to a low background pressure of only 2×10^{-6} mmHg, collisional effects were completely negligible.

Using sensitive measurement techniques, similar to a lock-in amplifier for very high frequency, Derfler and Simonen (1966) were able to study the propagation of electron plasma waves between the two grids shown in Fig. 3.11 over a frequency range of 37–115 MHz. The results of their measure-

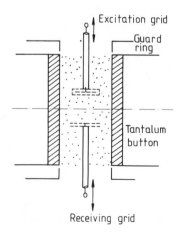

Fig. 3.11. Diagram of the sodium plasma tube showing the probe arrangement. (After Derfler and Simonen, 1966.)

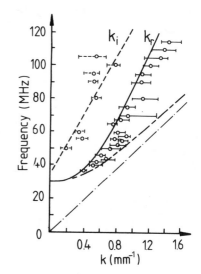

Fig. 3.12. Dispersion diagram: (○) experimental measurements; k_r and k_i, Landau Eq. (1); (- - -) thermal speed $\omega/k = (3kT/m)^{1/2}$; (- · -) Bohm and Gross, Eq. (2). (After Derfler and Simonen, 1966.)

ments are shown in Fig. 3.12. In this figure, the electron thermal speed is given by the slope of the dotted curve. The wave speed v_φ predicted by the Bohm–Gross dispersion relation [Eq. (3.31)] is given by the slope of the dashed–dotted curve. The open circles show the experimental measurements. At the lowest observable frequencies the experimental points for k_r ($2\pi/\lambda$, where λ is the measured wavelength) are seen to fall near the Bohm–Gross dispersion curve. At higher frequencies where Landau damping becomes strong, the experimental points deviate further and further from the Bohm–Gross curve (which assumes no wave damping—see Section 3.3.1). In observing the amplitude of the waves as a function of propagation distance, Derfler and Simonen (1966) were able to measure, also, the distance required for the wave amplitude to decrease by a factor of $1/e$. The reciprocal of this distance is denoted by k_i and is shown in Fig. 3.12 Also shown in Fig. 3.12 are two theoretical curves that result from a theoretical treatment in which Landau damping is taken into account. Agreement between the observed dispersion and the theoretically predicted dispersion is seen to be good.

ION-ACOUSTIC WAVES WITH
ION-NEUTRAL AND
ELECTRON-NEUTRAL COLLISIONS

4.1. INTRODUCTION

In Chapter 3 we studied the behavior of electrostatic waves propagating parallel to the applied magnetic field in warm plasmas. We explicitly eliminated any kind of wave damping or instability from our model so that both ω and k were real quantities. We want to continue the study of the preceding chapter, except that we now want to include in our model collisions between charged particles and neutrals, and calculate what effects these collisions have on wave motion in the plasma. Although we will consider only one type of wave explicitly, ion-acoustic waves, this case will be considered in sufficient detail that the reader should experience little difficulty in adapting the procedure to other types of waves in other kinds of plasmas.

As might be expected since there is a continual interchanging of energy between the wave and the charged particles in wave motion in plasmas, collisions between particles and neutrals can significantly interfere with the energy-exchange process and lead to strong damping of the wave. Not so obvious, perhaps, is the fact that the wave speed as well as the range of possible wavelengths can also be strongly modified. All of these changes manifest themselves in the theory as changes in the character of ω and k. In general, both ω and k can be complex at the same time, but when k is a complex quantity it is always possible, by an appropriate choice of a moving frame [with velocity $v_0 = \mathrm{Im}(\omega)/\mathrm{Im}(k)$], to reduce to the case of a real $- \omega$ value. Thus, we will consider the two following cases separately: First, we will allow ω to be complex, while simultaneously requiring that k be real. This situation characterizes the so-called *initial-value problem*. In the second case we will allow k to be complex, while keeping ω real. This situation characterizes the so-called *boundary-value problem*.

4.2. DISPERSION RELATION WITH COLLISIONS

With appropriate modifications, the equations that will be used here are the same as those in Chapter 3, i.e., the particle-conservation equations, the equations of motion, and Poisson's equation. The major modification occurs in the equations of motion, where we will introduce the effects of collisions via a term involving the collisional frequency. In the absence of a dc electric field, the presence of collisions with neutrals allows us to assume that the mean velocities of the charged particles are zero. This assumes, of course, that the mean velocity of the neutrals themselves is zero. With these modifications, these equations can be written as

$$\frac{\partial n_m}{\partial t} + \nabla \cdot (n_m v_m) = 0; \tag{4.1}$$

$$n_m m_m \left(\frac{\partial v_m}{\partial t} \right) + v_m \cdot (\nabla v_m) = n_m q_m E_1 - \nabla P_m - n_m m_m v_{mn} v_m; \tag{4.2}$$

$$\nabla \cdot E_1 = \frac{e}{\varepsilon_0} (n_{i1} - n_{e1}). \tag{4.3}$$

In Eq. (4.2) v_{mn} is the collision frequency of particles of specie m with the neutrals n. In this study, in keeping with our basic philosophy of keeping the physical situation as simple as possible without losing the physics of interest, we will consider wave propagation in only a two-component electron–ion plasma, as has already been assumed in Eq. (4.3). Thus, in Eqs. (4.1) and (4.2), $m = e,i$.

Since we have assumed charge neutrality throughout the plasma for the steady-state situation (no waves present), this means that $n_{i0} = n_{e0}$. Also, we have assumed that there are no dc electric fields, i.e., $E_0 = 0$. Thus, Poisson's equation, Eq. (4.3), already appears in linearized form (i.e., each term in the equation involves only one perturbed quantity to the first power). Since Eq. (4.1) is the same as in Chapter 3, its linearized form will be the same (except that here $v_{0m} = 0$). Thus, after linearization, Eq. (4.1) gives the perturbed velocity for each specie as

$$v_1 = \frac{\omega}{k} \frac{n_1}{n_0}. \tag{4.4}$$

After linearization, Eq. (4.2) gives for each specie

$$n_0 m \frac{\partial v_1}{\partial t} = n_0 q E_1 - \gamma K T \frac{\partial n_1}{\partial z} - n_0 m v v_1. \tag{4.5}$$

No vector signs appear in Eqs. (4.4) and (4.5) since our model requires that all perturbed quantities vary in only one direction, i.e., along the direction of the magnetic field (which in Eq. (4.5) we have assumed to be along the z axis of our coordinate system). Equation (4.3) can be rewritten without vectors as

$$ikE_1 = \frac{e}{\varepsilon_0}(n_{i1} - n_{e1}), \qquad (4.6)$$

where we have made a Fourier transform in space and have used the fact that $k//E_1$.

Making a Laplace transform in time and a Fourier transform in space, Eq. (4.5) gives

$$-i(\omega + iv)v_1 = \frac{q}{m}E_1 - \gamma\frac{KT}{m}ik\frac{n_1}{n_0}. \qquad (4.7)$$

Substituting the values of v_1 given by Eq. (4.4) into Eq. (4.7), substituting the resulting expressions for n_{i1} and n_{e1} into Eq. (4.6), and proceeding as in Chapter 3, gives a new dispersion relation that can be written as

$$\mathscr{D}(\omega,k) = 1 - \frac{\omega_{pe}^2}{\omega(\omega + iv_{en}) - k^2V_e^2} - \frac{\omega_{pi}^2}{\omega(\omega + iv_{in}) - k^2V_i^2} = 0. \qquad (4.8)$$

Equation (4.8) is, thus, the dispersion relation for purely electrostatic waves propagating parallel to the applied magnetic field in a warm, two-component, electron–ion plasma, including the effects of collisions of the ions and electrons with background neutrals. Except for collisions, Eq. (4.8) is identical with the dispersion relation given by Eq. (3.21) for a collisionless two-component plasma. Because of collisions, we see in Eq. (4.8) the appearance of two new "frequencies," $(\omega + iv_{en})$ and $(\omega + iv_{in})$, coming from the equations of motion; there is still an unmodified ω coming from the particle-conservation equations. Equation (4.8) differs from the corresponding equation for collisionless plasmas in another interesting mathematical sense: because of the appearance of the collision frequency terms and because ω and k are no longer required to be real, Eq. (4.8) is complex. In effect, Eq. (4.8) represents two equations since the real and imaginary parts of the left-hand side of the equation must separately be equal to zero.

Since we are interested only in ion-acoustic waves, in Eq. (4.8) we make the assumption that $\omega \ll \omega_{pi}$ (and, hence, also that $\omega \ll \omega_{pe}$, since $\omega_{pi} \ll \omega_{pe}$). In this approximation, the real parts of the second and third terms in Eq. (4.8) are much larger than the first term, so that the dispersion relation for ion-acoustic waves becomes

$$\frac{\omega_{pi}^2}{\omega(\omega + iv_{in}) - k^2 V_i^2} + \frac{\omega_{pe}^2}{\omega(\omega + iv_{en}) - k^2 V_e^2} = 0. \tag{4.9}$$

Equation (4.9) is a second-degree algebraic equation in ω, and can be expressed in the simple form

$$\omega^2 + i\omega v_c - k^2 Cs_i^2 = 0, \tag{4.10}$$

where v_c is an *average collision frequency* defined by

$$v_c = \frac{m_e v_{en} + m_i v_{in}}{m_e + m_i} \simeq v_{in}, \tag{4.11}$$

and Cs_i is the ion-acoustic wave velocity, taking into account the electron and the ion temperatures, given by

$$Cs_i = \left(\frac{m_i V_i^2 + m_e V_e^2}{m_e + m_i}\right)^{1/2} \simeq \left(\frac{kT_e + 3kT_i}{m_i}\right). \tag{4.12}$$

So far we have put no restrictions on the values that the collision frequencies may have. However, for reasons that will become clear shortly (related to our assumption that $\omega \ll \omega_{pi}$), we will now require that the ion-neutral and electron-neutral collision frequencies be restricted so that the average collision frequency given by Eq. (4.11) will be much smaller than the ion plasma frequency, i.e., $v_c \ll \omega_{pi}$. Thus, the results that will be found in the last two sections of this chapter, dealing with initial-value and boundary-value problems involving ion-acoustic waves, will be valid only when this condition is satisfied. It should be stated explicitly that this restriction does not represent a basic limitation with the theory of this chapter: the general dispersion relation given by Eq. (4.8) has no restrictions on the values of the ion-neutral and electron-neutral collision frequencies. It is only when we approximate Eq. (4.8) for low frequencies by Eqs. (4.9) and (4.10) that the restriction $v_c \ll \omega_{pi}$ becomes necessary. Ion-acoustic wave propagation at much higher frequencies (even approaching ω_{pi}), and with much higher collision frequencies, can be validly handled by Eq. (4.8).

Until now we have not specified the nature of ω and k. We will consider two cases. The first case will be that k is real and ω is complex, while the second case will be that ω is real and k is complex. For both of these cases we will assume, as in Chapter 3 [see Eq. (3.1)], that all of the perturbed physical quantities associated with the wave have the form

$$A_1(z,t) = A_1 \exp[i(kz - \omega t)], \tag{4.13}$$

where A_1 on the right-hand side of Eq. (4.13) is the amplitude of the physical quantity.

4.3. INITIAL-VALUE PROBLEM

We will study first what is known as an initial-value problem. That is to say, we will assume that at time $t = 0$, there is an ion-acoustic wave with k real propagating in the plasma. Therefore, ω must be complex in order to satisfy Eq. (4.10). This means [see Eq. (4.13)] that the wave will be damped in time, but not in space. Thus, we assume in Eq. (4.10) that ω has the form $\omega = \omega_r + i\omega_i$, where ω_r and ω_i are the real and imaginary parts, respectively, of ω. Thus ω_r is equal to $2\pi f$, where f is the real wave frequency, and ω_i is the so-called temporal damping rate. With this notation, Eq. (4.10) leads to two different equations, since both the real and imaginary parts of the equation must be separately equal to zero

$$\omega_r - \omega_i^2 - \omega_i \nu_c - k^2 C s_i^2 = 0; \tag{4.14}$$

$$2\omega_i + \nu_c = 0. \tag{4.15}$$

Equation (4.15) gives the damping rate directly

$$\omega_i = -\frac{\nu_c}{2}. \tag{4.16}$$

We can see that ω_i is a negative quantity, as we might expect for a damping rate [a *positive* ω_i would have indicated wave *growth*; see Eq. (4.13)].

Using Eq. (4.16) we see from Eq. (4.13) that all of the perturbed physical quantities associated with the wave for the initial-value problem will have the form

$$A_1(z,t) = A_1 \left[\exp\left(-\frac{\nu_c t}{2} \right) \right] \exp[i(kz - \omega_r t)], \tag{4.17}$$

which explicitly exhibits the damping effect due to collisions predicted by Eq. (4.16). This damping can modify the propagation properties of the wave in a number of other ways. For example, in contrast to collisionless propagation, the wave has a cutoff wavenumber, as we will now demonstrate. Substituting

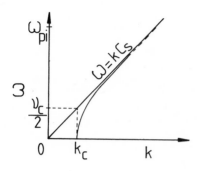

Fig. 4.1. Ion-acoustic wave dispersion in a collisional plasma for the initial-value problem. The theory is not valid for frequencies near ω_{pi}.

$\omega_i = -v_c/2$ into Eq. (4.14) gives

$$\omega_r^2 + \frac{v_c^2}{4} - k^2 Cs_i^2 = 0. \tag{4.18}$$

Figure 4.1 shows a plot of ω_r versus k for Eq. (4.18). From both the figure and the equation it can be seen that the range of possible wavelengths has been strongly modified. Because of collisions there is now a *cutoff* wavenumber k_c such that $k > k_c$, where k_c is given by

$$k_c = \frac{v_c}{2Cs_i}. \tag{4.19}$$

Thus, the wavelength of the wave is always smaller than a cutoff wavelength λ_c given by

$$\lambda_c = \frac{2\pi}{k_c} = \frac{4\pi Cs_i}{v_c}, \tag{4.20}$$

which one might note in passing is the same as the wavelength of an ion-acoustic wave of wave frequency $\omega_{rc} = v_c/2$ in a collision*less* plasma.

Collisions of the ions and electrons with neutrals produce another important change in the wave's properties. Whereas before the wave exhibited no dispersion at low frequencies (i.e., for $\omega \ll \omega_{pi}$), we now have a strong dispersion at low frequencies. To see this effect directly, we solve Eq. (4.18) for the phase velocity. This gives

$$v_\varphi = \frac{\omega_r}{k} = Cs_i \left[1 - \frac{1}{1 + (2\omega_r/v_c)^2} \right]. \tag{4.21}$$

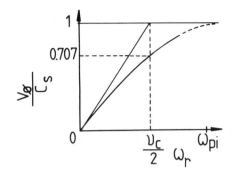

Fig. 4.2. Variation of the phase velocity of an ion-acoustic wave in a collisional plasma as a function of frequency for the initial-value problem.

The variation of phase velocity with wave frequency ω_r for a given collision frequency v_c is shown in Fig. 4.2. Both from Eq. (4.2) and from the figure we can see that the wave velocity is always less than the collisionless wave speed Cs_i, decreasing toward zero for wave frequencies below v_c. The upper part of the curves in both Fig. 4.1 and Fig. 4.2 is not validly predicted by the present theory since we have assumed that $\omega_r \ll \omega_{pi}$.

4.4. BOUNDARY-VALUE PROBLEM

We now study what is known as a boundary-value problem. That is, we assume we have applied for a very long time some real frequency ω in the plane $z = 0$ where we have some kind of wave exciter. Thus, a wave is continuously excited and propagates but due to collisions is spatially damped. Now k will be considered to be a complex quantity, with ω being real. We choose k to have the form

$$k = k_r + ik_i, \tag{4.22}$$

where k_r and k_i are the real and imaginary parts of the wave number, respectively. Substituting Eq. (4.22) into Eq. (4.10) and setting the real and imaginary parts of the resulting equation separately equal to zero gives the two equations

$$k_r^2 - k_i^2 = k_0^2 \tag{4.23}$$

$$k_r k_i = \frac{k_0^2}{2} \frac{v_c}{\omega} \tag{4.24}$$

where $k_0 = \omega/Cs_i$ would correspond to the real wave number of the same wave in a collision*less* plasma. Using this relationship and Eq. (4.23) we can

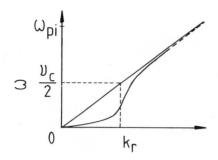

Fig. 4.3. Ion-acoustic wave dispersion in a collisional plasma for the boundary-value problem. The theory is not valid for frequencies near ω_{pi}.

eliminate k_i from Eq. (4.24). This gives the equation

$$\omega^2 = \frac{k_r^4 \, Cs_i^2}{k_r^2 + v_c^2/(4 \, Cs_i^2)}. \tag{4.25}$$

Since for a given set of conditions v_c and Cs_i are constants, Eq. (4.25) gives the functional relationship between ω and k_r. Thus, Eq. (4.25) is the dispersion equation for our boundary-value problem. A plot of Eq. (4.25) is shown in Fig. 4.3. From the figure it is seen that all wave numbers as well as all frequencies below the plasma frequency (i.e., all $\omega \ll \omega_{pi}$) are permitted. Thus, in contrast with the temporally damped case there is no cutoff wave number and, therefore, no upper limit to the wavelength of ion-acoustic waves propagating in a collisional plasma.

We are now interested in finding the phase velocity $v_\varphi = \omega/k_r$ and the collisional damping rate k_i/k_r. Thus, we need to know k_r and k_i. From Eqs. (4.23) and (4.24), we note that we know the sum and product of k_r^2 and $-k_i^2$. Solving for these two quantities, we easily find that

$$k_r^2 = \frac{k_0^2}{2}\left[\left(1 + \frac{v_c^2}{\omega^2}\right)^{1/2} + 1\right]; \tag{4.26}$$

$$k_i^2 = \frac{k_0^2}{2}\left[\left(1 + \frac{v_c^2}{\omega^2}\right)^{1/2} - 1\right]. \tag{4.27}$$

The phase velocity $v_\varphi = \omega/k_r$ is therefore given by

$$v_\varphi = \frac{(2)^{1/2}}{1 + (1 + v_c^2/\omega^2)^{1/2}} \, Cs_i. \tag{4.28}$$

Figure 4.4 shows the variation of phase velocity with wave frequency for a given value of v_c. As for the temporally damped case the high-frequency part of the curve is not valid. On the other hand the low-frequency part would

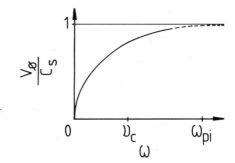

Fig. 4.4. Variation of the phase velocity of an ion-acoustic wave in a collisional plasma as a function of frequency for the boundary-value problem.

also seem not to be valid because the phase velocity approaches the ion thermal speed V_i, so that Landau damping (discussed in Chapter 7) can be expected to occur. However, as we will now show, the collisional damping will be much stronger than the Landau damping, so that the theory is still valid at these low frequencies, even though wave–particle interactions have not been included in the model.

The collisional damping rate is given simply by the ratio, $\mathrm{Im}(k)/\mathrm{Re}(k) = k_i/k_r$. Thus, using Eqs. (4.26) and (4.27)

$$k_i/k_r = \left[\left(1 + \frac{v_c^2}{\omega^2}\right)^{1/2} - 1\right]^{1/2}\left[\left(1 + \frac{v_c^2}{\omega^2}\right)^{1/2} + 1\right]^{-1/2}. \quad (4.29)$$

Figure 4.5 shows a plot of the collisional damping rate versus frequency. This figure shows that collisional damping is very strong for $\omega \ll v_c$, being even stronger than Landau damping in an isothermal ($T_i = T_e$) plasma, which has a damping rate given by $\mathrm{Im}(k)/\mathrm{Re}(k) \simeq 0.39$ (see Chapter 10). Thus, the present description is still approximately correct, even when the phase velocity of the wave is reduced to values close to the ion thermal speed. For lower frequencies, Landau damping dominates and the present description is not valid.

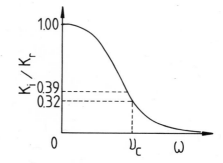

Fig. 4.5. Comparison of Landau damping and collisional damping of ion-acoustic waves for the boundary-value problem. For $\omega < v_c$, the collisional damping dominates.

4.5. SELECTED EXPERIMENT: BOUNDARY-VALUE PROBLEM FOR ION-ACOUSTIC WAVES IN A COLLISION-DOMINATED DISCHARGE PLASMA

In the preceding section we studied the boundary-value problem for ion-acoustic wave propagation in plasmas having significant collision rates between the plasma ions and the background neutral atoms for the conditions that ω, $v_c \ll \omega_{pi}$. In particular, we found that for $\omega \ll v_c$, the waves are both strongly dispersive and strongly damped. We now describe an experiment by Hatta and Sato (1962) in which they measured both the dispersion and damping of ion-acoustic waves propagating in a collision-dominated argon plasma.

Figure 4.6 shows the experimental facility used by Hatta and Sato (1962). The argon plasma between the hot cathode K and the grid G was formed from a mixture of electrons emitted from the hot cathode and ions which drifted from the plasma-generating region between A and G. A sinusoidally time-varying voltage was placed on the floating grid G to produce a continuous modulation of the ion density in the vicinity of G. This *ion density variation* then propagated away from the grid as an *ion-acoustic wave* toward K. The floating Langmuir probe P was used to measure both the amplitude and phase of the wave at the position of the probe.

Using the same theoretical model as used in Section 4.2 [see Eqs. (4.1)–(4.3)], Hatta and Sato (1962) obtained expressions for v_φ and k_i^2 that, for the purpose of the present discussion, differ from Eqs. (4.28) and (4.27), respectively, only by insignificant details (they assumed as *isothermal* compres-

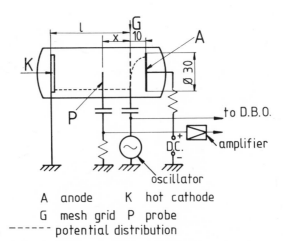

A   anode   K   hot cathode
G   mesh grid   P   probe
- - - - - potential distribution

Fig. 4.6. The test tube and the experimental circuit. (After Hatta and Sato, 1962.)

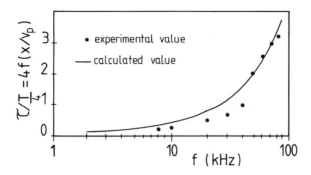

Fig. 4.7. Phase difference versus frequency (f). (After Hatta and Sato, 1962.)

sion of the ions, giving $\gamma_i = 1$, rather than 3; their effective ion-neutral collision frequency was only one-half that which we have used).

The first experiment of Hatta and Sato (1962) was to confirm the predicted dispersion of ion-acoustic waves at low frequencies ($f \ll f_{pi}$) in a collision-dominated plasma ($\nu_c \geqslant f$). The argon plasma density was chosen to be 2.5×10^9 cm^{-3}, giving $f_{pi} \sim 10^6$ Hz. The wave frequency was varied over the range of $f = 8$–80 kHz, and the results are presented for background argon gas pressure of 0.04 mmHg. The ion-neutral collision frequency for argon at pressure P(Torr) at room temperature is about $\nu_c \sim 1.6 \times 10^7 P$. Rather than measure v_φ as a function of f, these workers chose to measure the phase of the wave at the position of the probe P with respect to its phase at the grid G as a function of f, allowing them to calculate the phase shift $\Delta\varphi$ of the wave for each value of f.

Figure 4.7 gives the results of their experiment. To understand what has been plotted on the vertical axis, consider the following: if P is located a distance x ($= 1$ cm in the experiment) from G, and if the phase of the wave at G is arbitrarily chosen to be zero, then the phase at P will be simply $\varphi = 2\pi x/\lambda$, where λ is the wavelength. Using the general relationship that $v_\varphi = f\lambda$, we can write that the phase difference between P and G is given by

$$\Delta\varphi = \frac{2\pi f x}{v_\varphi}. \tag{4.30}$$

Thus, as stated in their figure caption, the vertical axis corresponds to phase difference.

From Figure 4.7, we see that $\Delta\varphi$ is indeed dispersive and, since v_φ and $\Delta\varphi$ are reciprocally related [see Eq. (4.30)], that v_φ is also dispersive. The solid curve is a plot of their theoretical equation corresponding to our Eq. (4.28). To *qualitatively* check the observed dispersion we can make the assumption that

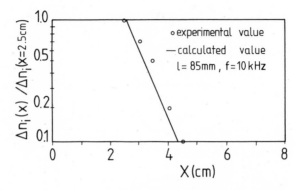

Fig. 4.8. Dependence of the attenuation of amplitude on the distance (x) from G. (After Hatta and Sato, 1962.)

$v_c \gg \omega$. Then Eq. (4.28) can be approximated by

$$v_\varphi = C s_i \left(\frac{2\omega}{v_c} \right)^{1/2}. \qquad (4.31)$$

Substituting Eq. (4.31) into Eq. (4.30) gives

$$\Delta\varphi = \frac{(2\pi v_c f)^{1/2} x}{C s_i}. \qquad (4.32)$$

Thus, we see that $\Delta\varphi$ should exhibit an approximate square-root dependence on f, as is observed to be closely the case.

In their second experiment, Hatta and Sato (1962) checked the predicted strong damping of ion-acoustic waves by ion-neutral collisions. In this experiment, the gas pressure was 0.15 mm Hg, giving $v_c \gg f$, f being chosen to be 10 kHz in the experiment. Figure 4.8 shows the results of this experiment. The amplitude of the wave $\Delta n_i(x)$, as measured by the probe at several positions, is compared to $\Delta n_i(x = 2.5$ cm$)$, the amplitude measured at the 2.5 cm position. Over a propagation distance of only 2 cm (from $x = 2.5$ cm to $x = 4.5$ cm), the amplitude is seen to decrease by a factor of 10, indicating strong damping indeed. (Since these waves are very nearly *planar*, no significant geometrical factor is involved in the measured change in amplitude). The straight line in Fig. 4.8 is the theoretical damping curve, calculated using their theoretical equation corresponding to our Eq. (4.27). The agreement between experiment and theory is seen to be good.

To understand why the theoretical damping curve in Fig. 4.8 should be a straight line, we need only remember that a planar wave can be represented by the equation

$$n(x,t) = n(0,0) \exp[i(kx - \omega t)]. \tag{4.33}$$

For the boundary-value problem, we replace k by $k_r + ik_i$. This gives

$$n(x,t) = n(0,0) \exp(-k_i x) \exp[i(k_r x - \omega t)]. \tag{4.34}$$

Equation (4.34) shows that the amplitude of the wave damps in space at a rate determined by $\exp(-k_i x)$. Thus, a semi-log plot of amplitude as a function of distance gives a straight line, as seen in Fig. 4.8. Furthermore, k_i can be determined directly from the slope of this line.

FINITE-SIZE-GEOMETRY EFFECTS

5.1. INTRODUCTION

In all the models we have studied so far we have assumed that the plasmas are infinite and homogeneous. It is found both theoretically and experimentally, however, that the properties of plasma waves predicted by such models can be strongly modified when studied in plasmas whose dimensions are comparable with the wavelength of the waves. These modifications are called *finite-size-geometry* effects. In this chapter, we study the propagation of two waves we are already familiar with—electron plasma waves and ion-acoustic waves—along a column of plasma infinite in length but finite in radius, which is supported by a strong magnetic field, and which may be contained inside a waveguide, for example.

For the electron plasma wave study, we study a cold plasma of constant density. For the ion-acoustic wave study, we study two cases: a warm plasma whose density is assumed in the first case to be constant throughout the column, but in the second case, to have a more realistic radial variation.

Electron plasma wave propagation inside a cold-plasma-filled waveguide differs significantly from what was found in Chapter 2 for an infinite cold plasma. For the infinite case, this mode was purely electrostatic, existed only at one frequency ($\omega = \omega_{pe}$), and did not propagate along the magnetic field. For the finite-plasma case, this mode is a propagating mode at frequencies both above and below ω_{pe}, and has an electric field that can vary from purely electrostatic to purely electromagnetic, depending on frequency and radius. Of interest also is the fact that the presence of the plasma column in the waveguide shifts the waveguide cutoff frequency to a higher value than what is observed for waves propagating in an empty waveguide. This can be understood as a waveguide filled with a substance having a relative dielectric constant less than unity, which is often the case for a plasma.

For the case of ion-acoustic wave propagation in a warm plasma column of assumed constant density, the finite-geometry theory predicts ion-acoustic waves that can propagate at speeds significantly greater than Cs (the infinite-plasma-model phase velocity) and that exhibit strong dispersion around ω_{ci}, the ion cyclotron frequency. If the finite-geometry theory is modified to include realistic density profiles for the plasma column and the density perturbation produced by the wave, and to assume a plasma column radius that is much larger than the actual radius (infinite-plasma model), then ion-acoustic waves are predicted whose phase velocity is Cs and which do not exhibit dispersion around ω_{ci}.

The general procedure for determining finite-size effects, both for electron plasma waves and ion-acoustic waves, consists of three steps. In the first step, starting with Maxwell's equations, we determine the general wave equation for $E_z(r)$ for each type of wave propagating in the plasma, assuming that the waves have an electric field of the form, $E_z(r,z,t) = E_z(r) \exp[i(kz - \omega t)]$, where z is in the direction of the supporting magnetic field. The second step is solving the general wave equations. This involves finding the solutions of these equations, taking into account the physical boundary conditions of the waves. This gives $E_z(r)$ for each type of wave in terms of ordinary Bessel functions. In the third step, using the well-known behavior of these Bessel functions, we investigate the functional behavior of $E_z(r)$. When r is chosen to be finite, for example, dispersion relations are found that describe the finite-size-geometry effects that occur. When r is allowed to become infinite, on the other hand, the dispersion relations found in Chapters 3 and 4 for the propagation of electron-plasma waves and ion-acoustic waves, respectively, in infinite plasmas are recovered.

By limiting the electric field E of the waves to the form described above we arbitrarily restrict out study to waves that propagate *along* the plasma column. That is, we eliminate the possibility of learning about finite-size effects on waves propagating in the radial direction (such effects have been observed for both electron plasma waves and ion-acoustic waves, resulting in radial standing waves, for example). We set this restriction merely to keep the model as simple as possible and still be able to demonstrate some finite-size-geometry effects of interest. (After studying this chapter, the reader should be able to treat other finite-size effects, such as standing radial waves in a waveguide, for example, without too much difficulty.) The form we have chosen for the electric field eliminates from consideration, also, any azimuthally asymmetric physical phenomena such as waves which are polarized, rotating, or otherwise azimuthally asymmetric. To treat such phenomena we could assume an electric field of the form $E_z(r,z,\theta,t) = E_z(r) \exp[i(kz - m\theta - \omega t)]$, for example, where θ allows for any azimuthal dependence around the z axis. Such a variation would be required for electro*magnetic* waves with rotating electric fields. In the present study we restrict ourselves to electro*static* waves in order to study the propagation of compressional waves along the z axis. For such modes the wave equations giving the radial dependence of $E_z(r)$ turn out to be zero-order Bessel differential equations of the first kind.

5.2. *ELECTRON PLASMA WAVES IN A COLD PLASMA SUPPORTED BY A STRONG MAGNETIC FIELD*

In this section we obtain, first, the general wave equation for azimuthally symmetric electron plasma waves in a cold, infinite, homogeneous plasma, propagating parallel to the supporting magnetic field. We then solve this equation and investigate the general properties of these waves when confined to propagate along a *finite*-radius plasma column enclosed in a waveguide. In particular, we study the behavior of these waves as a function of radius and frequency.

5.2.1. *General Wave Equation*

The model we assume is very similar to the cold-plasma model used in Chapter 2. That is, we assume no wave–particle interactions, no damping or growth of the waves, a constant plasma density, etc. One important difference is that we assume that the magnetic field is strong enough to restrict the electron motion to be only along the direction of the magnetic field. We also assume that we work only at frequencies $\omega \gg \omega_{pi}$, so the ions are unable to respond to the wave's electric field and can be considered to have no motion.

We start with Maxwell's equations for wave propagation in a vacuum (see Chapter 2)

$$\nabla \times E = -\frac{\partial B}{\partial t}; \tag{5.1}$$

$$\nabla \times H = j + \frac{\partial(\varepsilon_0 E)}{\partial t}; \tag{5.2}$$

$$\nabla \cdot B = 0; \tag{5.3}$$

$$\nabla \cdot \varepsilon_0 E = \rho, \tag{5.4}$$

where

$$B = \mu_0 H. \tag{5.5}$$

If n_0 is the background ion density and n is the electron density, then

$$\rho = -e(n - n_0), \tag{5.6}$$

where e is the magnitude of the charge of an electron.

Since we are interested only in wave propagation parallel to the magnetic field, whose direction we will choose to be along the z axis of our cylindrical coordinate system, we represent the wave-electric field by

$$E_z(r,z,t) = E_z(r)e^{i(kz - \omega t)}, \tag{5.7}$$

where the wave vector k has been chosen to be in the $+z$ direction. With this expression for the electric field, we find that it is advantageous to make Laplace transforms in time but not Fourier transforms in space. This will reduce our *partial*-differential equation to an *ordinary* differential equation. By taking advantage of the fact that the resulting differential equation is a zero-order Bessel equation, whose properties are well known, we are able to find very easily the finite-size-geometry effects in which we are interested (without having to make Fourier transforms in space).

We now go back to Maxwell's equations. From Eq. (5.1), using Eq. (5.5) and making a Laplace transform in time (assuming, for simplicity, that the initial value of the electric field is zero—see Section 1.3.2.), we write

$$\nabla \times (\nabla \times E) = i\omega\mu_0(\nabla \times H). \tag{5.8}$$

From Eq. (5.2), using Eq. (5.6) and, again, making a Laplace transform in time [and assuming, as we shall throughout this section, that $E_z(r,z,t=0) = 0$], we write

$$\nabla \times H = j - i\omega\varepsilon_0 E. \tag{5.9}$$

Substituting from Eq. (5.9) into Eq. (5.8) gives the general equation for wave propagation,

$$\nabla \times (\nabla \times E) = i\omega\mu_0 j + k_0^2 E, \tag{5.10}$$

where we defined k_0 as the wave number of an electromagnetic wave of the same frequency ω of interest, propagating in a vacuum

$$k_0 = \frac{\omega}{c} = \omega(\mu_0\varepsilon_0)^{1/2}. \tag{5.11}$$

We begin finding the solution of Eq. (5.10) by replacing the left-hand side of the equation by its mathematical equivalent

$$\nabla \times (\nabla \times E) = \nabla(\nabla \cdot E) - \nabla^2 E. \tag{5.12}$$

Then, since we are interested only in wave propagation parallel to the magnetic field, we take the z-component of each of the terms of the general wave equation to obtain the wave equation for propagation in the z direction

$$\nabla^2 E_z - \nabla_z(\nabla \cdot E) + i\omega\mu_0 j_z + k_0^2 E_z = 0, \tag{5.13}$$

∇_z being the component of the gradient operator parallel to z.

We need now to find an expression, in terms of E_z, for the current j_z produced in the plasma by the wave, and to find $\nabla \cdot E$. Since we have assumed that the ions are motionless and that the electrons move only along z, j_z will be equal to the total current j produced by the motion of the electrons. That is,

$$j_z = -en_0v_z, \tag{5.14}$$

where n_0 is the unperturbed electron (plasma) density, and v_z is the perturbed velocity of the electrons caused by the electric field E_z of the propagating wave. From the equation of motion v_z can be found for an electron which is, after a Laplace transform in time,

$$v_z = \frac{1}{i\omega}\left(\frac{e}{m}\right)E_z. \tag{5.15}$$

Substituting this expression for v_z into Eq. (5.14) to find j_z in terms of E_z allows us to rewrite Eq. (5.13) as

$$\nabla^2 E_z - \nabla_z(\nabla \cdot E) + k_0^2\left(1 - \frac{\omega_{pe}^2}{\omega^2}\right)E_z = 0, \tag{5.16}$$

where we have used Eq. (5.11) and the relationship $\omega_{pe}^2 = n_0e^2/\varepsilon_0 m$ (see Chapter 2) for the electron plasma frequency. To find $\nabla \cdot E$ we first combine Eqs. (5.4) and (5.6) to obtain Poisson's equation

$$\nabla \cdot E = \frac{\rho}{\varepsilon_0}, \tag{5.17}$$

where ρ is the total charge perturbation produced by the propagating wave. Thus, letting n_1 be the electron-density perturbation produced by the wave (no ion-density perturbation is produced), Poisson's equation becomes

$$\nabla \cdot E = \frac{-en_1}{\varepsilon_0}. \tag{5.18}$$

n_1 can be obtained easily from the particle-conservation equation for electrons

$$\frac{\partial n}{\partial t} + \nabla(nv) = 0, \tag{5.19}$$

where $n = n_0 + n_1$, and $v = v_1 = v_z$. After linearization of Eq. (5.19) and a Laplace transform in time, we find

$$n_1 = \frac{n_0(\nabla \cdot v_1)}{i\omega} \tag{5.20}$$

We replace v_1 in Eq. (5.20) in terms of E_z, using Eq. (5.15). Using the resulting expression for n_1 to substitute in Eq. (5.18), we obtain

$$\nabla \cdot E = ik \frac{\omega_{pe}^2}{\omega^2} E_z, \tag{5.21}$$

where again we have used the definition of the electron plasma frequency ω_{pe}. Substituting from Eq. (5.21) into Eq. (5.16) allows us to write

$$\nabla_\perp^2 E_z + (k_0^2 - k^2)\left(1 - \frac{\omega_{pe}^2}{\omega^2}\right)E_z, \tag{5.22}$$

where we have replaced ∇^2 in terms of the perpendicular Laplacian operator defined by

$$\nabla^2 = \nabla_\perp^2 - k^2. \tag{5.23}$$

Since E_z is not a function of θ [see Eq. (5.7)] we can replace $\nabla_\perp^2$ by only its r components, so that Eq. (5.22) can be written as

$$\frac{d^2 E_z(r)}{dr^2} + \frac{1}{r}\frac{dE_z(r)}{dr} + T^2 E_z(r) = 0, \tag{5.24}$$

where

$$T^2(\omega,k) = (k_0^2 - k^2)\left(1 - \frac{\omega_{pe}^2}{\omega^2}\right) \tag{5.25}$$

$T(\omega,k)$ is seen to contain all the ω-k information about the wave, as well as plasma-density information (through ω_{pe}). Since, in our model, we have assumed the plasma density to be constant, we see that T is not a function of r and can, thus, be treated as a constant in Eq. (5.24).

Equation (5.24) allows us to determine the dependence of E_z on r. As we shall see shortly, that fact will be sufficient for us to be able to obtain the dispersion relation for any electron plasma wave mode for which propagation along a cold-plasma-filled waveguide is possible. For example, if we assume that ω and k are real quantities (as we shall in this discussion), Eq. (5.24) reduces to a zero-order Bessel differential equation of the first kind, the general solution of which [see Abramowitz and Stegun (1972)] is given by

$$E_z(r) = A\mathcal{J}_0(Tr) + BY_0(Tr), \tag{5.26}$$

where $\mathcal{J}_0$ and Y_0 are ordinary Bessel functions, and A and B are arbitrary constants to be determined by the particular physical situation described by Eq. (5.26). Tr in Eq. (5.26) is the product of the constant T [see Eq. (5.25)]

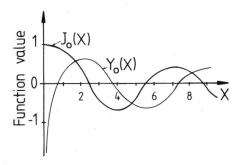

Fig. 5.1. Plots of the two Bessel functions
in Eq. (5.26).

and of the radial position r, measured from the z-axis of our assumed cylindrical
coordinate system. We show graphs of $J_0(x)$ and $Y_0(x)$ in Fig. 5.1. From this
figure we see that for finite T, $Y_0(Tr)$ will approach $-\infty$ for small values of
r. Thus, since $E_z(r)$ for waves of interest needs to remain finite, we see that
B in Eq. (5.26) must be chosen to be zero. Also, since $J_0(Tr)$ approaches
unity for small values of r, A in Eq. (5.26) must be chosen to be equal to
$E_z(0) = E_0$. Thus, the solution of Eq. (5.24), for our model, is given by

$$E_z(r) = E_0 J_0(Tr), \tag{5.27}$$

where E_0 is the amplitude of the electric field on the axis. From the form of
the differential equation given by Eq. (5.24) we see that the solution given by
Eq. (5.27) can be multiplied by any arbitrary constant and the result will also
be a solution of the differential equation. However, due to the requirement
of the linear theory that

$$n_1 \ll n_0, \tag{5.28}$$

we find that there is an upper limitation on the electric field strength which
the wave can have and still be described by our linear model. This upper
limit is here, as in the case of an infinite plasma, given by Eqs. (5.15) and
(5.20), which yield

$$|E_z| = \frac{n_1}{n_0} \frac{m}{e} \frac{\omega^2}{k} \ll \frac{m}{e} \frac{\omega^2}{k}, \tag{5.29}$$

where m and e are, respectively, the mass and magnitude of the charge of the
electron.

5.2.2. *Propagation in a Plasma-Filled Waveguide*

We now want to investigate the propagation of electron plasma waves
in a finite, magnetically supported, cold plasma contained in a cylindrical

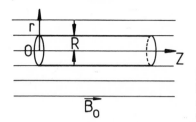

Fig. 5.2. An infinite, cold-plasma column contained in a circular waveguide of radius $r = R$. The boundary conditions at the surface of the plasma determine the character of waves propagating along the column.

waveguide, which is assumed to be infinitely long and to have a radius of $r = R$, as shown in Fig. 5.2. The density of the plasma is assumed to be uniform and constant.

Equation (5.27) gives the radial dependence of the E_z component of an electron plasma wave propagating in a magnetically supported plasma in the direction of the magnetic field. [The dependence of E_z on z and t was assumed, at the beginning, to be of the form given by Eq. (5.7).] For an infinite plasma the radial dependence of E_z is governed by $J_0(Tr)$, with Tr varying over its entire theoretical range: $0 \leqslant Tr \leqslant \infty$ (see Fig. 5.1). For a finite-plasma column, however, the values that Tr may have are limited by the boundary condition on E_z at the edge of the column. As we shall now see, it is this limitation that allows us to obtain the dispersion relation for electron plasma waves propagating along the plasma column.

5.2.2.1. Dispersion Relation and Possible Modes. Since the wall of a waveguide is a conducting surface, a boundary condition on any wave propagating along the plasma column is that, at the edge of the column, the electric field component of the wave parallel to the conducting surface must be zero. That is, $E_z(R) = 0$, since the conducting surface is along the z direction. From Eq. (5.27) we see that this requires that $J_0(TR) = 0$. From Fig. 5.1 we see that there are specific values of TR, called *roots*, for which $J_0 = 0$. If we call these roots P_{0n}, where $n = 1, 2, 3, \ldots$, then we can write that the boundary condition imposed on the wave by the conducting surface of the waveguide requires that

$$TR = P_{0n} \qquad (5.30)$$

or

$$T(\omega,k) = \frac{P_{0n}}{R}. \qquad (5.31)$$

Tabulated values of P_{0n} are easily found [see Abramowitz and Stegun (1972)]. The first three values, for example, are (see Fig. 5.1) $P_{01} = 2.40$, $P_{02} = 5.52$, and $P_{03} = 8.65$. As noted earlier, $T(\omega,k)$ contains all the ω-k information about the propagating wave. Thus, Eq. (5.31) gives the dispersion relation for the electron plasma waves propagating along the plasma column. Using Eq. (5.25) we can rewrite the dispersion relation as

$$T^2 = (k_0^2 - k^2)\left(1 - \frac{\omega_{pe}^2}{\omega^2}\right) = \left(\frac{P_{0n}}{R}\right)^2. \tag{5.32}$$

Equation (5.32) can be rewritten to give the classical dispersion relation for electron plasma waves in a cold-plasma-filled waveguide

$$k^2 = k_0^2 - \frac{P_{0n}^2}{R^2}\frac{1}{1 - \omega_{pe}^2/\omega^2}. \tag{5.33}$$

A cutoff (see Chapter 2) occurs when $k = 0$. The corresponding cutoff frequency $\omega = \omega_c$ is then found from Eq. (5.33) to be given by

$$\omega_c^2 = \omega_{pe}^2 + \frac{c^2 P_{0n}^2}{R^2}, \tag{5.34}$$

where we have used Eq. (5.11) to replace k_0 by ω/c. For an electromagnetic wave propagating in an empty waveguide, there is found to be a low-frequency cutoff given by cP_{01}/R. Thus, Eq. (5.34) shows that when a plasma is introduced into the waveguide, its cutoff frequency is shifted to a higher value.

When the wave frequency is much larger than the electron plasma frequency, the dispersion relation given by Eq. (5.33) becomes the classical dispersion relation for an electromagnetic wave traveling in an empty waveguide

$$\omega^2 = k^2 c^2 + \frac{P_{0n}^2 c^2}{R^2}, \tag{5.35}$$

where Eq. (5.11) has been used again to replace k_0^2 by ω^2/c^2.

As we now show, the introduction of a plasma in the waveguide not only modifies the cutoff frequency of the waveguide, but it also introduces the possibility of having electron plasma wave propagation at frequencies lower than the plasma frequency. If we look again at Eq. (5.33) we see that when $\omega < \omega_{pe}$ we can write

$$k^2 = k_0^2 + \frac{P_{0n}^2}{R^2}\frac{\omega^2}{\omega_{pe}^2 - \omega^2} > 0. \tag{5.36}$$

Since $k^2 > 0$, k is real and propagation can occur for $\omega < \omega_{pe}$. Note that when ω approaches ω_{pe} from below, $k \to \infty$ (i.e., there is a resonance at ω_{pe}, see Chapter 2). When k is much larger than k_0, the phase velocity of the wave is much smaller than the speed of light in a vacuum. For this case, we say that the electron plasma wave is a *slow wave*. For $k \gg k_0$, Eq. (5.36) can be written as

$$k^2 \simeq \frac{(P_{0n}^2/R^2)\omega^2}{\omega_{pe}^2 - \omega^2}. \tag{5.37}$$

Using the definition $v_\varphi = \omega/k$, Eq. (5.37) gives the phase velocity of the slow wave as

$$v_\varphi \simeq \frac{(\omega_{pe}^2 - \omega^2)^{1/2}}{T}. \tag{5.38}$$

For $\omega \ll \omega_{pe}$, v_φ approaches ω_{pe}/T. It must be remembered, however, that there is a low-frequency limit to our description, since we have assumed that the ions remain motionless during the wave motion. Thus, the slow waves described here must be at frequencies considerably higher than the ion plasma frequency in order for the description to remain valid.

The dispersion curves given by Eq. (5.33) are shown in Fig. 5.3. These curves show very clearly propagation of electron plasma waves at frequencies both above (the *fast wave*) and below (the *slow wave*) the electron plasma frequency, quite in contrast to the nonpropagating mode at $\omega = \omega_{pe}$ found in Chapter 2 for an *infinite* magnetically supported cold plasma.

5.2.2.2. Electrostatic and Electromagnetic Properties. We now look at the electrostatic and electromagnetic properties of electron plasma waves in a plasma-filled cylindrical waveguide. The character of the wave depends on the local

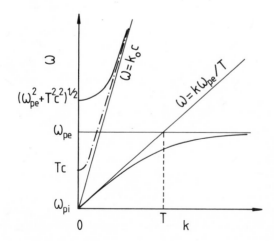

Fig. 5.3. Propagation of electron plasma waves in a cold-plasma-filled waveguide. The solid line represents propagation with a plasma, while the broken line shows electromagnetic wave propagation in the same waveguide without a plasma.

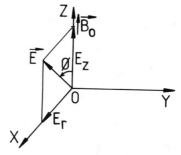

Fig. 5.4. Coordinate system showing the direction of the electric field of an electron plasma wave propagating along the axis of a magnetically supported cold-plasma column. When $\Theta = 0$, then $E = E_z$, $E_r = 0$, and the wave is purely electrostatic. When $\Theta = 90°$, the wave is purely electromagnetic.

direction of the electric field with respect to 0_z (see Section 2.3.3.). We first calculate

$$\tan \Phi = \frac{E_r}{E_z}, \tag{5.39}$$

where Φ is the angle that E makes with respect to B_0, see Fig. 5.4. To derive an expression for $\tan \Phi$, we start with Eq. (5.21)

$$\frac{1}{r}\frac{\partial(rE_r)}{\partial r} + \frac{\partial E_z}{\partial z} = ik\left(\frac{\omega_{pe}^2}{\omega^2}\right)E_z, \tag{5.40}$$

where we have written $\nabla \cdot E$ in cylindrical coordinates. Using the expression for E_z given by Eq. (5.27), and remembering that E_z has a z-dependence given by Eq. (5.7), Eq. (5.40) becomes

$$\frac{1}{r}\frac{\partial(rE_r)}{\partial r} = ik\left(\frac{\omega_{pe}^2}{\omega^2} - 1\right)E_0 J_0(Tr). \tag{5.41}$$

Integrating both sides of Eq. (5.41) from 0 to r gives

$$rE_r = ik\left(\frac{\omega_{pe}^2}{\omega^2} - 1\right)E_0 \int_0^r y J_0(Ty)\, dy. \tag{5.42}$$

The integral can be computed by making the change of variables

$$x = Ty, \tag{5.43}$$

and remembering that [see Abramowitz and Stegun (1972), for example]

$$\int_0^x x' J_0(x')\, dx' = x J_1(x). \tag{5.44}$$

Equation (5.42) thus becomes

$$E_r = ik\left(\frac{\omega_{pe}^2}{\omega^2} - 1\right)\frac{1}{T}\mathcal{J}_1(Tr)E_0.\tag{5.45}$$

Then, using Eq. (5.27) again, we obtain an equation for $\tan\Phi$

$$\tan\Phi = \frac{E_r}{E_z} = ik\left(\frac{\omega_{pe}^2}{\omega^2} - 1\right)\frac{1}{T}\frac{\mathcal{J}_1(Tr)}{\mathcal{J}_0(Tr)}.\tag{5.46}$$

Equation (5.46) can be used to determine qualitatively how the character of the wave for a given frequency varies with the radius. For example, for the fundamental mode corresponding to $TR = 2.40$ [see Eq. (5.30)], the variation of $\tan\Phi$ between $Tr = 0$ and $Tr = TR$ is shown in Fig. 5.5. For $Tr = 0$, i.e., for $r = 0$, $\mathcal{J}_1/\mathcal{J}_0 = 0$, which by Eq. (5.46) means that $E_r = 0$. Thus, on the axis of the plasma column, the wave has no radial component and is purely electrostatic. For $Tr = 2.40$, i.e., when $r = R$, $\mathcal{J}_1/\mathcal{J}_0 = \infty$. Thus, at the edge of the plasma column, the wave is purely electromagnetic (this we already knew, of course, because of the boundary condition at the conducting surface of the waveguide, which requires that $E_z = 0$). For any radial position between $r = 0$ and $r = R$, the electric field for the fundamental mode ($TR = 2.40$) will have both electrostatic and electromagnetic character.

The character of the electron plasma wave also depends on its frequency. For example (see Fig. 5.3), for the fast mode, when $\omega \gg \omega_{pe}$, Eq. (5.46) gives

$$\tan\Phi = \frac{E_r}{E_z} \simeq -i\frac{\omega}{cT}\frac{\mathcal{J}_1}{\mathcal{J}_0},\tag{5.47}$$

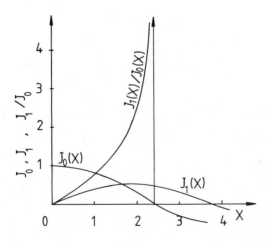

Fig. 5.5. Plots of $\mathcal{J}_0$, $\mathcal{J}_1$, and $\mathcal{J}_1/\mathcal{J}_0$ as a function of $x = Tr$. The behavior of $\mathcal{J}_1/\mathcal{J}_0$ shows that the electron plasma wave is purely electrostatic on the axis ($x = 0$) of the plasma column, but is purely electromagnetic at the edge ($x = 2.40$) of the column.

where k has been replaced by ω/c. Thus, for a given radius, the electromagnetic character increases with frequency. (In the limit of $\omega_{pe} \to 0$ we recover the purely electromagnetic wave in an empty waveguide.) When ω of the fast mode approaches the cutoff frequency, $k \to 0$ (see Fig. 5.3) and, by Eq. (5.46), $\tan \Phi \to 0$, so that the wave becomes purely electrostatic. For the slow wave, when ω approaches the resonance at the plasma frequency (i.e., $\omega \to \omega_{pe}$), Eq. (5.46) shows that $\tan \Phi \to 0$, so that the mode becomes purely electrostatic. At low frequencies, i.e., for $\omega_{pe}^2/\omega^2 \gg 1$, Eq. (5.46) shows that the slow wave becomes more and more electromagnetic with decreasing frequency.

5.3. ION-ACOUSTIC WAVES IN A WARM PLASMA SUPPORTED BY A STRONG MAGNETIC FIELD

In this section we study the propagation of ion-acoustic waves along a plasma column which is contained in a cylindrical waveguide and which is supported by a strong axial magnetic field. Since we are now dealing with ion-acoustic waves and no longer with electron plasma waves, we need to make different assumptions in our model. First, since ion-acoustic waves do not exist in a cold-plasma theory, we use a so-called *warm plasma theory*, with $T_e \gg T_i$. That is, we use a fluid theory in which thermal effects appear through pressure terms, as discussed in Chapter 3. A second important difference is that we are concerned with waves whose frequencies are below the ion plasma frequency, so that the ions can no longer be considered to be motionless. Otherwise, the same assumptions of no wave–particle interactions, no wave growth or damping mechanisms, etc., made in Section 5.2, are made again in this section.

As in Section 5.2, we first obtain the general wave equation for cylindrically symmetric ion-acoustic waves in a cylindrical plasma, propagating parallel to the supporting magnetic field. We then solve this equation and investigate the general properties of these waves when confined to propagate along a finite-radius plasma column contained in a waveguide. Two types of plasma columns are considered. In this first type, the plasma density is assumed to be uniform and constant, as it is for the electron plasma wave study. In the second type, however, the plasma column is considered to have a realistic density profile in the radial direction.

5.3.1. General Wave Equation

We assume that an ion-acoustic wave exists whose electric field does not depend on the azimuthal angle and which has the form [see Eq. (5.7)]

$$E_z(r,z,t) = E_z(r) \exp[i(kz - \omega t)], \qquad (5.48)$$

where ω and k will be chosen to be real. By choosing ω and k to be real, we ignore any growth or damping mechanisms.

Proceeding as in Section 5.2, we can determine a differential equation for $E_z(r)$, the wave-electric field in the axial direction

$$\nabla^2 E_z - \nabla_z(\nabla \cdot E) + i\omega\mu_0 j_z + k_0^2 E_z = 0, \tag{5.49}$$

where as in Eq. (5.13) j_z is the total current produced in the plasma by the electric field of the ion wave, and $\nabla \cdot E$ is the divergence produced in the electric field by the net charge perturbation ρ produced in the plasma by the wave. In form, Eqs. (5.13) and (5.49) are identical. The big difference is that, whereas in Section 5.2 j_z and ρ resulted only from the motion of the electrons, we must now consider both electron and ion motion. Thus, as in Section 5.2, we must now find expressions for j_z and $\nabla \cdot E$, in terms of E_z.

The linearized current in the axial direction due to the wave is given simply by

$$j_z = en_0(v_{iz} - v_{ez}), \tag{5.50}$$

where n_0 is the unperturbed plasma density, and v_{iz} and v_{ez} are, respectively, the perturbed ion- and electron-axial velocities caused by the E_z field of the propagating wave in the plasma-filled waveguide. We use the electron and ion equations of motion, respectively, to calculate v_{ez} and v_{iz}.

For the electrons we include a pressure-gradient term in the equation of motion

$$\frac{nm\partial v}{\partial t} = -ne(E + v \times B) - \nabla P, \tag{5.51}$$

where $P = nKT_e$ is the thermal pressure of the electrons. Since we have assumed that the electrons have motion only parallel to B, the $v \times B$ term is zero. After linearization, Eq. (5.51) becomes

$$\frac{\partial v_1}{\partial t} = -\frac{e}{m}E_1 - \frac{V_e^2 \nabla n_1}{n_0}, \tag{5.52}$$

where V_e is the thermal speed of the electrons, and we have taken into account the fact that T_e is constant in space (because of our assumption of *isothermal* compression of the electrons). ∇n_1 can be obtained from the particle-conservation equation for electrons

$$\frac{\partial n}{\partial t} + \nabla \cdot (nv) = 0. \tag{5.53}$$

After linearization and a Laplace transform in time (assuming, for simplicity, that the initial value of the density perturbation is zero), Eq. (5.53) can be solved to give

$$n_1 = \frac{n_0}{i\omega} \nabla \cdot v_1. \tag{5.54}$$

Then, ∇n_1 becomes

$$\nabla n_1 = \frac{n_0}{i\omega} \nabla(\nabla \cdot v_1). \tag{5.55}$$

But, since v_1 is assumed to be purely axial

$$\nabla \cdot v_1 = \nabla_z v_z = ikv_1, \tag{5.56}$$

where we have made a Fourier transform in space (see Section 1.3). Going back to the vectorial equation, Eq. (5.52), picking out only the z-component terms, using Eq. (5.56), it is easy to show that

$$v_1 = \frac{e}{m} \frac{1}{i\omega} E_z + \frac{V_e^2}{\omega^2} k^2 v_1. \tag{5.57}$$

Since we are interested only in slow waves at low frequencies, we assume that the wave speed is much less than the electron thermal speed, i.e.,

$$v_\varphi = \frac{\omega}{k} \ll V_e. \tag{5.58}$$

This corresponds to our neglecting the inertia of the electrons. With this assumption we can ignore the term on the left-hand side of Eq. (5.57) and write

$$v_{ez} = v_1 = \frac{i(e/m)\omega}{k^2 V_e^2} E_z \tag{5.59}$$

Equation (5.59) gives the v_{ez} needed for Eq. (5.50).

Before proceeding to find a corresponding expression for v_{iz}, it is of interest to note an important restriction put on the present model by the assumption made in Eq. (5.58), and that is that Eq. (5.58) prevents the dispersion relation that we derive in this section from being able to describe cutoffs $(k \to 0)$. This can be seen from the equation by observing that, as $k \to 0$, $v_\varphi \to \infty$ for any finite value of ω, thus violating the assumption that $v_\varphi \ll V_e$. At the end of this section we show that when $k \to 0$, the wave propagation becomes purely electro*magnetic*, which is not of interest in this chapter. Also, we note a simple modification that can be made, enabling the modified relation to describe cutoffs.

We now use the ion equation of motion to calculate v_{iz} in terms of E_z. To simplify the calculation, we assume that any thermal effects are due only to electrons, i.e., we assume that

$$T_e \gg T_i, \tag{5.60}$$

so that we can neglect the ion thermal pressure gradient in the equation of motion, giving

$$dv_i/dt = \frac{e}{m_i} (E + v_i \times B_0). \tag{5.61}$$

We now take only the terms in the z direction. After a Laplace transform in time this gives

$$-i\omega v_{iz} = \frac{e}{m_i} E_z. \tag{5.62}$$

Note that the $v_i \times B_0$ term gives no contribution, since B_0 is along z.

Substituting from Eqs. (5.59) and (5.62) into Eq. (5.50), we can now calculate the parallel component of the current j_z

$$j_z = ien_0 \left[\frac{e}{m_i} \frac{1}{\omega} - \frac{(e/m_e)\omega}{k^2 V_e^2} \right] E_z. \tag{5.63}$$

or

$$j_z = i \frac{\omega_{pi}^2}{\omega} \varepsilon_0 \left(1 - \frac{\omega^2}{k^2 C_s^2} \right) E_z, \tag{5.64}$$

where $C_s = V_e(m_e/m_i)^{1/2} = (KT_e/m_i)^{1/2}$ is the classical ion-acoustic wave velocity when $T_e \gg T_i$. From Eq. (5.64) the term containing j_z of Eq. (5.49) can be found to be

$$i\omega\mu_0 j_z = \frac{-\omega_{pi}^2}{c^2} \left(1 - \frac{\omega^2}{k^2 C_s^2} \right) E_z, \tag{5.65}$$

where c is the speed of light in vacuum.

We must now calculate the space charge ρ associated with the ion wave. The space charge involved in the wave is given simply by

$$\rho = \rho_i + \rho_e = e(n_i - n_e). \tag{5.66}$$

Thus, we must calculate ρ_i and ρ_e, the ion and electron space charge, respectively, due to the propagating wave. From Eqs. (5.54) and (5.59) we obtain for the electrons

$$n_e = i \frac{n_0 e}{m} \left(\frac{1}{k V_e^2} \right) E_z. \tag{5.67}$$

Thus

$$\frac{\rho_e}{\varepsilon_0} = -i \frac{1}{k} \frac{\omega_{pe}^2}{V_e^2} E_z. \tag{5.68}$$

The calculation of the *ion* space charge is somewhat more complicated. From Eq. (5.54) we have a relationship between n_1 and $\mathbf{V} \cdot v_1$. We must now calculate $\mathbf{V} \cdot v_1$ as a function of E_z, starting from the equation of motion for the ions

$$-i\omega v = \frac{e}{m} (E + v \times B_0). \tag{5.69}$$

As we pointed out in Chapter 2, this equation is of the form

$$v = a + v \times b, \tag{5.70}$$

where $a = i(e/m)(1/\omega)E$ and $b = i(e/m)(1/\omega)B_0$. The solution of Eq. (5.70) can be written as

$$v = \frac{1}{1 + b^2} [a + a \times b + b(a \cdot b)]. \tag{5.71}$$

One can check that the result given in Eq. (5.71) is identical to the result given in Eq. (2.45) in Chapter 2.
   Using Eq. (5.71), $\mathbf{V} \cdot v$ is given by

$$\mathbf{V} \cdot v = \frac{1}{1 + b^2} \{ \mathbf{V} \cdot a + \mathbf{V} \cdot (a \times b) + \mathbf{V} \cdot [b(a \cdot b)] \}. \tag{5.72}$$

We now look at the various terms on the right-hand side of Eq. (5.72). Since a is proportional to E, $\mathbf{V} \cdot a$ can be written immediately in terms of ρ. The second term on the right-hand side of Eq. (5.72) can be expanded as

$$\mathbf{V} \cdot (a \times b) = b \cdot (\mathbf{V} \times a) - a \cdot (\mathbf{V} \times b). \tag{5.73}$$

We can easily demonstrate that both terms on the right-hand side of Eq. (5.73) are zero. The last term is zero because b, which is proportional to B_0, is a constant. In the first term, a is proportional to $E(r,z)$, which has no azimuthal variation and no azimuthal component. Thus, $\nabla \times a$ is a purely azimuthal vector and is, thus, perpendicular to B_0 and to b. It follows that $b \cdot (\nabla \times a)$ is zero. Thus, $\nabla \cdot (a \times b)$, the second term on the right-hand side of Eq. (5.72), is zero.

We now look at the last term of Eq. (5.72). Since a is proportional to E, and b is proportional to B_0, the product $a \cdot b$ is proportional to $E_z B_0$. Thus, the vector $b(a \cdot b)$ is proportional to $E_z B_0 B_0$ which, in (x,y,z) vector notation, is the vector $(0,0,B_0^2 E_z)$. Thus, the divergence of $b(a \cdot b)$ is $(1/B_0)(\Omega_{ci}^3/\omega^3)kE_z$. The coefficient $1/(1 + b^2)$ of the right-hand side of Eq. (5.72) can be written as

$$\frac{1}{1 + b^2} = \frac{1}{1 - \Omega_{ci}^2/\omega^2} = \frac{\omega^2}{\omega^2 - \Omega_{ci}^2}, \tag{5.74}$$

where $\Omega_{ci} = eB_0/m$ is the ion cyclotron frequency. Thus, Eq. (5.72) can be written as

$$\nabla \cdot v = \frac{1}{B_0}\left(\frac{\omega^2}{\omega^2 - \Omega_{ci}^2}\right)\left(\frac{i\Omega_{ci}(\nabla \cdot E)}{\omega} + \frac{\Omega_{ci}^3 kE_z}{\omega^3}\right). \tag{5.75}$$

From Eq. (5.54) we have a relationship between n_1 and $\nabla \cdot v_1$, so that we can write an expression for the ion space charge as

$$\frac{\rho_i}{\varepsilon_0} = \frac{en_{i1}}{\varepsilon_0} = \frac{en_0 \nabla \cdot v_1}{i\varepsilon_0 \omega}. \tag{5.76}$$

Then, using Eq. (5.72) and taking into account Poisson's equation, we obtain for the ion space charge the equation

$$\frac{\rho_i}{\varepsilon_0} = \frac{\omega_{pi}^2}{\omega^2 - \Omega_{ci}^2}\left(\frac{\rho_i}{\varepsilon_0} + \frac{\rho_e}{\varepsilon_0} - \frac{ik\Omega_{ci}^2}{\omega^2}E_z\right). \tag{5.77}$$

As Eq. (5.68) gives ρ_e/ε_0, the electron space charge as a function of E_z, the equation for the ion space charge finally becomes

$$\frac{\rho_i}{\varepsilon_0} = \frac{\omega_{pi}^2}{\omega^2 - \Omega_{ci}^2 - \omega_{pi}^2}\left(\frac{\omega_{pe}^2}{k^2 V_e^2} + \frac{\Omega_{ci}^2}{\omega^2}\right)(-ikE_z). \tag{5.78}$$

Thus, substituting Eqs. (5.68) and (5.78) into Eq. (5.66), we obtain an equation for the total space charge of the wave in terms of E_z

$$\frac{\rho}{\varepsilon_0} = -ik\left[\frac{\omega_{pi}^2}{\omega^2 - \omega_{pi}^2 - \Omega_{ci}^2}\left(\frac{\omega_{pe}^2}{k^2 V_e^2} + \frac{\Omega_{ci}^2}{\omega^2}\right) + \frac{\omega_{pe}^2}{k^2 V_e^2}\right]E_z. \tag{5.79}$$

We now have everything we need to obtain a differential equation involving only E_z, the axial electric field of the propagating wave. First, we substitute Eq. (5.79) into Poisson's equation [Eq. (5.17)] to obtain $\nabla \cdot E$. We then substitute this result, plus Eq. (5.65), into Eq. (5.49). Finally, making a Fourier transform in space, we obtain the result

$$
\left[\nabla^2 + k_0^2 - \frac{\omega_{pi}^2}{c^2}\left(1 - \frac{\omega^2}{k^2 Cs^2}\right) \right.
$$
$$
\left. - \frac{k^2 \omega_{pi}^2}{\omega^2 - \omega_{pi}^2 - \Omega_{ci}^2}\left(\frac{\omega_{pe}^2}{k^2 V_e^2} + \frac{\Omega_{ci}^2}{\omega^2}\right) + \frac{\omega_{pe}^2}{V_e^2} \right] E_z = 0. \tag{5.80}
$$

As in Section 5.2, we can write Eq. (5.80) as a zero-order Bessel differential equation of the first kind. To do this, we replace ∇^2 by $\nabla_\perp^2 - k^2$ [see Eq. (5.23) and the following discussion] to obtain the desired wave equation for $E_z(r)$

$$
\frac{d^2 E_z(r)}{dr^2} + \frac{1}{r}\frac{dE_z(r)}{dr} + T^2 E_z(r) = 0, \tag{5.81}
$$

where

$$
T^2 = -k^2 \left[\frac{\omega_{pi}^2}{\omega^2 - \omega_{pi}^2 - \Omega_{ci}^2}\left(\frac{\omega_{pe}^2}{k^2 V_e^2} + \frac{\Omega_{ci}^2}{\omega^2}\right) + \frac{\omega_{pe}^2}{k^2 V_e^2} + 1 \right]
$$
$$
+ k_0^2 - \frac{\omega_{pi}^2}{c^2}\left[1 - \frac{\omega^2}{k^2 Cs^2}\right]. \tag{5.82}
$$

In interpreting Eq. (5.82) we remind the reader that ω^2 and k^2 are *real* quantities. This corresponds to the assumptions, in our model, of no temporal or spatial damping of the waves.

Equation (5.82) contains seven terms. We now want to demonstrate that the last three can be neglected (at the end of this section we show that these terms are associated with the electro*magnetic* character of the waves). Our basis of justification for neglecting these terms will be to show that each of them is much smaller than the $-k^2$ term in the equation.

We first form the ratio k_0^2/k^2. Dividing both the numerator and denominator by ω^2 gives $(k_0^2/\omega^2)/(k^2/\omega^2)$. But, from Eq. (5.11) we see that the numerator of this ratio is $1/c^2$, and from Eq. (5.58) we see that the denominator is $1/v_\varphi^2$ so we can write that $k_0^2/k^2 = v_\varphi^2/c^2 \ll 1$. Thus, $k_0^2 \ll k^2$ and the first of the last three terms is seen immediately to be negligible. Again, using Eq. (5.11), the second of the last three terms can be written as $k_0^2(\omega_{pi}^2/\omega^2)$. For comparison with k^2, we form the ratio

$$
\frac{k_0^2 \omega_{pi}^2}{k^2 \omega^2} = \frac{\omega_{pi}^2/\omega^2}{c^2/v_\varphi^2} \ll 1.
$$

Thus, $k_0^2(\omega_{pi}^2/\omega^2) \ll k^2$. We see that the third term differs in magnitude from the second term only by the ratio, $\omega^2/k^2 Cs^2 = v_\varphi^2/Cs^2$. But v_φ is never much larger than Cs (in fact, it is always smaller, except near the cutoff). Thus, the third of the last three terms is also negligible.

Neglecting the last three terms, Eq. (5.82) becomes

$$T^2 = k^2 \left[\frac{\omega_{pi}^2}{\omega_{pi}^2 - \Omega_{ci}^2 - \omega^2} \left(\frac{\omega_{pe}^2}{k^2 V_e^2} + \frac{\Omega_{ci}^2}{\omega^2} \right) - \frac{\omega_{pe}^2}{k^2 V_e^2} - 1 \right]. \tag{5.83}$$

Observation of Eq. (5.83) shows that in order for T to be independent of r, two conditions must be satisfied. The first condition is the same as that for the cold plasma of Section 5.2, and that is that the density be everywhere uniform and constant (in order that ω_{pi} and ω_{pe}, in Eq. (5.83), be constant in r). The second condition is that the electron temperature T_e (directly related to $V_e^2 = k T_e/m_e$) be everywhere uniform and constant. For T a constant in r, as discussed already in Section 5.2 for electron plasma waves, the solutions of interest for Eq. (5.81) can be written in the form

$$E_z(r) = E_0 \mathcal{J}_0(Tr), \tag{5.84}$$

where $E_0 = E_z(0)$ is the amplitude of the ion-acoustic wave on the axis of the cylindrical plasma column, and $\mathcal{J}_0$ is an ordinary Bessel function (see Fig. 5.1).

As discussed in Section 5.2, Eq. (5.84) multiplied by any constant will also be a solution of Eq. (5.81); however, linear theory requires that any solutions be such that [see Eq. (5.29)]

$$|E_z| \ll \frac{m_i}{q} \frac{\omega^2}{k}, \tag{5.85}$$

where m_i and q are, respectively, the mass and charge of the ion.

Following the derivation of Eq. (5.59) it was noted that when $k \to 0$, the wave propagation becomes purely electromagnetic. We now show this by showing that the total space charge of the wave goes to zero when $k \to 0$. The total charge of the wave is given by Eq. (5.79). Observation of this equation shows that in order for $\rho \to 0$ when $k \to 0$, it is necessary that the terms within the square brackets remain finite. This is not immediately obvious, since two of those terms contain k^2 in their denominator. Each of these two terms has the factor, $\omega_{pe}^2/k^2 V_e^2$. Multiplying both the numerator and denominator by ω^2, we can write this factor as $(\omega_{pe}^2/\omega^2)(\omega^2/k^2 V_e^2)$. We now want to show that $\omega^2/k^2 V_e^2$ remains finite as $k \to 0$. To do this, we take another look at the derivation of Eq. (5.59). As noted in the discussion immediately following that derivation, the assumption made in Eq. (5.58) automatically excludes the possibility of $k \to 0$. If we want to look at the case of $k \to 0$ (i.e., cutoffs), then we see from Eq. (5.57) that it is the last term in this equation that is negligible with respect to the first and not the reversed situation of the previous case. Thus, for this case, v_1 is given approximately by

$$v_1 = \frac{-i}{\omega} \frac{e}{m} E_z, \tag{5.86}$$

rather than the approximation given by Eq. (5.59). A comparison of Eqs. (5.59) and (5.86) shows that the v_1 of Eq. (5.59) [i.e., the v_1 actually used in the complete theoretical discussion up through Eq. (5.83)], is equal to the v_1 of Eq. (5.86) multiplied by the factor $-\omega^2/k^2 V_e^2$. Thus, the entire theoretical treatment up through Eq. (5.83), which describes wave propagation for all values of k, except $k \to 0$, can be converted into a theoretical treatment describing only the $k \to 0$ case by replacing the factor, $\omega^2/k^2 V_e^2$, everywhere it occurs in the formulation by -1! In particular, the total space charge of the propagating wave given by Eq. (5.79) can be written for the case of $k \to 0$ as,

$$\frac{\rho}{\varepsilon_0} = -ik \left[\frac{\omega_{pi}^2}{\omega^2 - \omega_{pi}^2 - \Omega_{ci}^2} \left(\frac{\Omega_{ci}^2}{\omega^2} - \frac{\omega_{pe}^2}{\omega^2} \right) - \frac{\omega_{pe}^2}{\omega^2} \right] E_z. \tag{5.87}$$

From Eq. (5.87) it is now clear that as $k \to 0$ and $\omega \to \omega_c$, where ω_c is some finite cutoff frequency, the terms in the bracket remain finite, so that $\rho \to 0$. Thus, as $k \to 0$ the electrostatic character of the wave disappears, leaving the wave completely electromagnetic.

Based on the preceding discussion the justification for stating earlier that the three terms in Eq. (5.82) that were dropped are associated with the electromagnetic character of the wave should now be clear: electrostatic behavior is associated with large wave numbers, while electromagnetic behavior is associated with small wave numbers. Thus, the three terms in Eq. (5.82) which are associated with k_0, and which were shown to make minuscule contributions to the total space charge ρ, arise from the electromagnetic character of the wave.

5.3.2. Propagation in a Plasma-Filled Waveguide

We are now ready to investigate the propagation of ion-acoustic waves in a finite-size, warm-plasma column of the type shown in Fig. 5.2. As in Section 5.2 for electron plasma waves, we first assume that the plasma density

is everywhere uniform and constant. We then discuss the ramifications introduced by assuming that the plasma density has a more realistic radial variation. Finally, we note that the *dispersion relation for ion-acoustic waves* in a
warm-plasma column can be modified to describe the propagation of *low-
frequency electron plasma waves* in a warm-plasma column.

 5.3.2.1. Finite-Size Model with Constant Density. As discussed in Section 5.2,
the parallel component of the electric field of any wave at the surface of the
conducting wall of the waveguide must be zero. Thus, Eq. (5.84) when
evaluated at $r = R$ must give zero. That is, $E_z(R) = 0$. As for the electron
plasma wave this gives the dispersion relation for ion-acoustic waves propagating parallel to the axis of a magnetically supported warm-plasma column as

$$T(\omega,k) = \frac{P_{0n}}{R}, \tag{5.88}$$

where T is given by Eq. (5.83) and again the P_{0n} are the roots of the
Bessel equation $\mathcal{J}_0(TR) = 0$.

 In most of the experiments where ion-acoustic wave propagation in a
plasma column has been studied, the plasma density and wave frequencies
used have been such that $\omega \ll \omega_{pi}$. We now want to introduce this approximation into Eq. (5.83) which, as it stands, has no restrictions on the relative
values of ω, ω_{pi}, and Ω_{ci}. To do this we make use of what is called a
small-parameter approximation, a general technique that is useful for reasonably
complicated algebraic equations such as Eq. (5.83), where the implications of
requiring that one variable be small with respect to another are not always
obvious. To implement this technique we first form a ratio of the two variables
involved. Thus, for this case, we define

$$p = \frac{\omega^2}{\omega_{pi}^2}. \tag{5.89}$$

Using this definition, we can rewrite Eq. (5.83) as

$$T^2 = k^2 \left[\frac{1}{1-p} \left(\frac{\omega_{pe}^2}{k^2 V_e^2} + \frac{\Omega_{ci}^2}{\omega^2} \right) - \frac{\omega_{pe}^2}{k^2 V_e^2} - 1 \right]. \tag{5.90}$$

Requiring that $p \ll 1$ and using the binomial expansion theorem, we can
write that $(1 - p)^{-1} = 1 + p + p^2 + p^3 + \cdots$. Keeping only the first two
terms in the expansion, we rewrite Eq. (5.90) as

$$T^2 \simeq k^2 \left[\frac{p\omega_{pe}^2}{k^2 V_e^2} + \frac{\Omega_{ci}^2(1 + p)}{\omega^2} - 1 \right]. \tag{5.91}$$

Ignoring p with respect to 1 in the second term on the right-hand side of Eq. (5.91) we obtain the small-parameter approximation of Eq. (5.83) as

$$T^2 \simeq k^2 \left[\frac{p\omega_{pe}^2}{k^2 V_e^2} + \frac{\Omega_{ci}^2}{\omega^2} - 1 \right], \tag{5.92}$$

where $p \ll 1$ is defined by Eq. (5.89). Using the definitions of ω_{pe}^2, V_e^2, and p, the first term in Eq. (5.92) becomes $\omega^2/k^2 Cs^2$, so that Eq. (5.92) can be written as

$$T^2 \simeq k^2 \left[\frac{\omega^2}{k^2 Cs^2} + \frac{\Omega_{ci}^2}{\omega^2} - 1 \right], \tag{5.93}$$

where the possible values of T are defined by Eq. (5.88). Thus, Eq. (5.93) is the dispersion relation for the propagation of ion-acoustic waves parallel to the axis of a magnetically supported, uniform, warm-plasma column contained in a cylindrical waveguide, subject to the restrictions that $\omega \ll \omega_{pi}$ and that we are not near the cutoff ($k \to 0$).

As a partial check on the validity of Eq. (5.93) we can let $\Omega_{ci} \to 0$ (i.e., zero magnetic field), and let the radius R of the plasma column go to ∞ [i.e., let $T \to 0$; see Eq. (5.88)], to see if we recover the dispersion relation of Chapter 3. Doing this, Eq. (5.93) becomes

$$\frac{\omega^2}{k^2 Cs^2} - 1 = 0. \tag{5.94}$$

That is, we have an ion-acoustic wave that propagates without dispersion, with a phase velocity $v_\varphi = \omega/k = Cs$, where Cs [see Eq. (3.23)] is defined as the speed of an ion-acoustic wave in an infinite plasma for $\omega \ll \omega_{pi}$. Thus, Eq. (5.93) contains the results of Chapter 3 as a special case.

Normally, we think of finite-size effects, i.e, plasma-boundary effects, as occurring when the wavelength is of the same order as the smallest dimension of the plasma. However, Eq. (5.93) shows that for strong magnetic fields the transverse boundary effects of the plasma, which manifest themselves through the T^2-term on the left-hand side of the equation, can be negligible. More specifically, if

$$\frac{\Omega_{ci}}{\omega} \gg \frac{T}{k}, \tag{5.95}$$

then the T^2-term in Eq. (5.93) can be neglected. Using the relationships: $T = P_{0n}/R$ [see Eq. (5.88)], $v_\varphi = \omega/k$, and $V_i = r_{ci}\Omega_{ci}$, where V_i is the ion thermal speed and r_{ci} is the ion cyclotron radius, Eq. 5.95 becomes

$$r_{ci} \ll \frac{V_i}{v_\varphi} \frac{R}{P_{0n}}. \tag{5.96}$$

Since $(V_i/v_\varphi) < 1$ and $(1/P_{0n}) < 1$, Eq. (5.96) shows that if the ion cyclotron radius r_{ci} is on the order of, or smaller than, the plasma radius R, then finite-size effects are negligible. More quantitatively, Wong (1966), using both fluid-theory and kinetic-theory models, has shown that, for the fundamental ion wave mode in a Q-machine plasma ($T_e = T_i$), finite-size effects become important only if

$$\frac{r_{ci}^2}{R^2} \gtrsim \frac{2}{P_{01}^2}. \tag{5.97}$$

Thus, we come to the conclusion that the transverse boundary conditions in a cylindrical plasma column are important if the cyclotron radius (rather than the wavelength!) is comparable to the plasma radius. This agrees with the experiments of Wong et al. (1964), where it was found that the propagation of ion-acoustic waves along a cylindrical plasma column supported by strong magnetic fields of 4–14 kG was not affected by the size and shape of the column, even though the plasma radius R was of the same order as the wavelength λ. On the other hand, Little (1961, 1962), whose experiment we will discuss in more detail in Section 5.4.1, has observed the effects of boundaries on ion-acoustic waves in a cylindrical plasma column supported by *weak* magnetic fields (less than 50 G) for $\lambda \simeq R$.

It should be pointed out that for sufficiently high frequencies such that $\lambda \ll R$, one expects on a physical basis that finite-size effects will be negligible. This is simply because, in this case, the wave in the interior region of the plasma does not "see" or "experience" the part of the plasma that is modified by the boundary. As far as the wave is concerned, the entire universe is filled with the type of plasma in which the wave finds itself. For the magnetically supported plasma we are considering here, however, we see that an *insulation* of the wave from the boundary occurs, mathematically, via the magnetic-field-dependent term Ω_{ci}^2/ω^2 [see Eq. (5.93)], which is seen to go to zero when $\omega \gg \Omega_{ci}$.

Figure 5.6 also shows clearly that the dispersive behavior due to finite-size effects in magnetically supported plasma disappears for $\omega \gg \Omega_{ci}$. A little further on in this section, we will discuss this *magnetic insulation* of the wave from the plasma boundaries, which can manifest itself even when the wavelength is on the order of or even larger than the dimensions of the plasma.

We now want to investigate the general characteristics of the dispersion relation represented by Eq. (5.93). In this equation, we choose to regard ω as

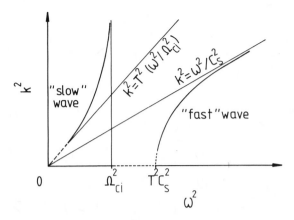

Fig. 5.6. Dispersion relation for ion-acoustic waves in a cylindrical plasma column for the case $(TCs > \Omega_{ci})$ where finite-size effects can be important. As indicated by the dashed parts of the curves, the model cannot validly describe the wave behavior for $k \to 0$.

an independent variable and to express k^2 in terms of this independent variable. Doing this, we can rewrite Eq. (5.93) as

$$k^2 = \left(\frac{T^2 Cs^2 - \omega^2}{Cs^2}\right)\left(\frac{\omega^2}{\Omega_{ci}^2 - \omega^2}\right). \tag{5.98}$$

Observation of Eq. (5.98) shows that there is a cutoff $(k \to 0)$ at $\omega = TCs$ and a resonance $(k \to \infty)$ at $\omega = \Omega_{ci}$. Since our model excludes cutoffs [see the discussion following Eq. (5.59)], we cannot use Eq. (5.98) to describe the behavior of ion-acoustic waves at the cutoff frequencies predicted by the equations; however, the description for frequencies reasonably *near* the cutoff frequency is valid.

From the form of Eq. (5.98) it is seen that a convenient way to display the functional relationship between ω and k is to plot k^2 versus ω^2. Before doing this, however, it is necessary to make a choice as to the relative values of TCs and Ω_{ci}. Since the wave behavior can be quite different, depending on which of these frequencies is higher, we consider the two cases separately. We first consider the case for $TCs \gg \Omega_{ci}$. Observing Eq. (5.98) we note first that in the frequency range $\Omega_{ci} < \omega < TCs$ the second factor on the right-hand side of Eq. (5.98) is negative, so that $k^2 < 0$. This corresponds to a purely imaginary k, so that no propagation is expected in this range of frequencies. This is shown in Fig. 5.6 by the hash marks along the ω^2 axis. Second, we note that as ω becomes small, $k^2 \to T^2 \omega^2/\Omega_{ci}^2$. That is, the phase velocity $v_\varphi \to \Omega_{ci}/T$, as indicated in Fig. 5.6 by the straight line, $k^2 = T^2(\omega^2/\Omega_{ci}^2)$, having a slope of $1/v_\varphi^2$. The upper curve in Fig. 5.6 shows the wave behavior for frequencies below the cyclotron resonance frequency, except for frequencies near zero, which our model excludes. As the frequency of this *slow* wave approaches Ω_{ci}, its speed decreases toward zero. The lower curve in Fig. 5.6 shows a plot of Eq. (5.98) for frequencies above the cutoff. At very high frequencies, the wave speed approaches the infinite-plasma wave speed Cs as shown by the straight line $k^2 = \omega^2/Cs^2$, whose slope is $1/Cs^2$. At frequencies near the cutoff, however, the speed of this *fast* wave can be much higher than Cs.

Observation of Fig. 5.6 shows that $\omega \geqslant TCs$ for the fast wave. If $TCs \gg \Omega_{ci}$ (Fig. 5.6 assumes only that $TCs > \Omega_{ci}$), then $\omega \gg \Omega_{ci}$ for the fast wave. For this case, the general dispersion relation given by Eq. (5.98) reduces to a particularly simple form, representing the dispersion relation for the fast wave, which can be written as

$$\omega^2 = (T^2 + k^2)Cs^2. \tag{5.99}$$

Equation (5.99) shows clearly, as does Fig. 5.6, that the fast wave has a low-frequency cutoff $(k \to 0)$ at $\omega = TCs$.

Figure 5.6 represents the general dispersion relation given by Eq. (5.98) for the case where $TCs > \Omega_{ci}$. It is of interest to see what this restriction means from a physical point of view. Proceeding as we did in the discussion of Eq. (5.95), we replace the quantities in the inequality by the following relationships:

$$T = \frac{P_{0n}}{R}, \qquad Cs = \left(\frac{KT_e}{m_i}\right)^{1/2}$$

$$\Omega_{ci} = \frac{V_i}{r_{ci}} \simeq \frac{(2KT_i/m_i)^{1/2}}{r_{ci}}.$$

Doing this, the inequality can be rewritten as

$$\frac{r_{ci}^2}{R^2} > \frac{2(T_i/T_e)}{P_{0n}^2}. \tag{5.100}$$

If, in Eq. (5.100), we choose the fundamental ion mode $(P_{0n} = P_{01})$ and let $T_i = T_e$, we see that Eq. (5.100) is the same as Eq. (5.97), which gives the criterion around the ion cyclotron frequency for finite-size effects to be important for the fundamental ion wave mode in a Q-machine plasma $(T_i = T_e)$. We shall see in Chapters 7–10 that because of wave–particle interactions, leading to so-called Landau damping, ion-acoustic waves in an isothermal $(T_i \simeq T_e)$ plasma cannot be treated adequately by the present simple fluid theory. It turns out, however, that, for finite-size effects, kinetic theory and fluid theory give very similar results [see Wong (1966)], so we will pursue our present argument. Thus, around the ion cyclotron frequency, Eq. (5.100) can be viewed as a generalized criterion for finite-size effects. Figure 5.6 therefore respresents the behavior of ion-acoustic waves in magnetically supported plasma columns where finite-size effects can be important. (This applies, of course, only to plasma columns having boundary conditions corresponding to our model.) As discussed earlier, and as observed in the figure, at sufficiently high frequencies, finite-size effects become unimportant, and the fast-wave behavior approximates that of planar ion waves in an infinite plasma.

We now want to examine the general dispersion relation given by Eq. (5.98) for the case where $TCs < \Omega_{ci}$. Figure 5.7 shows a graph of Eq. (5.98) for this case. Again, we see that there is a range of frequencies between the ion cyclotron frequency and the cutoff frequency for which ion wave propagation does not occur. And again we see that there is a slow wave and a fast wave. However, in contrast with Fig. 5.6, we see that the fast wave now occurs at frequencies *below* the ion cyclotron, whereas the slow wave occurs at frequencies *above* the cyclotron frequency. If we choose $TCs \ll \Omega_{ci}$, then $\omega \ll \Omega_{ci}$ for the fast wave and Eq. (5.98) reduces to the fast-wave dispersion relation, which can be written as

$$k^2 = \left(T^2 - \frac{\omega^2}{Cs^2} \right) \frac{\omega^2}{\Omega_{ci}^2}. \tag{5.101}$$

At low frequencies Eq. (5.101) shows that the phase velocity of the fast wave approaches a lower limit of $v_\varphi = \Omega_{ci}/T$, which is seen to be appreciably higher than Cs (the lower limit of the fast-wave speed for the $TCs > \Omega_{ci}$ case; see Fig. 5.6). As ω increases the phase velocity of the fast wave increases and,

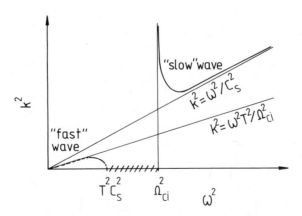

Fig. 5.7. Dispersion relation for ion-acoustic waves in a cylindrical plasma column for the case ($TCs < \Omega_{ci}$) where finite-size effects are negligible. As indicated by the dashed parts of the fast-wave curve, the model is not valid for $k \to 0$.

as before, there is a cutoff at $\omega = TCs$; however, the cutoff is now seen to repre-sent a high-frequency limit to fast-wave propagation, rather than a low-fre-quency limit.

As before (see Fig. 5.6), Fig. 5.7 shows that the slow wave has a resonance at the ion cyclotron frequency $(k \to \infty, v_\varphi \to 0)$. In the high-frequency limit $\omega \gg \Omega_{ci}$, $k^2 \to \omega^2/Cs^2$, i.e., the phase velocity of the slow wave approaches the speed Cs of ion-acoustic waves in an infinite warm plasma. This high-speed limit of the slow wave is seen to be appreciably higher than that observed for the $TCs > \Omega_{ci}$ case (see Fig. 5.6), and is seen to be a high-frequency limit rather than a low-frequency limit. If, as in the preceding discussion, we choose $TCs \ll \Omega_{ci}$, then $\omega \gg TCs$ for the slow wave, and the general dispersion rela-tion reduces to the slow-wave dispersion relation

$$k^2 = \frac{\omega^2}{Cs^2} \left(\frac{\omega^2}{\Omega_{ci}^2 - \omega^2} \right). \tag{5.102}$$

5.3.2.2. Infinite-Plasma Model with Realistic Density Profiles. In the preceding section we used a mathematical model that does not correspond to the physical reality of many experiments. Yet some of the predictions of the simple model are found to be in remarkably good agreement with some of the results of these experiments. Thus, it seems that the model is somewhat more general than one might *a priori* think. In this section we will give some "hand waving" arguments to try to explain why the model, in some cases, seems to accurately describe experiments not fulfilling the assumptions of the model.

In our simple model we have assumed that the density is uniform and constant. The motivation for this was simply so that Eqs. (5.24) and (5.81) would be simple to solve. For that case, these equations are simple Bessel differential equations having analytical solutions. For more realistic density profiles, the analytical solutions to these equations are either very difficult or nonexistent. (In practice, however, this does not present a problem because, even with modest computers the solutions of these equations for any desired density profile can be found without difficulty.) In our model, we have also assumed that there are conducting walls at the edges of the plasma. Thus, this allows us to assume that $E_z(R) = 0$, even for a constant-density plasma. However, even for plasmas enclosed by nonconducting walls we can still have $E_z(R) = 0$.

Figure 5.8 shows the plasma-density profile assumed by our simple model, as well as the perturbed-density profile predicted by this model for the fun-damental ion-acoustic wave mode. For typical experimental conditions, the plasma density is not constant with radius as shown, but rather is somewhat parabolic, having an almost flat central part and decreasing toward the edge of the plasma column. Also, the perturbed-density profile will be determined primarily by the experimental wave-excitation technique used and, for many

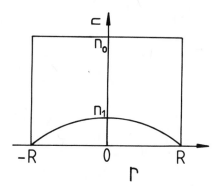

Fig. 5.8. The constant-density profile assumed by the model of Section 5.3.2.1, and the perturbed density profile predicted by this model for the fundamental ion-acoustic wave mode.

of the techniques used, will not greatly resemble the profile shown in Fig. 5.8. Figure 5.9 gives a more realistic approximation to both the unperturbed density profile and the density profile of a grid-excited wave.

It has been found that the same mathematical model that was used in the preceding section can be used here, if the value of R used in the equations is allowed to be much larger (but not infinite) than the actual plasma-column radius. This model is called the *infinite-plasma* model. With this modification the dispersion relation given by Eq. (5.88) can be written as

$$\left(1 - \frac{\omega^2}{k^2 C_s^2}\right)\left(k^2 \frac{\Omega_{ci}^2}{\omega^2} + \frac{\omega_{pi}^2}{c^2}\right) = 0. \tag{5.103}$$

From Eq. (5.103) we see that

$$\omega^2 = k^2 C_s^2. \tag{5.104}$$

Equation (5.104) is the dispersion relation for a wave propagating at the speed of an ion-acoustic wave in an infinite plasma, with no dispersion at the ion cyclotron frequency.

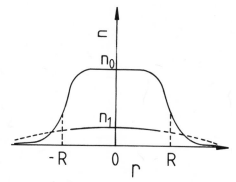

Fig. 5.9. Realistic approximation for both the unperturbed density profile in a plasma column and the perturbed density profile of a grid-excited ion-acoustic wave.

Thus, we have two models for ion-acoustic wave propagation in a plasma column: the finite-size model and the infinite-plasma model. The first model predicts ion-acoustic waves having $v_\varphi > Cs$, and having strong dispersion around the ion cyclotron frequency. The second model predicts ion-acoustic waves having $v_\varphi = Cs$, and having no dispersion at the ion cyclotron frequency. In fact, the modes predicted by both models have been observed in plasma columns immersed in strong magnetic fields.

5.3.2.3. Applicability of the Finite-Size Model to Electron Plasma Waves. In Section 5.2 we discussed finite-size effects on electron plasma waves in *cold* plasmas. What about finite-size effects on electron plasma waves in warm plasmas? In Section 5.3.2.1 we described a model for ion-acoustic waves in a warm plasma. That model, with certain restrictions, can be used to describe electron plasma waves because as far as the electrons are concerned only the electron inertia and the electron cyclotron motion have been neglected in the model. Thus, this same model can describe electron plasma waves below the electron plasma frequency as long as the phase velocity is smaller than the electron thermal speed. With these restrictions we can obtain the dispersion relation for electron plasma waves in a warm plasma column from Eq. (5.82) by letting the ion mass go to infinity (i.e., taking into account that $m_i \gg m_e$). Thus, we obtain

$$T^2 = \frac{\omega_{pe}^2}{V_e^2} - k^2 + k_0^2 = \frac{P_{0n}^2}{R^2}, \tag{5.105}$$

or, replacing k_0 by ω/c [see Eq. (5.11)] we can write

$$\omega^2 = k^2 c^2 + \left(\frac{\omega_{pe}^2}{V_e^2} - \frac{P_{0n}^2}{R^2} \right) c^2. \tag{5.106}$$

Thus, the thermal effect of the electrons modifies only the low-frequency part of the electron plasma wave in a plasma column.

5.4. SELECTED EXPERIMENTS

In this section we describe three experiments involving ion-acoustic wave propagation parallel to the axis of a magnetically supported warm-plasma column. In the first experiment by Little (1961, 1962), a mercury plasma was contained in a cylindrical glass tube. In the second and third experiments by Mills and Doucet (1968), a cesium plasma column was surrounded, in one experiment, by a cylindrical waveguide but, in the other experiment, only by vacuum. In all three experiments the measurements were made with

$\omega > \Omega_{ci}$. In two of the experiments good agreement was found with the theory of Section 5.3.2.1 for a waveguide filled with a constant-density plasma, although only one of the experiments closely approximates this model. In the other experiment the propagation was found to approximate that for an infinite plasma, with some slight dispersion near Ω_{ci}. This is a mode that is not described by our model.

5.4.1. Ion-Acoustic Wave Propagation in a Plasma-Filled Glass Tube

In these experiments (Little, 1961; 1962), ion-acoustic waves were propagated along a cylindrical column of mercury plasma contained in a glass tube and supported by a weak magnetic field. The plasma conditions, magnetic field, and geometry were such that $TCs \gg \Omega_{ci}$, and the wave frequencies were chosen such that $\omega \gg \Omega_{ci}$ (see Fig. 5.6). In a certain frequency range, as predicted by Fig. 5.6 for the fast ion-acoustic wave mode, it was found that the measured wavelength of the observed wave increased rapidly (i.e., $k = 2\pi/\lambda \to 0$) as the frequency was reduced. And below a critical frequency, accurately given by Eq. (5.100) for $k = 0$, no waves could be propagated. Near this lower limit, the phase velocity of the waves was found to be greater than the ion-acoustic wave speed in an infinite plasma, as predicted in Fig. 5.6 for the fast wave.

Figure 5.10 shows the experimental arrangement used by Little (1961, 1962). In the refrigerated, mercury-pool cathode C, a nickel cylinder N anchors the arc spot. The discharge is maintained between C and the water-cooled anode A. A_1 is the starting anode. The double Langmuir probe P consists of two tungsten wires sheathed in glass except for the extreme tips, which are ground flat to expose two plane-conducting disks, 1.5 mm in diameter and 2 mm apart. P measures the ion density, typically $10^{11} - 10^{12}$ ions/cc, representing $1\%-10\%$ ionization, in neutral-gas pressures of 1 μm or less. The magnetic-field coils M can provide up to 1500 G along the axis of the tube but, because of discharge instabilities that occurred at high fields, the experiments were conducted with magnetic fields of less than 100 G. The discharge tube is pumped at either end by mercury diffusion pumps. The whole system

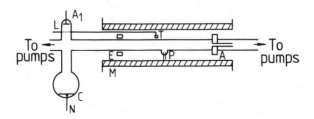

Fig. 5-10. Apparatus for studying ion-acoustic waves in a plasma column. (After Little, 1961.)

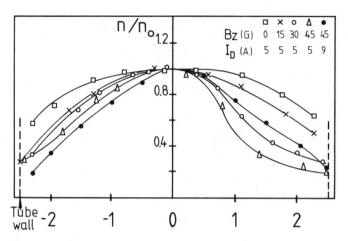

Fig. 5.11. Relative electron densities n/n_0 in the discharge as a function of radius, with different discharge parameters at 0.4 mmHg. (After Little, 1961.)

is assembled with O-rings that are shielded by inner glass tubes from the discharge, and the background pressure of noncondensible gases is less than 5×10^{-6} mmHg, with the discharge running.

Figure 5.11 shows the radial-density profiles measured with a Langmuir probe for various values of the magnetic field B_z and discharge current I_D. It is noted that as the magnetic field is made stronger diffusion of the plasma to the tube wall is decreased, resulting in sharper profiles as the plasma is confined more and more nearly toward the center of the tube. The assymmetry in the profiles was attributed to the presence of the probe, itself.

Two methods of wave excitation were used. In one method the current in a small coil E (see Fig. 5.10) around the plasma column was varied sinusoidally to produce a 10-G variation of the magnetic field in a small region of the plasma column centered at the plane of the coil. This time variation in the magnetic field produced in turn a time variation in the plasma density, which then propagated in both directions away from the plane of the coil as ion-acoustic waves. In the second method a large oscillating voltage was placed across two closely spaced grids (not shown) which were oriented perpendicular to the plasma column. The resulting large electric fields produced a time-varying ionization rate between the two grids, thereby producing again, a localized variation in the plasma density of the column.

Due to the excitation of background neutrals by plasma electrons, which leads to the emission of light as the excited neutrals radiate, a plasma such as the one produced by the apparatus shown in Fig. 5.10 is slightly luminous. Since the rate of excitation is a function of the plasma density, the slight per-

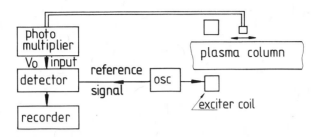

Fig. 5.12. Block diagram of the circuit. (After Little, 1961.)

turbations produced in the plasma density by a propagating ion-acoustic wave produces a slight variation in the light emitted by the plasma. This light variation at the same frequency as the wave can be detected by a sufficiently fast and sensitive light detector, such as a photomultiplier tube. Little (1961, 1962) used a photomultiplier tube which was collimated and which could be moved parallel to the discharge tube (see Figs. 5.10 and 5.12) to observe the propagation of the fast-ion mode along the plasma column. Because the variation in light intensity was so small, in order to achieve a tolerable signal-to-noise ratio it was necessary to use a phase-sensitive detector (see Fig. 5.12) which compared the photomultiplier signal V_0 with a reference signal from the oscillator driving the wave-exciter coil. The output from such a detector is proportional to $V_0 \cos \varphi$, where φ is the phase difference between the detected waveform and the reference waveform. Thus, as the photomultiplier tube was moved along the plasma column, both the amplitude and the phase change of the propagating wave could be measured.

Figure 5.13 shows a chart recorder graph for one such experiment. Since displacement along the graph was calibrated in terms of actual displacement of the photomultiplier along the plasma column, the wavelength, λ of the propagating ion wave could be determined directly from the graph by measuring the distance between any two points differing in phase by $\Delta\varphi = 2\pi$. Knowing λ, the speed of the wave for that particular frequency f could be found from the relationship, $v_\varphi = \lambda f$. Figure 5.14 shows a plot of $1/\lambda$ versus f for a series of measurements made in the manner just described. Since $k = 2\pi/\lambda$ and $\omega = 2\pi f$, we see that except for a scale factor of 2π, Fig. 5.14 can be viewed as a plot of k versus ω, which can then be compared qualitatively with the k^2 versus ω^2 graph given in Fig. 5.6 of our general dispersion relation for the case of $TCs > \Omega_{ci}$. Making this comparison one immediately concludes, as did Little (1961, 1962), that the experimentally observed wave is the fast ion-acoustic mode. As predicted by theory, the wave is seen to be very dispersive near the cutoff frequency: a factor of two change in frequency (from 50 kHz to 25 kHz) is seen to result in a factor of ten change in wavelength (from 3 cm to 30 cm). That is, the phase velocity of the wave, $v_\varphi = \lambda f$

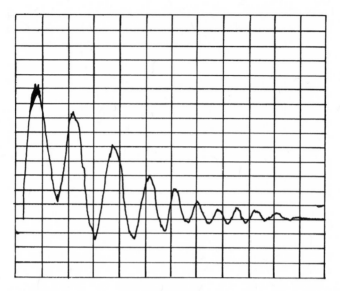

Fig. 5.13. Typical chart record of the output from the phase-sensitive detector when the photo-multiplier provides the input signal. The motion of the chart is synchronous with the motion of the photomultiplier along the plasma column. (1 in. on chart = 4.6 cm movement along column.) Discharge current 9 A; magnetic field 60 G: mercury vapor pressure 0.4 mmHg; frequency of signal 44 kHz. (After Little, 1961.)

increases by a factor of five as the frequency is decreased (if there were no dispersion, i.e., if $v_\varphi = $ const, then of course no change in wave speed would be expected). Also, as predicted by theory, the experimental points seem to be asymptotically approaching the straight line $k = \omega/Cs$, which has been

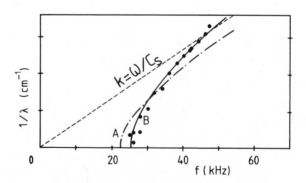

Fig. 5.14. The wavenumber $1/\lambda$ as a function of frequency. Anode-directed waves in a 9-A discharge with an axial field of 45 G. Experimental points, •; theoretical curve A for $R = 2.5$ cm, $T_e = 5.2 \times 10^4$ K; theoretical curve B for $R = 2$ cm, $T_e = 4 \times 10^4$ K. (After Little, 1961.)

drawn in Fig. 5.14 using the value of Cs that would be expected for an infinite plasma having the same electron temperature as that measured for the plasma column.

In order to compare *quantitatively* the observed cutoff frequency with the theoretically predicted cutoff frequency, one must know which one(s) of the modes that are predicted to be theoretically possible has, in fact, been observed. Going back to our original discussion in Section 5.2.2.1, where finite-boundary effects were discussed in terms of the roots of the zero-order Bessel functions, we see that in order to know which mode was observed we must know how the amplitude of the wave varied in the radial direction. Figure 5.15 shows such information obtained by Little (1961), using the double-Langmuir probe to measure the amplitude of the wave at several radial positions, all measurements being made, of course, in a diametral plane of the column. Comparison of the experimental profiles of Fig. 5.15 with the theoretically predicted perturbed-density profile in Fig. 5.8 shows that qualitatively they are similar. Although Little (1961) speculated that other modes were probably present, the overall similarity led him to assume, for the purposes of comparison with theory, that only the fundamental ion mode (i.e., that $P_{0n} = P_{01} = 2.40$, or $T = 2.40/R$) was present. Going back to Eq. (5.98) and setting $k = 0$ gives the theoretically predicted cutoff frequency for the fundamental mode as $\omega_c = TCs = 2.4Cs/R$, where R is the radius of the plasma column. From the experimentally measured value of $T_e = 0.46$ eV, Cs is calculated to be 1.5×10^5 cm/s. Using the diameter of the glass tube $R = 2.5$ cm as the diameter of the plasma column (Little, 1961), then gives

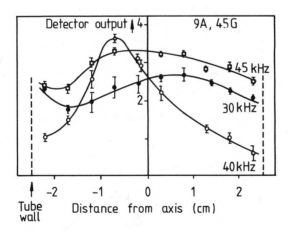

Fig. 5.15. The radial dependence of the in-phase signal at the Langmuir probe position for different frequencies in a discharge of 9 A with an axial field of 45 G at 0.4 mmHg. (After Little, 1961.)

$\omega_c/2\pi = 22$ kHz, a value that is seen to be in reasonable agreement with the cutoff frequency which can be deduced from the data of Fig. 5.14.

The data of Fig. 5.14 were obtained with an applied axial magnetic field of 45 G. From Fig. 5.11 we see that the radial density profile for this case more nearly approximates a Gaussian shape than the uniform density profile assumed in our theoretical model. Thus, at first glance, it might seem surprising that the results of Little's (1961) experiments should agree so well with the predictions of the model. However, as discussed in Section 5.3.2.2, although a nonconstant density profile can greatly complicate the solution of the general wave equation, the exact shape of the density profile cannot greatly alter the general characteristics of the fundamental mode: an amplitude that is finite on the axis, is approximately zero at the edge of the column, and is monotonically decreasing in between. So, perhaps, one should not be so surprised that the fundamental mode that *was* excited agrees fairly well with the simple theory but, rather, that it was the fundamental mode that was primarily excited (rather than, say, a mixture of modes).

5.4.2. Ion-Acoustic Wave Propagation in a Cylindrical Cesium Plasma

Figure 5.16 shows the experimental apparatus, called a Q-machine [see, for example, Motley (1975)], that was used by Mills and Doucet (1968) to study the propagation of ion-acoustic waves along a cesium plasma column supported by a weak magnetic field. In this device the plasma is generated by having a beam of cesium vapor from an oven incident upon a hot (2200° C)

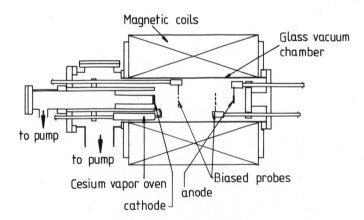

Fig. 5.16. Q-machine used to study the propagation of ion-acoustic waves along a magnetically supported cesium plasma column. Contact ionization of the cesium vapor incident on the hot cathode followed by diffusion along the weak axial magnetic field creates a highly ionized cylindrical plasma column between the anode and cathode. Wave excitation and detection was via the two biased grids.

tungsten cathode. On contact the cesium vapor is ionized and diffuses along the weak ($\leqslant 50$ G) axial magnetic field to produce a plasma column that terminates on an anode of the same size and shape as the cathode. Since great care is exercised to insure that the surface temperature of the cathode is very uniform, and that the entire surface is uniformly exposed to the vapor beam, this type of machine produces a plasma column that has a highly uniform, constant density, as assumed in our theoretical model. That is, the radial density profile very closely approximates the upper curve shown in Fig. 5.8. In Q-machine plasmas, however, $T_e \simeq Ti$, so that wave–particle interactions, which we have ignored in our model by assuming $T_e \gg Ti$ (see Chapter 3), can occur and cause strong damping of the waves.

Mills and Doucet (1968) saw two types of ion-acoustic waves, depending on the boundary conditions of the plasma. When the plasma column was enclosed in a metallic cylindrical tube (not shown) whose inner wall just touched the surface of the plasma, the fast wave (see Fig. 5.6) was observed. When the plasma column was surrounded by vacuum (in this case, several centimeters of high vacuum separated the column from the glass walls of the vacuum chamber), an ion-acoustic wave was seen whose speed, at high frequencies, approached the speed Cs of an ion wave in an infinite plasma, but whose speed exhibited some dispersion as the ion cyclotron frequency was approached. Under some conditions, it sometimes appeared that both types of waves were present simultaneously. However, definitive measurements could be made only when the two different boundary conditions were imposed separately. Because of the weakness of the detected signals due to strong damping, phase-sensitive detection similar to that used by Little (1961, 1962) was necessary. Both generation and detection of the ion waves was via biased planar grids oriented perpendicular to the axis of the plasma column, see Fig. 5.16.

Figure 5.17 shows plots of phase velocity v_φ versus the frequency f for the two waves observed. The upper curve shows the fast wave, which exhibits strong dispersion at low frequencies, with an extrapolated cutoff frequency of about 13 kHz. The cutoff frequency of this fast wave is closely predicted by Eq. (5.99) when the fundamental mode (i.e., $TR = 2.40$) is assumed and when R is taken to be the radius of the plasma column. As found by Little (1961, 1962), the speed of the fast wave approaches that of an ion wave in an infinite plasma as the wave frequency increases. In agreement with the predictions of our theoretical model, no propagation of the fast wave was observed for frequencies below TCs. The slow wave observed by Mills and Doucet (1968) is shown by the bottom curve in Fig. 5.17. As noted earlier some slight dispersion is exhibited by this wave as the cyclotron frequency is approached. Also, we note that this wave was observed to propagate in the frequency range of $\Omega_{ci} \leqslant \omega \leqslant TCs$, a range for which no ion wave propagation is predicted by our simple model.

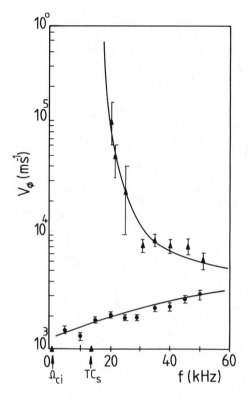

Fig. 5.17. Fast and slow ion-acoustic waves observed to propagate along a cylindrical plasma column supported by a weak magnetic field. The fast wave could be observed clearly only when the column was enclosed by a cylindrical waveguide. The slow curve was observed clearly only when the column was surrounded by a vacuum. (After Mills and Doucet, 1968.)

The plasma conditions used for studying the fast wave, except for $T_e \simeq T_i$, represent a very close approximation to our assumed theoretical model of a magnetically supported uniform plasma in a waveguide. Thus, the close agreement between experiment and theory is not surprising. As noted, however, our theory does *not* predict the observed slow wave. One can wonder why this slow wave was not also observed by Little (1961, 1962). Also, since $\lambda \gg R$ for all frequencies studied, it is somewhat surprising that this wave apparently experiences no finite-size effects. The explanation of this wave is not presently known.

ION-ACOUSTIC WAVES IN A SMALL DENSITY GRADIENT

6.1. INTRODUCTION

A plasma that is infinite in extent and everywhere uniform and constant in density is a theoretical approximation to reality. Although for all practical purposes the interior regions of some plasmas of interest closely approximate the conditions assumed in such theoretical models, all plasmas have regions of strong density variation in the vicinity of their surfaces. As we show in this chapter, the associated density gradients can produce strong modifications of the wave motion occurring in plasmas.

As an example of density-gradient effects, we discuss the propagation and reflection of ion-acoustic waves in nonuniform, nonmagnetically supported plasmas. One theoretical method that has been used to study waves in density gradients involves the so-called Wentzel–Kramers–Brillouin (WKB) approximation. This technique was introduced in quantum mechanics for solving Schrodinger's wave equation (Wentzel, 1926; Kramers, 1926; Brillouin, 1926). For application of this technique to wave propagation in plasmas see Yeh and Liu (1972). Unfortunately, the approximation is not valid when the characteristic length of density variation L defined by the equation $1/L = (1/n_0)(dn_0/dx)$ is of the order of, or smaller than, the wavelength, as is the case of interest here. For the present study we find and solve the wave equation directly.

A density gradient in a plasma is always associated with the variation of other plasma quantities such as electric field, mean velocities of the particles, and charge separation. Ishihara *et al.* (1978) have conveniently divided the density-gradient transition region between the main body of the plasma and its surface into three parts: a subsonic-flow region in which the ion drift speed (i.e., the plasma diffusion speed toward its surface) is less than the ion-acoustic wave speed; a supersonic-flow region in which shock formation and other nonlinear phenomena can occur; and an exterior region where sheaths form and charge neutrality breaks down. We limit our discussion to a study of the wave propagation and reflection in the first region. That is, we are interested only in *small*-density-gradient effects. As a point of practical interest, we note that although reflection in the supersonic-flow region may occur and, perhaps, produce a modification of the local density gradient itself, such a reflection will not result in the wave being reflected back into the main body of the plasma. This is because the flow speed of the plasma toward the surface in the supersonic-flow region is, by definition, greater than the ion wave speed, so the wave will be Doppler-shifted toward the surface of the plasma even if reflection should occur in the rest frame of the diffusing plasma.

6.2. WAVE EQUATION

Here we are interested in the propagation of slow, low-frequency waves. By this we mean that the wave speed is much slower than the electron thermal speed V_e, and the wave frequency is much smaller than the ion plasma frequency. The first restriction allows us to neglect the electron's inertia or, equivalently, to describe the electrons by a Boltzmann distribution function. The second restriction guarantees that the ion- and electron-density perturbations produced by the wave will be equal—a condition that greatly simplifies the wave equation.

We begin by showing that the electrons will be Boltzmann. This property can be demonstrated by use of the equation of motion for the electrons, including in the equation a pressure-gradient term which arises from the assumed density gradient. This equation can be written as

$$nm_e \frac{dv}{dt} = -ne(E + v \times B) - \nabla P_e \qquad (6.1)$$

Since we assume no applied magnetic field, $B_0 = 0$ and $v \times B = 0$. Therefore, for equilibrium Eq. (6.1) becomes

$$n_0 m_e v_0 \nabla v_0 = -n_0 e E_0 - \gamma_e K T_e \nabla n_0, \qquad (6.2)$$

where $\nabla P_e = \gamma_e K T \nabla n_0$ has been used. For first-order perturbations Eq. (6.1) gives

$$\frac{\partial v_1}{\partial t} + v_0 \nabla v_1 = \frac{-eE_1}{m_e} - \frac{\gamma_e K T_e}{m_e} \frac{\nabla n_1}{n_0}. \qquad (6.3)$$

If ω/k and v_0 are both $\ll V_e$, then the two terms on the left-hand side of Eq. (6.3) are much smaller than the two terms on the right-hand side. We now consider this case by limiting our description to slow waves in small-density gradients. With this restriction the mean velocity v_0 of the particles given by Eq. (6.2) is much smaller than the electron thermal speed and, for slow waves such as ion-acoustic waves, $\omega/k \ll V_e$. Assuming that $\gamma_e = 1$ and replacing E_1

by $-\nabla V_1$, where V_1 is the local potential, Eq. (6.3) can then be written as

$$e\nabla V_1 - KT_e \frac{\nabla n_1}{n_0} = 0. \tag{6.4}$$

The one-dimensional solution of Eq. (6.4) is

$$n_1(x,t) = n_0(x = 0) \exp\left[\frac{eV_1(x)}{KT_e}\right]. \tag{6.5}$$

We note that Eq. (6.5) includes both the unperturbed density variation n_0 given by

$$n_0(x) = n_0(x = 0) \exp\left[\frac{eV_0(x)}{KT_e}\right], \tag{6.6}$$

and the perturbed density n_1 defined by

$$n(x,t) = n_0(x) + n_1(x,t). \tag{6.7}$$

Using the usual linearization procedure, Eqs. (6.5) and (6.7) can be used to find the electron density perturbation

$$n_1(x,t) = n_0(x)\left[\frac{eV_1(x,t)}{KT_e}\right]. \tag{6.8}$$

Equation (6.8) shows that an increase in the wave potential $V_1(x,t)$ leads to an exponential increase in the size of the electron density perturbation, $n_1(x,t)$. Because of this, the electrons are said to exhibit Boltzmann behavior.

We now show that when we are interested in low frequencies we can assume complete charge neutrality. From Poisson's equation

$$\nabla \cdot E_1 = \frac{e}{\varepsilon_0}(n_i - n_e), \tag{6.9}$$

we can write

$$-\nabla^2 V_1 = \frac{n_0 e}{\varepsilon_0}\left[\left(\frac{n_1}{n_0}\right)_i - \left(\frac{n_1}{n_0}\right)_e\right]. \tag{6.10}$$

Since

$$\left(\frac{n_1}{n_0}\right)_e = \frac{eV_1}{KT_e},$$

see Eq. (6.8), the left-hand side of Eq. (6.10) can be written as

$$-\nabla^2 V_1 = k^2 V_1 = \frac{KT_e}{e} k^2 \left(\frac{n_1}{n_0}\right)_e.$$

Thus, Eq. (6.10) becomes

$$\left(\frac{n_1}{n_0}\right)_i - \left(\frac{n_1}{n_0}\right)_e \simeq \left(\frac{n_1}{n_0}\right)_e \frac{\omega^2}{\omega_{pi}^2}. \tag{6.11}$$

From Eq. (6.11) we see that, for $\omega^2 \ll \omega_{pi}^2$, we can write

$$n_{1i} = n_{1e}. \tag{6.12}$$

Thus, from Eq. (6.8) we can write for the ions

$$n_{1i} = n_0(x) \frac{eV_1(x,t)}{KT_e}, \tag{6.13}$$

and we see that in the limit of low frequency the ions exhibit anti-Boltzmann behavior, i.e., an increase in V_1 leads to an exponential *increase* in n_{1i}. Therefore, we can write the differential equations of low-frequency waves, using only the equation of particle conservation and the *linearized* equation of motion for the ions, as

$$\frac{\partial n}{\partial t} + \nabla \cdot (nv) = 0; \tag{6.14}$$

$$\frac{\partial v}{\partial t} = \frac{eE_1}{m_i}. \tag{6.15}$$

We assume that we produce a density perturbation in the plasma, at $x = 0$, of the form

$$n(x = 0, t) = n(x = 0) \exp(-i\omega t). \tag{6.16}$$

At these low frequencies, all quantities $A(x,t)$ will, in fact, have the same general behavior, i.e.,

$$A(x,t) = A(x) \exp(-i\omega t). \tag{6.17}$$

Thus, we can replace the time operator everywhere by

$$\frac{\partial}{\partial t} = -i\omega. \tag{6.18}$$

For plane-wave propagation, which we are assuming, the ∇ operator can be replaced by

$$\nabla = \frac{\partial}{\partial x}. \tag{6.19}$$

Making these substitutions in Eqs. (6.13)–(6.15) and linearizing gives

$$\frac{n_1}{n_0} = \frac{eV_1}{KT_e}; \tag{6.20}$$

$$-i\omega n_1 + \left(\frac{\partial}{\partial x}\right)(n_0 v_1) = 0; \tag{6.21}$$

$$-i\omega v_1 = -\left(\frac{e}{m_i}\right)\left(\frac{\partial V_1}{\partial x}\right). \tag{6.22}$$

Defining the ratio of densities given by Eq. (6.20) as

$$\varepsilon_n = \frac{n_1}{n_0} = \frac{eV_1}{KT_e}, \tag{6.23}$$

(which can be thought of as either the relative density perturbation or as the reduced wave potential), reduces Eqs. (6.21) and (6.22) to the single differential equation

$$\frac{d^2 \varepsilon_n}{dx^2} + \frac{1}{n_0(x)}\frac{dn_0(x)}{dx}\frac{d\varepsilon_n}{dx} + k_0^2 \varepsilon_n = 0, \tag{6.24}$$

where k_0 is the usual wave number in a uniform plasma $k_0 = \omega/Cs$ with

$$Cs = \left(\frac{KT_e}{m_i}\right)^{1/2} \tag{6.25}$$

being the usual ion-acoustic wave velocity. [Rather than define the variable ε_n and obtain Eq. (6.24), we could equally well have replaced Eqs. (6.20)–(6.22) by a single differential equation in v_1. However, since v_1 and ε_n have a derivative relationship, as shown by Eqs. (6.21) and (6.23), no new information would have been gained. In the present discussion, we use only Eq. (6.24)].

In principle we can solve Eq. (6.24) for any density profile $n_0(x)$. The solution may be expressed in terms of Bessel functions, Neuman functions, Hankel functions, hypergeometric functions, etc., depending on the density profile chosen.

6.3. WAVE PROPAGATION IN A NONUNIFORM PLASMA HAVING A GAUSSIAN DENSITY PROFILE

We now describe the propagation of ion-acoustic waves in a plasma having a Gaussian profile (a density profile that is closely approximated in many experimental situations) given by

$$n_0(x) = n_0(0) \, \exp\left(\frac{-x^2}{\alpha^2}\right). \tag{6.26}$$

Differentiation of Eq. (6.26) gives

$$\frac{1}{n_0(x)} \frac{dn_0(x)}{dx} = \frac{-2x}{\alpha^2}. \tag{6.27}$$

Substitution of Eq. (6.27) into Eq. (6.24) yields

$$\frac{d^2\varepsilon_n}{dx^2} - \frac{2x}{\alpha^2} \frac{d\varepsilon_n}{dx} + k_0^2\varepsilon_n = 0. \tag{6.28}$$

This equation reduces to the classical parabolic cylinder equation [see, for example, Abramowitz and Stegun (1972)]

$$\frac{d^2y}{dz^2} = -\left(\frac{z^2}{4} + a\right), \tag{6.29}$$

where

$$a = -\frac{1 + k_0^2\alpha^2}{2};$$

$$y = \varepsilon_n \exp\left(\frac{-x^2}{\alpha^2}\right); \tag{6.30}$$

$$z = \frac{2x}{\alpha}.$$

The so-called "even solution" of Eq. (6.29) is given by

$$n_1(x) = An_0(0) \exp\left(\frac{-x^2}{\alpha^2}\right) M\left(\frac{-k_0^2\alpha^2}{4}, \frac{1}{2}, \frac{x^2}{\alpha^2}\right), \tag{6.31}$$

where $A \ll 1$ is a constant, and M is called the confluent hypergeometric function. For small wavelengths ($k_0\alpha \gg 1$), we can use an expansion for $M(a,b,x)$,

Fig. 6.1. Theoretical behavior of the amplitude of a linear ion-acoustic wave propagating in a plasma whose density profile is Gaussian. This wave is a plot of Eq. (6.32), where we have chosen k and α values typical of those occurring in many experiments: $\lambda = 2$ cm; $\alpha = 4$ cm. The dashed curve shows the corresponding square root of the unperturbed density for a Gaussian profile. Here the solution $k(x)$ of Eq. (6.37) is a constant, and the amplitude of the solution $n_1(x)$ of Eq. (6.38) is proportional to $n_0^{1/2}(x)$.

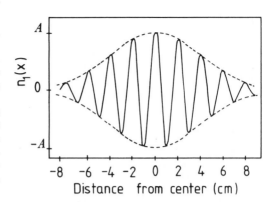

where $a \to -\infty$, b is bounded, and x is real, which gives

$$n_1(x) \simeq A n_0(0) \exp\left(\frac{-x^2}{2\alpha^2}\right) \cos kx, \tag{6.32}$$

where k is the wave number given by

$$k = k_0[1 + (k_0^2 \alpha^2)^{-1/2}] \simeq k_0. \tag{6.33}$$

Figure 6.1 is a plot of Eq. (6.32) for values of k and α that are typical of those used in many experiments.

6.4. WAVE PROPAGATION IN A NONUNIFORM PLASMA HAVING AN ARBITRARY DENSITY PROFILE

Density profiles other than the Gaussian profile used in the preceding section can be used, but all are found to demonstrate qualitatively the behavior seen in Fig. 6.1. Thus, we will use here a method (Doucet and Feix, 1975) that not only demonstrates this general behavior, but has the advantage of describing both wave propagation and reflection. For this purpose, we define a new variable v which is related to our previous variable ε_n by

$$v = \frac{\varepsilon_n}{[n_0(x)]^{1/2}}. \tag{6.34}$$

This permits us to write Eq. (6.24) as

$$\frac{d^2 v}{dx^2} + f(x)v = 0, \tag{6.35}$$

where

$$f(x) = k_0^2 - \left[\frac{1}{n_0(x)}\right]^{1/2} \frac{d^2 n_0^{1/2}(x)}{dx^2}.\tag{6.36}$$

We can make some general observations and predictions for the wave motion described by Eqs. (6.35) and (6.36). First, we note that $f(x)$ has the general character of being the square of a wave number. Second, we note that the mathematical form of Eq. (6.35) is the same as that of a harmonic oscillator. If we look at the form of $f(x)$, we see that for most experimental situations (except for periodic structures), $f(x)$ will be constant in sign over a wide range of space, say $a < x < b$, which might include, for example, the subsonic part of the density-gradient transition region at the edge of a plasma. As can be observed from Eq. (6.36), whether $f(x)$ is >0 or <0 depends, for a given wave, on the properties of the density variation in this non-uniform region of the plasma. If $f(x) < 0$, we can think of this as corresponding to a wave number that is purely imaginary or, alternatively, to a strongly damped harmonic oscillator. For either interpretation we expect the solution $v(x)$ of Eq. (6.35) to be a monotonic function in the region of space (a,b). That is, we expect that the wave will not exhibit oscillatory behavior in this nonuniform region of the plasma. Such a wave is sometimes said to be evanescent. Since we have not included any damping mechanism in our simple model, this corresponds to *wave reflection* in our case.

If $f(x) > 0$ in (a,b), then we expect $v(x)$ (which is directly related to n_1 the quantity of interest here) to be oscillatory in nature. In other words the *wave propagates*. In this case we can define a local wave number $k(x)$ by writing

$$k^2(x) = f(x) = k_0^2 - \left[\frac{1}{n_0(x)}\right]^{1/2} \frac{d^2 [n_0(x)]^{1/2}}{d^2 x},\tag{6.37}$$

and the density perturbation produced by the propagating wave can be shown to be given by the equation

$$n_1(x) = A n_0(x) [n_0(x)]^{1/2} \exp[ik(x)x],\tag{6.38}$$

where $k(x)$ is given by Eq. (6.37).

Equation (6.38) exhibits the $[n_0(x)]^{1/2}$ variation shown in Fig. 6.1 for the case of a Gaussian profile, but appears here as a general result, being independent of the exact shape of the density profile in the region (a,b) where Eq. (6.38) assumes that $f(x) > 0$.

From Eqs. (6.37) and (6.38), we point out explicitly that both the amplitude and the wave number of the propagating wave depend on the shape $n_0(x)$ of the density profile in the region of propagation.

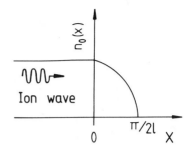

Fig. 6.2. Ion-acoustic wave propagation in a uniform plasma having a density profile at its edge as shown, and as given by Eq. (6.39).

6.5. WAVE PROPAGATION IN A UNIFORM PLASMA HAVING A SUBSONIC DENSITY GRADIENT AT ITS EDGE

Any steady-state or quasi-steady-state plasma having a well-established, diffusive-loss process at its boundaries necessarily has a density gradient with a subsonic part. Normally this is the region between the main body of the plasma and the supersonic-flow region near the surface of the plasma. We consider here the case of ion-acoustic wave propagation in a plasma whose density profile is as shown in Fig. 6.2, where

$$n_0(x) = n_0(0) = \text{const}, \qquad \text{for } x \leqslant 0;$$

$$n_0(x) = n_0(0)(\cos lx)^2, \qquad \text{for } 0 \leqslant x \leqslant \frac{\pi}{2l}; \qquad (6.39)$$

$$n_0(x) = 0, \qquad \text{for } x > \frac{\pi}{2l}.$$

For $x \leqslant 0$, the plasma is uniform, and we have classical ion-acoustic wave propagation, with $k = \omega/Cs = k_0$, for all frequencies $\omega \ll \omega_{pi}$. *In the density gradient*, we can write $f(x)$ [see Eq. (6.37)] as

$$f(x) = k_0^2 - l^2, \qquad (6.40)$$

where

$$l^2 = \left(\frac{1}{n_0}\right)^{1/2}\left[\frac{d^2(n_0)^{1/2}}{dx^2}\right], \qquad (6.41)$$

and

$$n_0(x) = n_0(0)\cos^2(lx). \qquad (6.42)$$

Letting $f(x) = k^2$ in Eq. (6.40) and replacing k_0 by ω/Cs, we see that there is a particular frequency $\omega_c = lCs$ at which cutoff $(k \to 0)$ occurs. Thus, for

$\omega < \omega_c, f(x) < 0$ (i.e., k is purely imaginary) and the wave is reflected. For $\omega > \omega_c, f(x) > 0$ and the wave propagates without energy loss to the plasma edge at $x = \pi/(2l)$.

6.6. SELECTED EXPERIMENTS

Doucet *et al.* (1974) studied the behavior of short-wavelength ion-acoustic waves in small density gradients both theoretically and experimentally. Parts of that theoretical study appear in the preceding sections of this chapter. Jones *et al.* (1976) studied experimentally the behavior of both short-wavelength and long-wavelength ion-acoustic waves in small density gradients. We now briefly summarize the experimental methods and results of these two studies.

6.6.1. Short-Wavelength Ion-Acoustic Waves in Small Density Gradients

The predictions of our simple theory for linear ion-acoustic waves in a density gradient have been checked by Doucet *et al.* (1974) using the experimental arrangement shown schematically in Fig. 6.3. The so-called diffusion plasma was produced by ionizing collisions between the background atoms and energetic electrons from the anode–hot-cathode assembly. The plasma density was limited to the range of 1×10^8 to 5×10^8 cm^{-3} in order to obtain the desired density gradient. The electron temperature was about 1 eV. To avoid significant collisional damping, the argon background neutral pressure was kept below 10^{-3} Torr. These parameters led to a plasma which, for the purposes of the experiment, could be considered to be collisionless. The density profile was shaped by means of the two 30-cm-diam parallel copper plates shown in Figure 6.3. By use of variable resistors to ground, the potential of these plates could be varied from approximately plasma potential to approximately floating potential, allowing a wide range of density gradients to be produced.

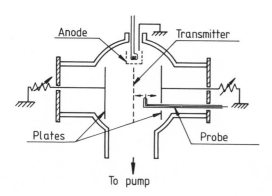

Fig. 6.3. Experimental apparatus used to study the behavior of linear ion-acoustic waves in density gradients. The magnitudes of the density gradients along the direction of propagation were controlled by means of the two 30-cm-diam copper plates. (After Doucet *et al.*, 1974.)

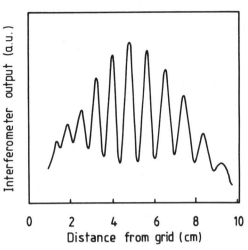

Fig. 6.4. Experimentally observed behavior of linear ion-acoustic waves in a discharge plasma whose density profile was approximately Gaussian and centered 5 cm from the grid. Here $C_s \sim 1.5 \times 10^5$ cm/s, $\lambda \sim 0.9$ cm, $\alpha \sim 2$ cm [see Eq. (6.26)], $T_e \sim 1.1$ eV, $n_e \sim 2 \times 10^8$ cm^{-3}, and the excitation voltage is ~ 0.5 V, peak-to-peak, at $f = 170$ kHz. (After Doucet *et al.*, 1974.)

The wave-amplitude measurements were made for continuous wave (cw) excitation. Excitation voltages having amplitudes of only a few tenths of a volt were used. A small probe, biased slightly positive with respect to the plasma (in order to measure density fluctuations) was used as the ion wave detector. The detected signal was processed by a lock-in amplifier. Figure 6.4 shows some typical propagation data. For these data, the copper plate was located approximately 15 cm from the grid, while the position of maximum

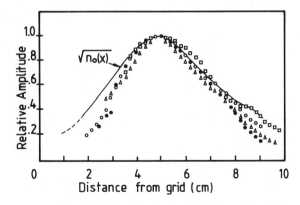

Fig. 6.5. A comparison of the experimentally measured amplitude behavior of linear ion-acoustic waves with the square root of the experimentally measured density profile: $\triangle$, $\bullet$, $\bigcirc$, and $\square$ correspond to excitation frequencies of 150, 170, 180, and 200 kHz, respectively. The systematic deviation of the amplitude data points from the square-root density curve near the grid represents, we believe, a difficulty in making an accurate measurement of the density profile in the vicinity of the sheath near the grid (the dashed portion of the profile). (After Doucet *et al.*, 1974.)

density was observed to be approximately 5 cm from the grid. Thus, the amplitude of the wave increases for propagation in the direction of increasing density and decreases for propagation in the direction of decreasing density.

Comparison of Fig. 6.4 with Fig. 6.1 shows a strong qualitative agreement between experimental observation and the predictions of the theory. In Fig. 6.5, a more quantitative comparison is made. Here, the observed spatial variation of the wave amplitude for several frequencies is compared with the square root of the measured density profile. Very good agreement is seen in the region of wave damping, and reasonably good agreement is seen in the region of wave growth. An attempt to show systematically that the wave behavior was independent of wave frequency, as predicted by theory, was only partially successful. At low frequencies, the comparison was limited by the interferometric technique, which requires that a number of wavelengths be present in order to accurately observe a rapidly varying amplitude. At high frequencies, the comparison was limited by the ion plasma frequency (which was kept low by the fact that the desired density gradients could be achieved only at low plasma densities). Also in agreement with the theory, the phase velocity of the wave was found to be approximately constant in the density gradient.

6.6.2. *Reflection of Long-Wavelength Ion-Acoustic Waves in Small Density Gradients*

Jones *et al.* (1976) performed an experiment in which it was found that the density variation in the subsonic region of the density gradient was very closely exponential. Defining a gradient scale length $l_p = 1/k_p$, the distance over which an *e*-fold variation occurs in the density, and using the theory of Doucet *et al.* (1974), a simple criterion was found for reflection: If $|k_p/k| \geqslant 1$, corresponding to ion-acoustic waves having wavelengths longer than the gradient scale length, then reflection is predicted. If $|k_p/k| \ll 1$, corresponding to short wavelengths, then transmission is predicted. The reflected wave is predicted to be 180° out of phase with the incident wave when the incident wave is propagating in the direction of decreasing density, i.e., when $k_p < 0$, which was the case for the experiment.

The method of plasma production was similar to that of Doucet *et al.* (1974). To minimize unwanted density gradients at the surface, the 50-cm-long, 50-cm-diam cylindrical plasma was surrounded by small confining magnets, producing so-called magnetic multipole confinement. Argon plasma having densities of $10^9 - 10^{10}$ cm^{-3} and $T_e \sim 3-4$ eV were typical. Wave generation and reflection was via two 20-cm-diam, planar wire grids. Wave detection was via a negatively biased cylindrical probe.

The experiment was divided into two parts. In the first part, a time-of-flight technique was used to study the propagation of an ion-acoustic wave

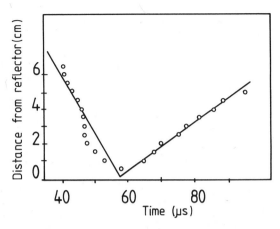

Fig. 6.6. Plot of probe position versus time of passage of two ion-acoustic waves with the reflector present. The line on the left is for the incident wave, and the slope gives a speed of 3×10^5 cm/s. (After Jones et al., 1976.)

generated by applying a short voltage pulse to the transmitter grid. In the second part, a cw sine wave voltage signal was applied to the transmitter, and a lock-in-amplifier was used to provide both frequency and phase information about the incident and reflected waves. Since wave generation by pulsing produces a wave packet consisting of a mixture of frequencies and phases, no frequency or phase information could be learned from the pulse experiment.

Figure 6.6 shows the result of making time-of-flight measurements of ion wave packets, in which the position of the packets with respect to the reflector was measured at different times after excitation. The slope of the line on the left, corresponding to incident waves, gives a speed, $Cs = 3 \times 10^5$ cm/s, which is in good agreement with the theoretical value [see Eq. (6.25)] calculated using an experimentally measured value of $T_e = 3.8$ eV. The line on the right, corresponding to reflected waves, has a slope of only one-half that of the line on the left, and is of opposite sign. The change in sign indicates propagation in the opposite direction, while the factor of one-half is because the position of the detector was measured with respect to the *reflector* (to reach the detector position, the reflected wave must traverse the distance between that position and the reflector twice, once going toward the reflector and once more after reflection). Taking this factor of 2 into account, the incident and reflected waves have the same speed.

Figure 6.7 shows the results of applying cw sinewaves of different frequencies to the transmitter and looking at the wave patterns observed between the transmitter and reflector by a lock-in amplifier. Careful analysis, including comparison of these wave patterns with wave patterns obtained under the same conditions, but without a reflector, revealed that the peak in the 30-kHz curve, centered at approximately the 9-cm position, corresponded to a reflected

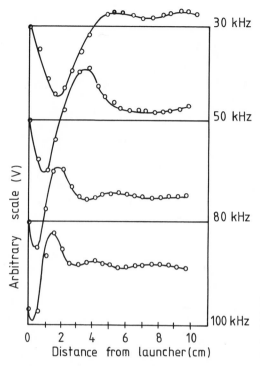

30 kHz

50 kHz

80 kHz

100 kHz

Arbitrary scale (V)

0   2   4   6   8   10
Distance from launcher (cm)

Fig. 6.7. Plot of lock-in amplifier output versus distance from launcher with a screen reflector located at a distance of 11 cm from the launcher. (After Jones *et al.*, 1976.)

ion-acoustic wave that was 180° out of phase with respect to the incident wave. For the data shown the gradient-scale length in the subsonic region in front of the reflector was found to be $l_p = 1.9$ cm (giving $k_p = 0.55$ cm^{-1}). The wave number $k(= \omega/Cs)$ for the 30-kHz wave was $k = 0.67$ cm^{-1}. When the potential on the reflector grid was varied to produce $k_p = 0.45$ cm^{-1}, reflection of the 30-kHz signal no longer occurred. Although neither criterion was exactly met ($k_p/k \geqslant 1$ for reflection, and $k_p/k \ll 1$ for no reflection), the experiment did demonstrate that: (1) There was a critical k_p (for the 30-kHz signal used, somewhere in the range 0.45–0.55 cm^{-1}), which separated the condition favorable for reflection from the condition unfavorable for reflection; moreover, the ratio of this critical k_p to k was within a factor of 2 of what was predicted theoretically. (2) As the ratio, k_p/k, decreased, one went from a reflecting situation to a nonreflecting situation, as predicted theoretically.

LANDAU DAMPING
AN INITIAL-VALUE PROBLEM

7.1. *INTRODUCTION*

As we have seen in the preceding five chapters, many wave phenomena in plasmas can be studied using only the fluid-theory model in which the plasma is considered as two or more interpenetrating fluids (one for the electrons and one for each ion species). The main disadvantage of this model, as has been pointed out several times, is that only the velocity *average* of each plasma specie can be taken into account, so that any velocity-dependent effects, such as the Landau damping to be discussed in this chapter, are not predicted by the theory. From a mathematical point of view, the main advantage of the model is that it is a much simpler model than the kinetic-theory model to be used in the present and remaining chapters of the book. Because of this relative simplicity, unless velocity-dependent phenomena are of specific interest, the fluid-theory model is always used. For *non*-velocity-dependent effects, the fluid-theory model and the kinetic-theory model give identical results. As we employ the kinetic theory in the remaining chapters of the book, we shall see that not only are velocity-dependent effects predicted, but that we will recover many of the fluid-theory wave phenomena already found.

The reason that the two models can give the same results is that they are very closely related. As will be recalled from the preceding chapters, the fluid theory consists of the continuity equation, Newton's equation of motion for a fluid, and Maxwell's equations. The kinetic theory consists of Vlasov's equation and Maxwell's equations. But the continuity equation and Newton's equation of motion for a fluid are obtained by taking the first two velocity moments of the Vlasov equation. It is for this reason that the fluid theory misses any velocity-dependent effects; when the integrations on velocity of the Vlasov equation are made, most of the velocity information contained in the particle distribution functions of the Vlasov equations is lost. For the reader who would like to see a comprehensive discussion of the various theoretical models used in plasma physics, many excellent books, such as the recent one by Nicholson (1983), are available.

Except for the fact that the plasma will now be described by the kinetic theory—with the use of distribution functions for both the ions and electrons—the description of wave properties in this chapter is done in a way similar to the methods used in Chapter 3, where the plasma properties were introduced through local charge and current densities in Maxwell's equations.

Before beginning our discussion of collisionless "Landau damping" in plasmas, we want to first show that, due to the spread of particle velocities inherent in the distribution function, any density perturbation in *any* gaseous medium will be naturally "damped" *without* collisions due to the phase mixing caused by the normal diffusion of particles. Landau damping in a plasma will be seen then to be just a modification of "free-streaming" damping. In fact, due to the effect of the coherent electric field associated with the evolution of the density perturbation, the free-streaming damping caused by the diffusion of particles is strongly reduced, so that by comparison Landau damping is much smaller than damping due to free streaming.

There are primarily two techniques for studying waves in plasmas: the initial-value-problem technique and the boundary-value-problem technique. In the initial-value problem, we assume that, at time $t = 0$, we know the exact state of the plasma perturbation. That is, we assume that we know the mathematical form of the initial perturbed electron-density distribution function $f_1(x, v, t = 0)$ and, thus, also, the corresponding moments, such as the initial density perturbation $n_1(x, t = 0)$, the initial velocity perturbation $v_1(x, t = 0)$, and the resulting initial electric field perturbation $E_1(x, t = 0)$. Solving the initial-value problem consists of finding $f_1(x,v,t)$ and, thus, also, $n_1(x,t)$, $v_1(x,t)$, and $E_1(x,t)$, for all x and t.

In the boundary-value problem, we assume that for any time $t > 0$ we have some information on the perturbation produced at some position, say $x = 0$. For example, we may assume that we know $f_1(x = 0, v, t > 0)$ and, thus, the corresponding moments. Solving the boundary-value problem consists then of finding $f_1(x,v,t)$, $n_1(x,t)$, $v_1(x,t)$, and $E_1(x,t)$ for all x and all $t > 0$.

The boundary-value problem is somewhat better suited for the description of real experimental situations than is the initial-value problem. The initial-value problem in purely mathematical in the sense that we arbitrarily assign some initial perturbed state to the plasma, which we assume can exist and which we assume we know *throughout all space*. Such a model may or may not be very realistic. On the other hand, for the boundary-value problem, we assume we know the perturbed state of the plasma in some *localized region* of the plasma. This perturbed state can be produced, for example, by a perturbation that is produced locally in the plasma by a repetitive voltage on a grid, by an ion beam, etc. At $t = 0$, the perturbing disturbance is introduced in the plasma at $x = 0$. After such a disturbance has been present long enough to create a steady-state perturbation in the plasma (theoretically, one would have to wait infinitely long; in practice, one need wait a time equal only to several times the longest characteristic time scale of the plasma), one turns off the perturbing disturbance at some $t > 0$ and uses the boundary-value technique to calculate the evolution in time and space of the perturbation present in the plasma at turnoff. One may be able to calculate reasonably accurately the perturbation in the distribution function that is produced for such a perturbation or, based on experience, one may be able to make a reasonably accurate guess as to what the perturbation was at the time of turnoff.

The initial-value-problem technique is the more widely used of these two techniques, and is the technique that we will use in this chapter for studying both free-streaming and Landau damping.

7.2. COLLISIONLESS DAMPING DUE TO FREE STREAMING

To demonstrate collisionless damping in the absence of any electric field, we look at the evolution of a given density perturbation, or particle distribution function, in a simple situation in which the electric charge of each particle is zero. When there is no electric field the evolution of the distribution function of the particles $f(r,v,t)$ can be described by Liouville's equation

$$\frac{df(r,v,t)}{dt} = 0. \tag{7.1}$$

To permit a simple and complete calculation we restrict ourselves here to the case of a one-dimensional, uniform, and infinite medium of neutral particles characterized by the Maxwellian equilibrium distribution function

$$f_0(v) = (a\pi^{1/2})^{-1} \exp\left(\frac{-v^2}{a^2}\right), \tag{7.2}$$

where a is the thermal velocity characterizing the particles.

At each time t the distribution function $f(x,v,t)$ can be written as

$$f(x,v,t) = f_0(v) + f_1(x,v,t), \tag{7.3}$$

where f_1 is the perturbation of the distribution function associated with the wave or the density perturbation.

The evolution of f_1 is then given by Eq. (7.1) which, with Eq. (7.3), gives

$$\frac{\partial f_1}{\partial t} + v\frac{\partial f_1}{\partial x} = 0. \tag{7.4}$$

Any function of $y = (x - vt)$ is a solution of Eq. (7.4). Thus, for example if we want to describe the time and space evolution of the initial perturbation

$$f_1(x,v,0) = bf_0(v) \cos(k_0 x), \tag{7.5}$$

the corresponding perturbation $f_1(x,v,t)$ is obtained simply by remembering that, from Liouville's theorem, f_1 is a constant along a trajectory, which means that the particles that are at position x at time t were at the position, $x - vt$, at $t = 0$. Thus,

$$f_1(x,v,t) = f_1(x - vt, v, 0), \tag{7.6}$$

so that

$$f_1(x,v,t) = bf_0(v) \cos(k_0 x - k_0 vt), \tag{7.7}$$

where b is the amplitude of the perturbation.

The density perturbation is then easily obtained by the integration of $f_1(x,v,t)$ with respect to velocity:

$$n_1(x,t) = \int_{-\infty}^{+\infty} f_1(x,v,t) \, dv. \tag{7.8}$$

Substituting for f_1 from Eq. (7.7) and for f_0 from Eq. (7.2), then integrating, gives

$$n_1(x,t) = b(\cos k_0 x) \exp\left(\frac{-k_0^2 a^2 t^2}{4}\right). \tag{7.9}$$

Thus, we see that the perturbation of density is strongly attenuated in time and does not propagate. The corresponding behavior is just a free-streaming effect and shows us that we should not be surprised when a density perturbation, such as a wave, is damped without collisions.

This free-streaming effect can give rise to surprising *short*-time behavior (only the *long*-time-scale behavior is usually of interest in a typical damping experiment): on a time scale of the order of $\tau = (k_0 a)^{-1}$, the density variations depend strongly on the exact nature of the initial perturbation of the distribution function. As an example, we show that the initial density perturbation can actually increase on a short-time scale. Let us assume that in our medium of neutral particles we can produce the initial perturbation [see Eq. (7.5)] given by

$$f_1(x,v,0) = bf_0'(v) \cos(k_0 x), \tag{7.10}$$

where $f'_0(v) = df_0(v)/dv$. The corresponding *initial* density perturbation is zero since

$$n_1(x,0) = \int_{-\infty}^{+\infty} bf'_0(v) \cos(k_0 x) \, dv = b \cos(k_0 x) \, [f_0(v)]_{-\infty}^{+\infty} = 0. \quad (7.11)$$

Using the same arguments used to obtain Eqs. (7.6) and (7.7), we can write, respectively, that

$$f_1(x,v,t) = f_1(x - vt, v, 0), \quad (7.12)$$

and

$$f_1(x,v,t) = bf'_0(v) \cos(k_0 x - k_0 vt). \quad (7.13)$$

Substituting from Eq. (7.13) into Eq. (7.8), then integrating, gives the density perturbation as a function of time as

$$n_1(x,t) = -b \sin(k_0 x - k_0 at) \exp\left(\frac{-k_0^2 a^2 t^2}{4}\right). \quad (7.14)$$

Thus, $n_1(x,t)$ first grows approximately linearly with t, goes through a maximum at $t = 2^{1/2}/(ka)$, and decreases to zero as $t \to \infty$. The initial growth of n_1 does not mean that there is an instability but, rather, that the impulse given to the particles at $t = 0$ (even producing no density perturbation at that time) is kept in the particles' velocity distribution. The subsequent *phase mixing* gives rise to a nonzero density perturbation which initially grows and then vanishes as $t \to \infty$. In fact, there is no exchange of energy here between the particles and a wave, the energy given to the medium being restricted to the kinetic energy of the particles. Thus, the reader should not be surprised that a density perturbation can be damped without collisions when later in the chapter we introduce collisionless interactions between the electrons of a plasma and an electron plasma wave (via the electric field of the wave). We show, in fact, that such an interaction does not introduce damping that is additional to the free-streaming damping we have just studied but, instead, for the case of long wavelengths, it can strongly reduce the free-streaming damping and give rise to wave propagation.

7.3. LONGITUDINAL OSCILLATIONS IN AN INFINITE, HOMOGENEOUS PLASMA WITH NO APPLIED FIELDS— THE ELECTRON PLASMA WAVE

We consider an infinite, collisionless, homogeneous plasma which basically is a gas of electrons neutralized by cold, *motionless* ions. Moreover, we assume that the plasma has no dc electric or magnetic fields. The electrons are described by their distribution function $f(r,v,t)$, which obeys the Vlasov and Poisson equations

$$\frac{\partial f}{\partial t} + v \frac{\partial f}{\partial r} - \frac{e}{m} \frac{\partial f}{\partial v} E = 0, \tag{7.15}$$

$$\nabla \cdot E = \frac{-e}{\varepsilon_0} \left[\int_{-\infty}^{+\infty} f \, dv - n_0 \right] + \frac{\rho_{ex}}{\varepsilon_0}, \tag{7.16}$$

where n_0 is the equilibrium plasma density, i.e., the density of the electrons in the absence of any perturbation and also of the neutralizing, motionless ions. The term $\rho_{ex}(r,t)$ represents any external charge variations associated with the wave (excitation, for example).

We write

$$f(r,v,t) = f_0(v) + f_1(r,v,t), \tag{7.17}$$

where $f_0(v)$ is the equilibrium distribution function. The equilibrium density n_0 is then given by

$$n_0 = \int_{-\infty}^{+\infty} f_0(v) \, dv. \tag{7.18}$$

Using Eq. (7.17), the Vlasov equation can be written without any approximations as

$$\frac{\partial f_1}{\partial t} + v \frac{\partial f_1}{\partial r} - \frac{e}{m} \left[\frac{df_0(v)}{dv} + \frac{\partial f_1}{\partial v} \right] E = 0. \tag{7.19}$$

We linearize this equation by making the assumption that

$$\frac{df_0(v)}{dv} \gg \frac{\partial f_1}{\partial v} \tag{7.20}$$

for all values of v. Doing this gives

$$\frac{\partial f_1}{\partial t} + v \frac{\partial f_1}{\partial r} - \frac{e}{m} \frac{df_0(v)}{dv} E = 0. \tag{7.21}$$

Again, using Eq. (7.17) and keeping only first-order terms, the linearized Poisson equation is given by

$$\nabla \cdot E = -\frac{e}{\varepsilon_0} \int_{-\infty}^{+\infty} f_1(r,v,t) \, dv + \frac{\rho_{ex}}{\varepsilon_0}. \tag{7.22}$$

Note that in Eqs. (7.21) and (7.22) E is a first-order quantity (the electric field produced in the plasma by the perturbation), since we have assumed that there are no electric fields in the plasma except those produced by the perturbation. Equations (7.21) and (7.22) are said to be *coupled* by the electric field E.

7.3.1. Using Fourier–Laplace Transforms to Solve the Coupled Poisson and Vlasov Equations

We note that the linearized Vlasov equation is a partial-differential equation and that the linearized Poisson equation is an integro-partial-differential equation. We now use the classical Fourier–Laplace transform procedures discussed in Chapter 1 to simplify these equations. At this point, it is suggested that the reader review that part of Chapter 1 in order to re-familiarize himself with the assumptions and conditions that must be fulfilled in order to legitimately apply these techniques. Since we are dealing here with an initial-value problem, we must pay particular attention, for example, to the mathematical properties of the initial perturbation that we will choose later when we apply the theory to specific problems.

We begin by doing Fourier transforms in space for both Eqs. (7.21) and (7.22). This eliminates the spatial derivatives, leaving us with two equations involving time derivatives. Both equations involve the perturbed electric field E which we eliminate between them to give one ordinary differential for $f_1(k,v,t)$, the Fourier image of the perturbed distribution of the electrons. We then do a Laplace transform in time for this equation.

Using the properties of Fourier transforms, we Fourier-transform each of the terms in Eq. (7.21) to give the Fourier-transformed, linearized, Vlasov equation

$$\frac{\partial}{\partial t} f_1(k,v,t) + ik \cdot v f_1(k,v,t) - \frac{e}{m} \frac{df_0(v)}{dv} E(k,t) = 0, \qquad (7.23)$$

where we have used the same symbol f_1 to represent the Fourier image as is used to represent the real perturbation. As usual, the presence of k inside the parentheses indicates that $f_1(k,v,t)$ in Eq. (7.23) is the Fourier-transformed quantity and not the quantity in real space.

Since only the component of the electron's velocity that is parallel to the direction of propagation of the perturbation is of primary importance for the problem we are considering, we express the electron's velocity as

$$v = v_{//} + v_{\perp}, \qquad (7.24)$$

where $v_{//}$ is the component of v that is parallel to k (the direction of k being the same as the direction of motion of the perturbation), and $v_\perp$ is the component of v that is perpendicular to k. Thus,

$$k \cdot v_{//} = kv_{//},$$
$$k \cdot v_\perp = 0. \tag{7.25}$$

In order to give us a perturbed electron distribution function f_1, that is a function only of $v_{//}$, we make an integration on $v_\perp$. This is indicated as

$$f_1(k,v_{//},t) = \iint f_1(k,v,t)\, d^2 v_\perp, \tag{7.26}$$

where the integration is done from $-\infty$ to $+\infty$ on the two components of $v_\perp$ perpendicular to $v_{//}$. Similarly, although we don't show it explicitly, we do the same integration on f_0 in order to have an equilibrium electron-distribution function that is also a function only of $v_{//}$. Doing this, our Fourier-transformed, linearized, Vlasov equation, Eq. (7.23), becomes a *one-dimensional*, Fourier-transformed, linearized, Vlasov equation given by

$$\frac{\partial f_1}{\partial t}(k,v_{//},t) + ikv_{//}f_1(k,v_{//},t) - \frac{e}{m}\frac{df_0(v_{//})}{dv_{//}}E(k,t) = 0, \tag{7.27}$$

and the corresponding Poisson equation, Eq. (7.22) becomes

$$ikE(k,t) = -\frac{e}{\varepsilon_0}\int_{-\infty}^{+\infty} f_1(k,v_{//},t)\, dv_{//} + \frac{\rho_{ex}(k,t)}{\varepsilon_0}. \tag{7.28}$$

Equations (7.27) and (7.28) can be combined, by eliminating the electric field $E(k,t)$, to give an integro-differential equation for $f_1(k,v_{//},t)$

$$\frac{\partial f_1}{\partial t} + ikv_{//}f_1 - i\frac{\omega_p^2}{kn_0}\frac{df_0(v_{//})}{dv_{//}}\int_{-\infty}^{+\infty} f_1\, dv_{//} + i\frac{e}{m}\frac{df_0(v_{//})}{dv_{//}}\frac{\rho_{ex}(k,t)}{k\varepsilon_0} = 0, \tag{7.29}$$

where ω_p is the electron plasma frequency $\omega_p = (ne^2/\varepsilon_0 m)^{1/2}$. The initial-value problem consists of finding the solutions $f_1(k,v_{//},t)$ of Eq. (7.29) corresponding to the initial conditions of the perturbations

$$f_1(k,v_{//},0) = g(k,v_{//}),$$
$$\rho_{ex}(k,0) = 0, \tag{7.30}$$

where $f_1(k,v_{//},0)$ and $\rho_{ex}(k,0)$ are, respectively, the Fourier images of the initial perturbation in the distribution function, $f_1(r,v_{//},0) = g(r,v_{//})$, and any initial external charge perturbation introduced in the plasma $\rho_{ex}(r,0)$.

It is possible to calculate any macroscopic quantity associated with the wave, such as the density $n_1(k,t)$, the current density $j_1(k,t)$, and the electric field $E_1(k,t)$ and then, using the inversion Fourier techniques discussed in Chapter 1, to find the real density $n_1(r,t)$, the real current density $j_1(r,t)$, and the real electric field $E_1(r,t)$. In fact, as we shall see, we are actually more interested in calculating these macroscopic perturbations than in calculating f_1, per se. However, once f_1 is calculated, the calculation of the macroscopic quantities of interest is relatively easy and straightforward.

We now want to do a *Laplace* transform (see Chapter 1) in time of Eq. (7.29) in order to effectively replace the time-differential operator by a constant. Using the properties of Laplace transforms we make a Laplace transform on each term in Eq. (7.29) to obtain

$$-i\omega f_1 + ikv_{//}f_1 - i\frac{\omega_p^2}{kn_0}\frac{df_0}{dv_{//}}\int_{-\infty}^{+\infty} f_1\,dv_{//} + i\frac{e}{m}\frac{df_0}{dv_{//}}\frac{\rho_{ex}(k,\omega)}{k\varepsilon_0} = g(k,\omega), \quad (7.31)$$

where f_1 is now $f_1(k,v_{//},\omega)$, a function of k, $v_{//}$, and ω. Thus, except for the integral term, which represents the Fourier-and-Laplace-transformed electron-density perturbation produced by the initial perturbation

$$n_1(k,\omega) = \int_{-\infty}^{+\infty} f_1(k,v_{//},\omega)\,dv_{//}, \quad (7.32)$$

we see that we have transformed the differential equation, Eq. (7.29), into an algebraic equation, Eq. (7.31). Equation (7.31) can now be solved for $f_1(k,v_{//},\omega)$ to give

$$f_1 = \frac{1}{\omega - kv_{//}}\left[-\frac{\omega_p^2}{k}\frac{df_0}{dv_{//}}\frac{n_1(k,\omega)}{n_0} + \frac{e}{m}\frac{df_0}{dv_{//}}\frac{\rho_{ex}(k,\omega)}{k\varepsilon_0}\right] + i\frac{g(k,v_{//})}{\omega - kv_{//}}. \quad (7.33)$$

Equation (7.33) gives the Fourier-and-Laplace-transformed, one-dimensional distribution function for the electron-density perturbation. We are interested now to use f_1 to find the Fourier-and-Laplace-transformed macroscopic density and macroscopic electric field perturbations produced by this perturbation of the electron density. From Eq. (7.32), we see that the macroscopic density perturbation can be obtained from f_1 simply by doing an integration over $v_{//}$. Thus, for the macroscopic density perturbation we can write

$$n_1(k,\omega) = \frac{\dfrac{e}{m}\dfrac{\rho_{ex}(k,\omega)}{k\varepsilon_0}\displaystyle\int_{-\infty}^{+\infty}\frac{df_0/dv_{//}}{\omega - kv_{//}}\,dv_{//} + \displaystyle\int_{-\infty}^{+\infty} i\frac{g(k,v_{//})}{\omega - kv_{//}}\,dv_{//}}{1 + \dfrac{\omega_p^2}{kn_0}\displaystyle\int_{-\infty}^{+\infty}\frac{df_0/dv_{//}}{\omega - kv_{//}}\,dv_{//}}. \quad (7.34)$$

Equation (7.34) is of the form

$$n_1(k,\omega) = \frac{\mathcal{N}(k,\omega)}{1 + \chi(k,\omega)} = \frac{\mathcal{N}(k,\omega)}{\varepsilon(k,\omega)}, \tag{7.35}$$

where

$$\mathcal{N}(k,\omega) = \frac{e}{m} \frac{\rho_{ex}(k,\omega)}{k\varepsilon_0} \int_{-\infty}^{+\infty} \frac{df_0/dv_{//}}{\omega - kv_{//}} dv_{//} + \int_{-\infty}^{+\infty} i \frac{g(k,v_{//})}{\omega - kv_{//}} dv_{//}, \tag{7.36}$$

χ is the susceptibility, and $\varepsilon(k,\omega)$ is given by

$$\varepsilon(k,\omega) = 1 + \chi = 1 + \frac{\omega_p^2}{kn_0} \int_{-\infty}^{+\infty} \frac{df_0/dv_{//}}{\omega - kv_{//}} dv_{//}. \tag{7.37}$$

$\varepsilon(k,\omega)$ is the relative dielectric constant related to the displacement vector D [see Eq. (2.12)], defined by

$$D(k,\omega) = \varepsilon_0 \varepsilon(k,\omega) E(k,w). \tag{7.38}$$

The numerator of Eq. (7.35), $\mathcal{N}(k,\omega)$, is a function of the wave excitation via any external charge variation $\rho_{ex}(k,\omega)$ and/or the initial perturbation of the distribution function $g(k,v_{//})$. The functional dependence of the denominator $\varepsilon(k,\omega)$ is determined by χ, the plasma susceptibility, which is independent of the excitation, depending only on the unperturbed distribution function $f_0(v_{//})$, which characterizes the equilibrium plasma.

It is important to note also that the numerator $\mathcal{N}(k,\omega)$ often depends on $f_0(v_{//})$, either through the integral

$$\int_{-\infty}^{+\infty} \frac{f_0'(v_{//})}{\omega - kv_{//}} dv_{//}, \tag{7.39}$$

when the excitation involves some external charge variation $\rho_{ex}(k,\omega)$, or through $g(k,v_{//})$, the initial perturbation of the distribution function, $g(k,v_{//})$ [$g(k,v_{//})$ being, usually, the plasma response to a rapidly pulsed external field].

Making a Laplace transform in time of Poisson's equation and of Eqs. (7.28) and (7.32), we see that we can write an equation similar to Eq. (7.35) for the electric field as

$$E(k,\omega) = i\frac{e}{k} \frac{\mathcal{N}(k,\omega)}{\varepsilon_0[1 + \chi(k,\omega)]} - i\frac{\rho_{ex}(k,\omega)}{k\varepsilon_0}. \tag{7.40}$$

For simplicity we assume that the external charge $\rho_{ex}(k,\omega)$ is zero, the perturbation of the distribution function being the only excitation. Equation (7.40) then becomes

$$E(k,\omega) = -\frac{e}{k} \frac{1}{\varepsilon_0(1 + \chi)} \int_{-\infty}^{+\infty} \frac{g(k,v_{//})}{\omega - kv_{//}} \, dv_{//}. \qquad (7.41)$$

Before continuing our discussion of Landau damping, we want to point out that Eq. (7.41) is the plasma equivalent to Ohm's law for a resistor. In order to do this, we must show that the right-hand side of Eq. (7.41) has the form of a current density divided by a conductivity.

In Chapter 2 [see Eq. (2.21)], we saw that the susceptibility of the electron gas is related to the conductivity σ, a relationship that can be expressed as $1 + \chi = -(\sigma - i\omega\varepsilon_0)/i\omega\varepsilon_0$. Equation (7.41) can then be written as

$$(\sigma - i\omega\varepsilon_0)E(k,\omega) = i\frac{e\omega}{k} \int_{-\infty}^{+\infty} \frac{g(k,v_{//})}{\omega - kv_{//}} \, dv_{//}. \qquad (7.42)$$

We show first that the right-hand side of Eq. (7.42) can be identified as the free-streaming current-density perturbation produced by the initial perturbation of the distribution function. To do this we use the simple case of an initial perturbation $f_1(x,v,t=0)$ associated with a zero initial-density perturbation

$$n_1(x,t=0) = \int_{-\infty}^{+\infty} f_1(x,v_{//},t=0) \, dv_{//} = 0. \qquad (7.43)$$

For this purpose, we look first at the so-called ballistic, or free-streaming, solution—defined as $f_1^*(x,v_{//},t)$—of the Vlasov equation for the case where there are no electric fields in the plasma [see Eq. (7.15)]

$$\frac{\partial f_1^*}{\partial f} + v\frac{\partial f_1^*}{\partial x} = 0. \qquad (7.44)$$

Making a Fourier transform in space and a Laplace transform in time gives

$$f_1^*(k,v_{//},\omega) = i\frac{g(k,v_{//})}{\omega - kv_{//}}. \qquad (7.45)$$

The corresponding density perturbation (which we have defined to be zero) is given by

$$n^*(k,\omega) = i\int_{-\infty}^{+\infty} \frac{g(k,v_{//})}{\omega - kv_{//}} \, dv_{//}, \qquad (7.46)$$

while the current-density perturbation produced by f_1^* is given by

$$j^*(k,\omega) = -ie \int_{-\infty}^{+\infty} v_{//} \frac{g(k,v_{//})}{\omega - kv_{//}} \, dv_{//}. \tag{7.47}$$

Upon replacing the $v_{//}$ term in the integrand of Eq. (7.47) by $[-(-\omega + \omega - kv_{//})/k]$, and remembering that we have defined the initial-density perturbation [see Eq. (7.43)] to be zero, it is not difficult to show that this equation can be written as

$$j^*(k,\omega) = -i\frac{e\omega}{k} \int_{-\infty}^{+\infty} \frac{g(k,v_{//})}{\omega - kv_{//}} \, dv_{//}. \tag{7.48}$$

Thus, Eq. (7.42) becomes

$$(\sigma - i\omega\varepsilon_0)E(k,\omega) = -j^*(k,\omega). \tag{7.49}$$

In a vacuum or when the plasma susceptibility is negligible, in general, $\sigma \to 0$, and Eq. (7.49) becomes

$$-i\omega\varepsilon_0 E^*(k,\omega) = -j^*(k,\omega), \tag{7.50}$$

giving the electric field $E^*(k,\omega)$ produced by this current-density perturbation. We can see that Eq. (7.50) is the Fourier and Laplace transform of

$$\varepsilon_0 \frac{\partial E^*}{\partial t} + j^* = 0, \tag{7.51}$$

which comes from the following Maxwell equation for the case of electrostatic waves $(H = 0)$

$$\nabla \times H = j^* + \varepsilon \frac{\partial E^*}{\partial t} = 0. \tag{7.52}$$

For the case of an electrostatic wave in a plasma, Eq. (7.52) can be written as

$$\nabla \times H = j_p + \varepsilon_0 \frac{\partial E_p}{\partial t} + j_{ex} = 0, \tag{7.53}$$

the Fourier and Laplace transform of which is given by

$$j_p(k,\omega) - i\omega\varepsilon_0 E_p(k,\omega) = -j^*(k,\omega), \tag{7.54}$$

where E_p is the electric field induced in the plasma by the initial perturbation.

The free-streaming term $j^*(k,\omega)$ in this case is due to an external excitation and thus plays the role of the external current j_{ex}.

A comparison between Eqs. (7.54) and (7.49) gives

$$j_p(k,\omega) = \sigma E(k,\omega), \tag{7.55}$$

which is Ohm's law.

Taking Eqs. (7.48) and (7.50) into account we can express the total electric field $E(k,\omega)$ given in Eq. (7.41) as a function of the free-streaming electric field E^* and of the plasma susceptibility χ by

$$E(k,\omega) = \frac{E^*(k,\omega)}{1 + \chi(k,\omega)} = E^*(k,\omega)\left[1 - \frac{\chi(k,\omega)}{1 + \chi(k,\omega)}\right], \tag{7.56}$$

or

$$E(k,\omega) = E^*(k,\omega) + E_{col}(k,\omega), \tag{7.57}$$

where $E_{col}(k,\omega)$ is the collective part of the electric field

$$E_{col}(k,\omega) = -E^*(k,\omega)\frac{\chi(k,\omega)}{1 + \chi(k,\omega)}, \tag{7.58}$$

and $\chi(k,\omega)$ is the susceptibility

$$\chi(k,\omega) = \frac{\omega_p^2}{kn_0}\int_{-\infty}^{+\infty}\frac{df_0(v_{//})/dv_{//}}{\omega - kv_{//}}\,dv_{//}. \tag{7.59}$$

It is interesting to note that this susceptibility is a function of e^2/m, rather than of the parameter e/m coming from the usual particle dynamics, such as from the equation of motion

$$\frac{m\,dv_{//}}{dt} = eE(r,t). \tag{7.60}$$

7.3.2. Free-Streaming and Collective Contributions to the Electric Field

We see that if we use heavy particles with small charges q instead of electrons, the free-streaming motion just described is important, and the corresponding total electric field can be approximated by the free-streaming electric field, since $\chi \to 0$ as a consequence of $q^2/m \to 0$. On the other hand, for lighter charged particles, and for the electrons considered here, we can treat the free-streaming term of Eq. (7.57) as the first term of an expansion, the other terms representing the collective effects of the particles.

7.3.3. Time Evolution of the Electric Field

We now want to do an inverse Laplace transform on $E(k,\omega)$—see Eqs. (7.40) and (7.41)—in order to find the variation of the electric field as a function of time. In order to do this, we must choose not only the form of the initial perturbation, but also the equilibrium distribution characterizing the plasma. In this section we investigate plasmas characterized by cold and Lorentzian electron equilibrium distribution functions. For the Lorentzian case we derive Landau damping valid for all wavelengths. In Section 7.3.4., we derive Landau damping in the long-wavelength approximation valid for arbitrary distribution functions. In Section 7.3.5., we treat Landau damping in detail for Maxwellian plasmas.

7.3.3.1. Inversion Procedure. In order to have tractable, simple solutions, we continue to assume that no external charges $\rho_{ex}(k,\omega)$ are involved in the wave excitation at time $t = 0$. Thus, we make an inverse Laplace transform of Eq. (7.41). To do this, we use the procedure given in Section 1.3.3.

$$E(k,t) = \frac{1}{2\pi} \int_c \exp(-i\omega t) E(k,\omega)\, d\omega, \qquad (7.61)$$

where c is a path lying above all of the singularities of $E(k,\omega)$. It is necessary that $\omega_i = \text{Im}(\omega) > v$, v being a positive number.

From Eqs. (7.37) and (7.40) we see that the functional dependence of $E(k,\omega)$ is determined by $N(k,\omega)$ and $\varepsilon(k,\omega)$. From Eqs. (7.36) and (7.37) we see that $N(k,\omega)$ and $\varepsilon(k,\omega)$ have the same *form* as the Hilbert transform of $f(v_{//})$ (see Section 1.5)

$$H(z) = \int_{-\infty}^{+\infty} \frac{f(v_{//})}{v_{//} - z}\, dv_{//}. \qquad (7.62)$$

This integral represents two different analytical functions, $H^+(z)$ and $H^-(z)$, which are related by the equation

$$H^-(z) = H^+(z) - 2i\pi f(z). \qquad (7.63)$$

By analytical continuation these two functions are analytical in the whole complex plane and can be represented by the two Hilbert transforms

$$H^\pm(z) = \int_{L^\pm} \frac{f(v_{//})}{v_{//} - z}\, dv_{//}, \qquad (7.64)$$

having paths of integration L^+ and L^- as discussed in Chapter 1.

The integral involved in an inverse Laplace transform requires that $\text{Im}(\omega) > \mu$, where μ is positive. If $z = \omega/k$, then when $k > 0$ we must choose $H^+(z)$, while for $k < 0$ we must choose $H^-(z)$.

Thus, $E(k,\omega)$ is the ratio of two Hilbert transforms, that is to say, of two functions that are analytical in the whole complex plane, in ω or in ω/k, except at the singularities of $\mathcal{N}(k,\omega)/\varepsilon(k,\omega)$. Among these singularities special attention will be given to the zeros of the denominator $\varepsilon(k,\omega)$. These zeros are given by $\mathcal{D}(k,\omega) = \varepsilon(k,\omega) = 0$, which is the dispersion relation. While it is not always possible, we try to use either Jordan's lemma or theorem (see Chapter 1) to find a contour that includes the poles and gives a zero integral on the half-circle when $|\omega| \to \infty$ (see the example discussed in Chapter 1). We will than compute the integral using the residues theorem

7.3.3.2. A Cold Plasma. The one-dimensional (see Sec. 7.3.1.) distribution function for a cold plasma can be written as

$$f_0(v_{//}) = \delta(v_{//})n_0, \tag{7.65}$$

where $\delta(v_{//})$ is the Dirac delta function. Substituting from Eq. (7.65) into Eq. (7.37) and setting the result equal to zero gives the dispersion relation for a cold plasma

$$\mathcal{D}(k,\omega) = 1 + \frac{\omega_p^2}{k} \int_{-\infty}^{+\infty} \frac{d[\delta(v_{//})]}{dv_{//}} \frac{dv_{//}}{\omega - kv_{//}} = 0. \tag{7.66}$$

Looking only at the integral term of Eq. (7.66) we can write

$$\int_{-\infty}^{+\infty} \frac{d[\delta(v_{//})]}{dv_{//}} \frac{dv_{//}}{\omega - kv_{//}} = -\int_{-\infty}^{+\infty} \delta(v_{//}) \frac{k}{(\omega - kv_{//})^2} \, dv_{//} = -\frac{k}{\omega^2}. \tag{7.67}$$

Thus

$$\mathcal{D}(k,\omega) = 1 - \frac{\omega_p^2}{\omega^2} = 0, \tag{7.68}$$

which is the well-known dispersive relation found for a cold plasma in Chapter 2.

Using the inversion procedure discussed in the preceding section we can write $E(k,t)$ for the perturbed electric field in a cold plasma as

$$E(k,t) = \frac{1}{2\pi} \int_{-\infty+iv}^{+\infty+iv} E(k,\omega)e^{-i\omega t} \, d\omega, \tag{7.69}$$

for any $v > 0$. If $E(k,\omega) \to 0$ as $|\omega| \to \infty$, we can use Jordan's lemma. For $t < 0$ we choose a closed contour using a half-circle in the upper half-plane, $\mathrm{Im}\,\omega > 0$ [see Fig. 1.4(a)]. Then, for $t < 0$

$$\int_{-\infty+iv}^{+\infty+iv} E(k,\omega)e^{-i\omega t} \, d\omega = 2\pi i \sum \text{Residues}. \tag{7.70}$$

There are no poles in the upper half-plane Im $\omega > v$, because the Laplace transform of $E(k,t)$ exists; therefore, the integral is zero and we find, in agreement with the causality principle, that the electric field is zero everywhere before the excitation was applied (at $t = 0$).

For $t > 0$, we choose a contour closed by a half-circle in the lower-half plane, Im$(\omega) < 0$ [see Fig. 1.4(b)]. To calculate the residues we must at this point choose the form of the initial perturbation $f_1(r, v_{//}, t = 0)$, of the electron distribution function. We see later that f_1 is usually chosen to be proportional to the derivative $df_0(v_{//})/dv_{//}$ of the unperturbed distribution function. Thus, we choose

$$g(x,v_{//}) = f_1(r, v_{//}, t = 0) = A\delta'(v_{//}) \cos k_0 x. \qquad (7.71)$$

We must now compute $E(k,\omega)$. Observation of Eqs. (7.35) and (7.40) shows that for the case where $\rho_{ex} = 0$, $E(k,\omega)$ and $n_1(k,\omega)$ differ only by a coefficient term in front of the integral. Thus, for simplicity we will compute $n_1(k,\omega)$, rather than $E(k,\omega)$. Observation of Eqs. (7.35) and (7.36) shows that we need to know $g(k,v_{//})$, not $g(x,v_{//})$, as given in Eq. (7.71). Fourier transforming Eq. (7.71) in space gives the Fourier image $g(k,v_{//})$ as

$$g(k,v_{//}) = \int_{-\infty}^{+\infty} g(x,v_{//})e^{-ikx}\,dx = A\delta'(v_{//})[\pi\delta(k - k_0) + \pi\delta(k + k_0)]. \qquad (7.72)$$

Substituting from Eq. (7.72) into Eq. (7.35) gives

$$n_1(k,\omega) = \frac{iA\pi[\delta(k - k_0) + \delta(k + k_0)]}{1 - \omega_p^2/\omega^2} \int_{-\infty}^{+\infty} \frac{\delta'(v_{//})}{\omega - kv_{//}}\,dv_{//}. \qquad (7.73)$$

But from Eq. (7.67) we have found that the integral term is Eq. (7.73) is given by

$$\int_{-\infty}^{+\infty} \frac{\delta'(v_{//})}{\omega - kv_{//}}\,dv_{//} = -\frac{k}{\omega^2}. \qquad (7.74)$$

Thus, using Eq. (7.74), Eq. (7.73) becomes

$$n_1(k,\omega) = \frac{-ikA\pi[\delta(k - k_0) + \delta(k + k_0)]}{\omega^2 - \omega_p^2}. \qquad (7.75)$$

We want now to find $n_1(x,t)$. First, doing an inverse Laplace transform in time we obtain

$$n_1(k,t) = -\frac{ikA\pi[\delta(k - k_0) + \delta(k + k_0)]}{2\pi} \int_{-\infty}^{+\infty} \frac{e^{-i\omega t}}{\omega^2 - \omega_p^2}\,d\omega. \qquad (7.76)$$

Since the function, $1/(\omega^2 - \omega_p^2)$, $\to 0$ for $|\omega| \to \infty$, Jordan's lemma can be used, and the integral can be evaluated at the two first-order poles $\pm\omega_p$ (see

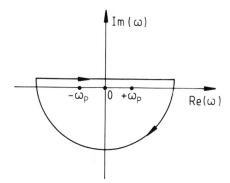

Fig. 7.1. Path of integration of Eq. (7.76).

Fig. 7.1). Using the residues theorem the integral in Eq. (7.76) is given by

$$\int_{-\infty}^{+\infty} \frac{e^{-i\omega t}}{\omega^2 - \omega_p^2} \, d\omega = -2\pi i \sum \text{Res}\left[\frac{e^{-i\omega t}}{\omega^2 - \omega_p^2}, \omega = \pm\omega_p\right] = -\frac{2\pi}{\omega_p} \sin \omega_p t.$$

(7.77)

Substituting from Eq. (7.77) into Eq. (7.76) gives

$$n_1(k,t) = iA\pi k[\delta(k - k_0) + \delta(k + k_0)] \frac{\sin \omega_p t}{\omega_p}.$$

(7.78)

Taking the inverse Fourier transform in space of Eq. (7.78) gives

$$n_1(x,t) = \frac{1}{2\pi} \int_{-\infty}^{+\infty} n_1(k,t) e^{ikx} \, dk = iA \frac{\sin \omega_p t}{\omega_p} k_0 \sin k_0 x.$$

(7.79)

Equation (7.79) shows that in a cold plasma any perturbation produced at time $t = 0$, with any wave number k_0, will oscillate at the plasma frequency, in agreement with the results in Chapter 2 of the fluid-theory treatment of a cold-electron plasma.

7.3.3.3. A Lorentzian Plasma. We now look in some detail at the case of a neutral plasma consisting of cold, motionless ions and electrons characterized by a one-dimensional Lorentzian distribution

$$f_0(v_{//}) = \frac{n_0}{\pi} \frac{a}{a^2 + v_{//}^2}.$$

(7.80)

For this case it is still possible to compute all the quantities calculated in the preceding cold-plasma example, but this case also contains most of the physics involved in ballistic and collective (wave-particle) behavior. Before calculating these quantities, however, we make some comments that are useful for distribution functions in general.

When calculating the plasma susceptibility χ or the relative dielectric constant ε we must evaluate the integral

$$\varepsilon(k,\omega) = 1 - \frac{\omega_p^2}{n_0 k^2} \int_{-\infty}^{+\infty} \frac{f'_0(v_{//})}{v_{//} - \omega/k} \, dv_{//}. \qquad (7.81)$$

Assuming that a is a characteristic velocity of the distribution, let us define the following *normalized* quantities

$$u = \frac{v_{//}}{a}; \qquad \xi = \frac{\omega}{ka}; \qquad \varphi_0(u) = \frac{f_0(v_{//})}{n_0}; \qquad k_d = \frac{\omega_p}{a}. \qquad (7.82)$$

Then, Eq. (7.81) can be written as

$$\varepsilon(k,\omega) = 1 - \frac{k_d^2}{k^2} \int_{-\infty}^{+\infty} \frac{\varphi'_0(u)}{u - \xi} \, du, \qquad (7.83)$$

involving the Hilbert transform,

$$Q'(\xi) = \int_{-\infty}^{+\infty} \frac{\varphi'_0(u)}{u - \xi} \, du. \qquad (7.84)$$

Since a Hilbert transform is a convolution, we can write

$$Q'(\xi) = \frac{dQ(\xi)}{d\xi}. \qquad (7.85)$$

In general, it turns out to be simpler to first compute $Q(\xi)$ and then the derivative $dQ/d\xi$.

Using the definitions given in Eq. (7.82) and the Lorentzian distribution given in Eq. (7.80) we can write the normalized one-dimensional Lorentzian distribution function as

$$\varphi_0(u) = \frac{1}{\pi} \frac{1}{u^2 + 1}, \qquad (7.86)$$

so that

$$Q^+(\xi) = \int_{L^+} \frac{\varphi_0(u)}{u - \xi} \, du = \frac{1}{\pi} \int_{L^+} \frac{du}{(u - i)(u + i)(u - \xi)}. \qquad (7.87)$$

The integral in Eq. (7.87) has three poles: $u = \xi$, $\pm i$. Since $Q^+(\xi)$ is normally given for $\text{Im}(\xi) > 0$, we can choose L^+ as the contour shown in Fig. 7.2. Thus, only the pole at $u = -i$ is inside the closed contour. Applying the residues

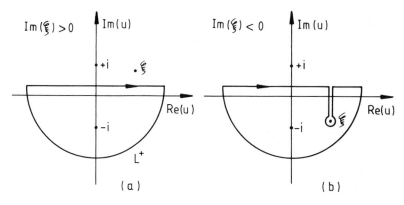

Fig. 7.2. Path of integration of Eq. (7.87): (a) for $\text{Im}(\xi) > 0$; (b) for $\text{Im}(\xi) < 0$.

theorem gives

$$\int_{L^+} \frac{du}{(u^2 + 1)(u - \xi)} = -2\pi i \, \text{Res}\left[\frac{1}{(u^2 + 1)(u - \xi)}, u = -i\right] = \frac{-\pi}{\xi + i}.$$

(7.88)

Therefore

$$Q^+(\xi) = -\frac{1}{\xi + i},$$

(7.89)

and by direct differentiation

$$Q^{+\prime}(\xi) = \frac{1}{(\xi + i)^2}.$$

(7.90)

Thus, the susceptibility and the relative dielectric constant are both very simple functions. For $k > 0$

$$\varepsilon^+(k,\omega) = 1 - \frac{k_d^2}{k^2} Q^{+\prime}(\xi) = 1 - \frac{\omega_p^2}{(\omega + ika)^2}.$$

(7.91)

For $k < 0$, we have

$$Q^-(\xi) = \int_{L_-} \frac{\varphi_0(u)\,du}{u - \xi} = \frac{1}{\pi} \int_{L_-} \frac{du}{(u - i)(u + i)(u - \xi)}.$$

(7.92)

There are still three poles, but only the pole at $+i$ is inside the L_- contour (see Fig. 7.3). Thus, using the residues theorem we find

$$\int_{L_-} \frac{du}{(u^2 + 1)(u - \xi)} = +2\pi i \, \text{Res}\left[\frac{1}{(u^2 + 1)(u - \xi)}, u = +i\right] = -\frac{\pi}{\xi - i}.$$

(7.93)

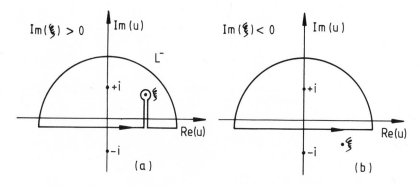

Fig. 7.3. Path of integration of Eq. (7.92): (a) for $\text{Im}(\xi) > 0$; (b) for $\text{Im}(\xi) < 0$.

Thus

$$Q^-(\xi) = -\frac{1}{\xi - i}, \tag{7.94}$$

and

$$Q^{-\prime}(\xi) = \frac{1}{(\xi - i)^2}. \tag{7.95}$$

Thus

$$\varepsilon^-(k,\omega) = 1 - \frac{k_d^2}{k^2} Q^{-\prime}(\xi) = 1 - \frac{\omega_p^2}{(\omega - ika)^2}. \tag{7.96}$$

We can now compute $E(k,\omega)$ from Eq. (7.41)

$$E^\pm(k,\omega) = \frac{-e}{k\varepsilon_0} \frac{\int_{L\pm} g(k,v_{//})/(\omega - kv_{//})\, dv_{//}}{\varepsilon^\pm(k,\omega)} \tag{7.97}$$

Using the same excitation as used for the cold-plasma example [see Eqs. (7.71) and (7.72)], we obtain

$$\int_{L\pm} \frac{f_0'(v_{//})\, dv_{//}}{v_{//} - \omega/k} = \int_{L\pm} \frac{\varphi_0'(u)\, du}{u - \xi} = \frac{dQ^\pm(\xi)}{d\xi} = \begin{cases} = \dfrac{1}{(\xi + i)^2}, & \text{for } k > 0 \\[3mm] = \dfrac{1}{(\xi - i)^2}, & \text{for } k < 0 \end{cases} \tag{7.98}$$

and

$$\int_{L\pm} \frac{g(k,v_{//})}{\omega - kv_{//}}\, dv_{//} = -A\pi[\delta(k - k_0) + \delta(k + k_0)]\frac{ka^2}{(\omega \pm ika)^2}. \tag{7.99}$$

Therefore

$$E^{\pm}(k,\omega) = \frac{(eA\pi/\varepsilon_0)\,[\delta(k - k_0) + (k + k_0)]\,[a^2/(\omega \pm ika)^2]}{1 + \chi}, \quad (7.100)$$

or, since $\chi = -\omega_p^2/(\omega + ika)^2$, we can write

$$E^{\pm}(k,\omega) = \frac{eA\pi}{\varepsilon_0}\,a^2\,\frac{\delta(k - k_0) + \delta(k + k_0)}{(\omega \pm ika)^2 - \omega_p^2}. \quad (7.101)$$

We could immediately make an inverse Laplace transform on $E(k,\omega)$ to obtain $E(k,t)$. Instead, we want to look separately at the free-streaming and collective contributions to the electric field in order to make a comparison with the free-streaming observed for the case of *neutral* particles having a Maxwellian distribution (see Section 7.2). Looking at Eqs. (7.100) and (7.101), if $\chi \to 0$ or, alternatively, if $\omega_p^2 \to 0$ (as for neutral particles), then $E(k,\omega)$ is a free-streaming electric field $E^*(k,\omega)$. For $k > 0$, for example,

$$E^{+}*(k,\omega) = \alpha(k)\,\frac{1}{(\omega + ika)^2}, \quad (7.102)$$

where

$$\alpha(k) = \frac{eA\pi}{\varepsilon_0}\,a^2[\delta(k - k_0) + \delta(k + k_0)]. \quad (7.103)$$

We use the procedure given in Chapter 1 to write the inverse Laplace transform of Eq. (7.102) as

$$E^{+}*(k,t) = \frac{1}{2\pi} \int_{-\omega+iv}^{+\infty+iv} \alpha(k)\,\frac{e^{-i\omega t}}{(\omega + ika)^2}\,d\omega. \quad (7.104)$$

Using the contour shown in Fig. 7.4, we can evaluate the integral in Eq. (7.104) by means of the residues theorem to obtain

$$\int_{L^+} \frac{e^{-i\omega t}}{(\omega + ika)^2}\,d\omega = -2\pi i\,\text{Res}\left[\frac{e^{-i\omega t}}{(\omega + ika)^2},\,\omega = ika\right] = -2\pi t\,e^{-kat}, \quad (7.105)$$

so that the free-streaming electric field for $k > 0$ becomes

$$E^{+}*(k,t) = -\alpha(k)t\,e^{-kat}. \quad (7.106)$$

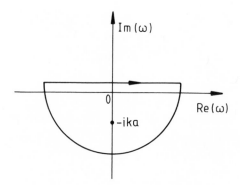

Fig. 7.4. Path of integration of Eq. (7.104).

Thus, as found for the case of a neutral gas having a Maxwellian distribution, the free-streaming contribution for a Lorentzian plasma first rises and then vanishes in time without oscillation.

We now look at the collective effect, which is represented here by the susceptibility $\chi = -\omega_p^2/(\omega + ika)^2$. We start again with Eq. (7.100) and assume that $|\chi| < 1$, allowing us to make an expansion so that the electric field can be written as

$$E^+(k,\omega) = \frac{\alpha(k)}{(\omega + ika)^2} \left[1 - \chi + \chi^2 + \cdots + (-1)^n\chi^n + \cdots\right]. \quad (7.107)$$

For $\omega_p^2 \to 0$, $\chi \to 0$, and as already found, the first term is the free-streaming electric field [see Eq. (7.102)], so that we can write

$$E^+(k,\omega) = E^{+*}(k,\omega) + \sum_{n=1}^{\infty} E_{n\,\text{col}}^+(k,\omega), \quad (7.108)$$

where the successive $E_{n\,\text{col}}^+(k,\omega)$ are the collective contributions of order n

$$E_{n\,\text{col}}^+(k,\omega) = \alpha \frac{\omega_p^{2n}}{(\omega + ika)^{2n+2}}. \quad (7.109)$$

We make an inverse Laplace transform of Eq. (7.108) to obtain

$$E^+(k,t) = E^{+*}(k,t) + \sum_{n=1}^{\infty} E_{n\,\text{col}}^+(k,t), \quad (7.110)$$

where

$$E^{+}_{n\,\text{col}}(k,t) = \frac{1}{2\pi} \int_{-\infty+iv}^{+\infty+iv} E^{+}_{n\,\text{col}}(k,\omega)e^{-i\omega t}\,d\omega$$

$$= \frac{\alpha}{2\pi}\,\omega_{p}^{2n} \int_{-\omega+iv}^{+\omega+iv} \frac{e^{-i\omega t}}{(\omega+ika)^{2n+2}}\,d\omega. \tag{7.111}$$

The integral in Eq. (7.111) can be calculated in the same manner as that used for the free-streaming term except that now $\omega = -ika$ is a pole of order $2n + 2$. Thus, using the residues theorem

$$\text{Res}\left[\frac{e^{-i\omega t}}{(\omega+ika)^{2n+2}}, \omega = -ika\right] = \frac{1}{(2n+1)!}\,\frac{d^{2n+1}}{d\omega^{2n+1}}\,e^{-i\omega t}\bigg|_{\omega=ika}. \tag{7.112}$$

Since

$$\frac{d^{2n+1}}{d\omega^{2n+1}}\,e^{-i\omega t} = -i(-1)^{n}t^{2n+1}\,e^{-i\omega t}, \tag{7.113}$$

the integral in Eq. (7.111) is given by

$$\int_{L^{+}} \frac{e^{-i\omega t}\,d\omega}{(\omega+ika)^{2n+2}} = -2\pi\,\frac{(-1)^{n}}{(2n+1)!}\,t^{2n+1}\,e^{-kat}, \tag{7.114}$$

so that the collective electric field can be written

$$En^{+}_{\text{col}}(k,t) = -\frac{\alpha}{\omega_{p}}\,e^{-kat}\left[(-1)^{n}\frac{(\omega_{p}t)^{2n+1}}{(2n+1)!}\right]. \tag{7.115}$$

Thus Eq. (7.110) shows that when ω_p increases from 0 to any finite value an expansion of $E(k,t)$ can be obtained by using a finite number of collective terms, valid for a finite time t

$$E^{+}(k,t) \simeq E^{+}*(k,t) + E^{+}_{1\,\text{col}}(k,t) + \cdots + E_{n\,\text{col}}(k,t). \tag{7.116}$$

The number of terms needed increases with $\omega_p t$ so that, eventually, all the terms are required. Then, using the series expansion of $\sin \omega_p t$

$$\sin \omega_p t = \sum_{n=1}^{+\infty} (-1)^{n}\frac{(\omega_p t)^{2n+1}}{(2n+1)!}, \tag{7.117}$$

we can write $E^+(k,t)$ as

$$E^+(k,t) = -\alpha(k)e^{-kat}\frac{\sin \omega_p t}{\omega_p}, \tag{7.118}$$

or using Eq. (7.103) we can write finally

$$E^+(k,t) = -\frac{eA}{\varepsilon_0}\pi a^2 [\delta(k - k_0) + \delta(k + k_0)]\frac{\sin \omega_p t}{\omega_p}e^{-kat}, \tag{7.119}$$

where $k > 0$, as assumed at the outset.

Comparing Eqs. (7.119) and (7.106) we observe that: (a) for $\omega_p t \ll 1$, the collective solution is identical to the free-streaming solution. This is because the plasma is unable to exhibit collective electron behavior on a time scale which is much smaller than $1/\omega_p$; (b) for $\omega_p t \gg 1$, both solutions are exponentially damped; however, because of the collective electron response, the collective solution exhibits oscillations at the electron plasma frequency, as shown in Fig. 7.5.

We leave it as an exercise for the reader to show that Eq. (7.119) can be found by a direct Laplace transform inversion of Eq. (7.100) *without* making an expansion in χ. The expansion was introduced here to enable an explicit

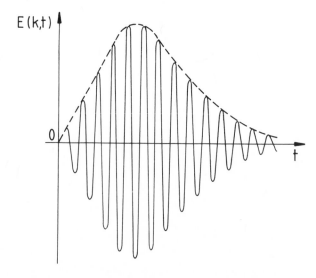

Fig. 7.5. Variations of the electric field $E^+(k,t)$ versus time in a Lorentzian plasma. The effect of the collective terms is the oscillatory behavior shown by the solid line. The amplitude of these oscillations is damped by the same term $\exp(-kat)$ that governs the long-time behavior of the free-streaming contribution of the perturbed electric field.

comparison to be made between the free-streaming and collective effects but, clearly, such an equivalent mathematical procedure should not give a result different from the direct inversion of $E^+(k,\omega)$.

The expansion made in Eq. (7.107) can be interpreted from a physical point of view in the following way: The first term is the free-streaming term, the second term is a second-order free streaming induced by the first-order free-streaming response, etc.. Thus, the reaction of the plasma may be viewed as an iterative effect. Such an interpretation, while interesting, does not, perhaps, prepare the interpreter for the net result of the iteration, i.e., a coherent sinusoidal oscillation at the plasma frequency.

We want now to complete the Fourier–Laplace inversion of Eq. (7.100) by making an inverse Fourier transform on $E(k,t)$. Using the procedure given in Chapter 1 we can write

$$E(x,t) = \frac{1}{2\pi} \int_{-\infty}^{+\infty} E(k,t)e^{-ikx}\, dk. \tag{7.120}$$

At this point, we must remember that the Hilbert transforms involved in Eq. (7.41) represent two different functions. For the same reason (see Section 8.3. for a detailed discussion) that the plasma dielectric function $\varepsilon(k,\omega)$ must be written as

$$\varepsilon(k,\omega) = \varepsilon^+(k,\omega)\,\Upsilon(k) + \varepsilon^-(k,\omega)\,\Upsilon(-k), \tag{7.121}$$

where ε^+ and ε^- are given, respectively, by Eqs. (7.91) and (7.96), we must express $E(k,\omega)$ as

$$E(k,\omega) = E^+(k,\omega)\,\Upsilon(k) + E^-(k,\omega)\,\Upsilon(-k), \tag{7.122}$$

where

$$E^+(k,\omega) = \frac{N^+(k,\omega)}{\varepsilon^+(k,\omega)}, \tag{7.123}$$

and

$$E^-(k,\omega) = \frac{N^-(k,\omega)}{\varepsilon^-(k,\omega)}, \tag{7.124}$$

with N^+ and N^- being given, respectively, by

$$N^+(k,\omega) = \frac{e}{k^2\varepsilon_0} \int_{L^+} \frac{g(k,v_{//})}{v_{//} - \omega/k}\, dv_{//}, \qquad \text{for } k > 0,$$

$$N^-(k,\omega) = \frac{e}{k^2\varepsilon_0} \int_{L^-} \frac{g(k,v_{//})}{v_{//} - \omega/k}\, dv_{//}, \qquad \text{for } k < 0. \tag{7.125}$$

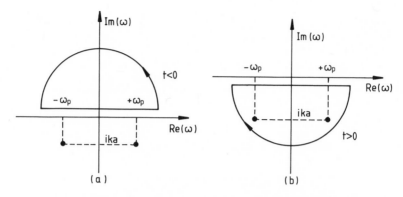

Fig. 7.6. Path of integration of Eq. (7.127): (a) for $t < 0$; (b) for $t > 0$.

We need, first, to make an inverse Laplace transform on $E^-(k,\omega)$ to obtain $E^-(k,t)$. From Eqs. (7.101) and (7.103), we can write

$$E^-(k,\omega) = \alpha(k)\,\frac{1}{(\omega - ika)^2 - \omega_p^2}. \tag{7.126}$$

Thus, using the Chapter 1 recipe, we can write

$$E^-(k,t) = \frac{\alpha(k)}{2\pi} \int_{-\omega+iv}^{+\omega+iv} \frac{e^{-i\omega t}\,d\omega}{(\omega - ika)^2 - \omega_p^2}. \tag{7.127}$$

There are now two poles, $\omega = ika \pm \omega_p$, and the integration is for $v > ka$. Using the contour indicated (see Fig. 7.6), we can write

$$E^-(k,t) = \frac{\alpha(k)}{2\pi} \left\{ -2\pi i\, \text{Res}\left[\frac{e^{-i\omega t}}{(\omega - ika)^2 - \omega_p^2},\ \omega = ika \pm \omega_p \right] \right\}. \tag{7.128}$$

It is left as an exercise for the reader to show that Eq. (7.128) yields

$$E^-(k,t) = -\alpha(k)e^{kat}\,\frac{\sin \omega_p t}{\omega_p}, \qquad \text{with } k < 0. \tag{7.129}$$

We see that since $E^+(k,t)$ and $E^-(k,t)$ are different, we must separate the two cases, $k > 0$ and $k < 0$. Thus, the inversion of the Fourier transform in k generally is not possible by the technique used for the inverse Laplace transform. However, we can still write

$$E(x,t) = \frac{1}{2\pi} \int_{-\infty}^{+\infty} E(k,t)e^{ikx}\,dk, \tag{7.130}$$

on the real axis in the k plane but, it will not be possible to evaluate this integral by the residues theorem because, even if $E^+(k,t)$ and $E^-(k,t)$ are analytical, the total field

$$E(k,t) = E^+(k,t)\,\Upsilon(k) + E^-(k,t)\,\Upsilon(-k),\qquad(7.131)$$

is not an analytical function.

Substituting from Eq. (7.131) into Eq. (7.130), we can write for all values of k

$$E(x,t) = \frac{1}{2\pi}\left[\int_{-\infty}^{+\infty} E^+(k,t)\,\Upsilon(k)e^{ikx}\,dk + \int_{-\infty}^{+\infty} E^-(k,t)\,\Upsilon(-k)e^{ikx}\,dk\right],\quad(7.132)$$

which can be written also as

$$E(x,t) = \frac{1}{2\pi}\left[\int_{-\infty}^{0} E^-(k,t)e^{ikx}\,dk + \int_{0}^{+\infty} E^+(k,t)e^{ikx}\,dk\right],\qquad(7.133)$$

where

$$E^+(k,t) = -\frac{eA}{\varepsilon_0}\pi a^2[\delta(k-k_0) + \delta(k+k_0)]\frac{\sin \omega_p t}{\omega_p}e^{-kat},$$

$$E^-(k,t) = -\frac{eA}{\varepsilon_0}\pi a^2[\delta(k-k_0) + \delta(k+k_0)]\frac{\sin \omega_p t}{\omega_p}e^{+kat},\qquad(7.134)$$

and $k > 0$.

Equation (7.133) can be written in alternative form as

$$E(k,t) = E_0(t)\pi[\delta(k-k_0) + \delta(k+k_0)][e^{-kat}\Upsilon(k) + e^{+kat}\Upsilon(-k)],\quad(7.135)$$

where

$$E_0(t) = -\frac{eA}{\varepsilon_0}a^2\frac{\sin \omega_p t}{\omega_p}.\qquad(7.136)$$

Substituting from Eq. (7.135) into Eq. (7.130) gives

$$E(x,t) = \frac{E_0(t)}{2}\left[\int_{-\infty}^{+\infty} \delta(k-k_0)e^{-kat}\Upsilon(k)e^{ikx}\,dk\right.$$

$$\left. + \int_{-\infty}^{+\infty} \delta(k+k_0)e^{+kat}\Upsilon(-k)e^{ikx}\,dk\right],\qquad(7.137)$$

from which we can write

$$E(x,t) = -\frac{eA}{\varepsilon_0} a^2 \frac{\sin \omega_p t}{\omega_p} e^{-k_0 a t} \cos k_0 x. \tag{7.138}$$

7.3.4. Long-Wavelength Oscillations

We have seen that

$$E(k,\omega) = \frac{N(k,\omega)}{\varepsilon(k,\omega)}, \tag{7.139}$$

and that the inverse image $E(k,t)$ will include the contributions of the various poles of $N(k,\omega)/\varepsilon(k,\omega)$. Assuming, for simplicity and in contrast with the preceding example that $N(k,\omega)$ has no poles corresponding to the zeros of $\varepsilon(k,\omega)$, we can write

$$E(k,t) = E_{\mathbf{exi}}(k,t) + \sum_j E_j(k,t), \tag{7.140}$$

where the first term $E_{\mathbf{exi}}(k,t)$ is the contribution of the excitation given by the poles of $N(k,\omega)$, when they exist, and $\sum_j E_j(k,t)$ is the sum of the contributions of the various zeros of the distribution function $\varepsilon(k,\omega)$. The poles can be finite or infinite in number. However, in many cases only one pole (or, only a few poles) is close enough to the real axis to give a strong contribution to the electric field. {Since a pole contribution to the electric field attenuates like $\exp[-\mathrm{Im}(\omega)t]$, the contribution of a pole close to the real axis—after a time which depends upon the relative damping rates and excitation coefficients—will dominate the contributions from poles not close to the real axis.}

We assume for simplicity that only one pole is very close to the real axis, and we write the dispersion relation

$$\mathscr{D}^+(k,\omega) = 1 - \frac{k_d^2}{k^2} Q'(\xi) = 0, \tag{7.141}$$

using the dimensionless parameters given in Eq. (7.82). Equation (7.141) is valid for ξ having any value in the ξ plane. For ξ anywhere on the real axis we write

$$Q'(\xi_r) = P \int_{-\infty}^{+\infty} \frac{\varphi_0'(u)\,du}{u - \xi_r} + i\pi\varphi_0'(\xi_r), \tag{7.142}$$

the integral being understood to be Cauchy's principal part, as discussed in Chapter 1.

We now want to make an expansion of $\varepsilon^+(k,\omega)$, in terms of $\omega = \varepsilon$, around the real axis so that the expansion $\varepsilon(k,\omega)$ is analytical. Then, because of its analyticity and by proper choice of $\text{Re}(\varepsilon) = \varepsilon_r$, we can make a Taylor expansion $\varepsilon(k,\varepsilon_0)$ around ε_r, where ε_0 is a pole near the real axis. From this expansion we can then find the phase velocity and damping rate of the wave associated with the pole ε_0.

Since $Q'(\xi)$ represents two different functions, we first concentrate on finding an expansion of $Q^{+\prime}(\varepsilon)$ around the real axis for $k > 0$. Now

$$Q^+(\xi) = \int_{L^+} \frac{\varphi_0(u)}{u - \xi} \, du. \qquad (7.143)$$

Thus, on the real axis we can write

$$Q^+(\xi_r) = P \int_{-\infty}^{+\infty} \frac{\varphi_0(u) \, du}{u - \xi} + i\varphi_0(\xi_r), \qquad (7.144)$$

the integral being understood to be Cauchy's principal part of the integral. In the long-wavelength approximation $(|k| \to 0)$, $|\xi = \omega/(ka)| \to \infty$, so that we can make an asymptotic expansion of $Q^+(\xi)$ of the form

$$Q^+(\xi) \simeq \sum_{n=1}^{\infty} a_n \frac{1}{\xi_n} + i\pi\varphi_0(\xi), \qquad (7.145)$$

which will be everywhere analytical except for $\xi = 0$. For that, we can make an expansion of $\varphi_0(u)/(u - \xi_r)$

$$\frac{\varphi_0(u)}{u - \xi_r} = -\frac{1}{\xi_r} \frac{\varphi_0(u)}{1 - u/\xi_r} = -\frac{\varphi_0(u)}{\xi_r} \left[1 + \frac{u}{\xi_r} + \frac{u^2}{\xi_r^2} + \cdots + \frac{u^n}{\xi_r^n} + \cdots \right],$$
$$(7.146)$$

where the series is convergent for $|u| < |\xi|$.

Despite the fact that we are not allowed to make an integration on u from $-\infty$ to $+\infty$ for each term of this series, we can note that $\varphi_0(u)$ is very small for large values of u, so that for large ξ_r

$$\int_{-\infty}^{+\infty} \varphi_0(u)u^n \, du \simeq \int_{-\xi_r}^{+\xi_r} \varphi_0(u)u^n \, du. \qquad (7.147)$$

Thus, for large ξ_r

$$\int_{-\infty}^{+\infty} \frac{\varphi_0(u)}{u - \xi_r} \, du \simeq -\frac{1}{\xi_r} \left[1 + \frac{\langle u \rangle}{\xi_r} + \frac{\langle u^2 \rangle}{\xi_r^2} + \cdots + \frac{\langle u^n \rangle}{\xi_r^n} \right]. \qquad (7.148)$$

The corresponding expansion is not a real expansion as an entire series, but is instead an asymptotic expansion (see Whitaker and Watson, 1965). In fact, the series is not really convergent. For example, we can form the ratio of two successive terms and apply d'Alembert's rule to test the convergence of the above series. Let us consider, for example, an even distribution function that has a Maxwellian-like shape for large values of u. For this case

$$\langle u \rangle = \langle u^3 \rangle = \cdots = \langle u^{2n+1} \rangle = 0, \tag{7.149}$$

and

$$\langle u^{2n} \rangle = \int_{-\infty}^{+\infty} u^{2n} e^{-u^2} \, du = \frac{1 \cdot 3 \cdot 5 \cdot \ldots \cdot (2n-1)}{2^{2n}} (\pi)^{1/2}, \tag{7.150}$$

so that the ratio of two successive terms in Eq. (7.148) is given by

$$\frac{1}{\xi^2} \frac{\langle u^{2n+2} \rangle}{\langle u^{2n} \rangle} \simeq \frac{1}{\xi^2} \frac{2n+1}{4}. \tag{7.151}$$

Thus, for a given value of ξ, even if very large, there is a value of $\mathcal{N}(\xi)$ (see Section 1.5.2.4.) such that if $n > \mathcal{N}(\xi)$, the series

$$1 + \frac{\langle u^2 \rangle}{\xi^2} + \cdots + \frac{\langle u^{2n} \rangle}{\xi^{2n}} + \cdots \tag{7.152}$$

diverges. When

$$\int_{-\infty}^{+\infty} \frac{\varphi_0(u)}{u - \xi_r} \, du + \frac{1}{\xi_r} \left[1 + \frac{\langle u \rangle}{\xi_r} + \cdots + \frac{\langle u^n \rangle}{\xi_r^n} \right] = O(\xi_r^{-n-2}) \tag{7.153}$$

i.e., is of the order of $1/\xi_r^{n+2}$, which can be very small for large ξ_r values, the asymptotic expansion can then be a good approximation of Cauchy's principal part of the integral [see Eq. (7.144)].

Let us now consider the function

$$F(\xi) = -\frac{1}{\xi} \left[1 + \frac{\langle u \rangle}{\xi} + \frac{\langle u^2 \rangle}{\xi^2} + \cdots + \frac{\langle u^n \rangle}{\xi^n} \right] + i\pi\varphi_0(\xi). \tag{7.154}$$

This function is analytical everywhere except for $\xi = 0$. In view of the preceding discussion, $F(\xi)$ is a good approximation of the integral appearing in $Q^+(\xi)$ [see Eq. (7.144)] for all real values of ξ, except $\xi = 0$, so that *near the* real axis we can write

$$Q^+(\xi) \simeq -\frac{1}{\xi} - \frac{\langle u \rangle}{\xi^2} - \frac{\langle u^2 \rangle}{\xi^3} - \cdots - \frac{\langle u^n \rangle}{\xi^{n+1}} + i\pi\varphi_0(\xi), \tag{7.155}$$

n being of the order of the minimum error between this approximation and the exact function. Straightforward differentiation of both sides of Eq. (7.155) with respect to ξ gives a similar expression for $Q^{+\prime}(\xi)$

$$Q^{+\prime}(\xi) \simeq \frac{1}{\xi^2} + \frac{2\langle u \rangle}{\xi^3} + \frac{3\langle u^2 \rangle}{\xi^4} + \cdots + \frac{(n+1)\langle u^n \rangle}{\xi^{n+2}} + i\pi\varphi_0'(\xi). \quad (7.156)$$

We now arbitrarily limit the expansion in Eq. (7.156) to the first three terms. If $\varphi_0(u)$ is an even function, this gives an error of the order of ξ^{-5} or ξ^{-6}. We then obtain an asymptotic expansion for $\varepsilon^+(k,\xi)$—see Eq. (7.141)—given by

$$\varepsilon^+(k,\xi) = 1 - \frac{k_d^2}{k^2}\left[\frac{1}{\xi^2} + \frac{2\langle u \rangle}{\xi^3} + \frac{3\langle u^2 \rangle}{\xi^4}\right] - i\pi\varphi_0'(\xi)\frac{k_d^2}{k^2}. \quad (7.157)$$

Since $\varepsilon^+(k,\xi)$ is an analytic function, we can make a Taylor expansion around a point $\xi = \xi_r$, where ξ_r is a real number. If

$$\varepsilon^+(k,\xi) = \varepsilon_r^+(k,\xi) + i\varepsilon_i^+(k,\xi), \quad (7.158)$$

ε_r^+ and ε_i^+ being the real and the imaginary parts, respectively, of $\varepsilon^+(k,\xi)$, then Eq. (7.157) becomes

$$\varepsilon^+(k,\xi_r) = 1 - \frac{k_d^2}{k^2}\left[\frac{1}{\xi_r^2} + \frac{2\langle u \rangle}{\xi_r^3} + \frac{3\langle u^2 \rangle}{\xi_r^4}\right] - i\pi\varphi_0'(\xi_r)\frac{k_d^2}{k^2}. \quad (7.159)$$

As noted in the last paragraph of Section 7.3.3.1, we are interested in the zeros of the dispersion relation $\mathscr{D}^+(k,\omega) = 0$. Since $\mathscr{D}^+ = 0$, the real and complex parts of $\mathscr{D}^+$ must separately be zero. Thus, we can write

$$\mathscr{D}_r^+(k,\xi_r) = \mathrm{Re}\ \mathscr{D}^+(k,\xi_r) = 1 - \frac{k_d^2}{k^2}\left[\frac{1}{\xi_r^2} + \frac{2\langle u \rangle}{\xi_r^3} + \frac{3\langle u^2 \rangle}{\xi_r^4}\right] = 0. \quad (7.160)$$

Obviously, since Eq. (7.160) does not represent the entire dispersion relation, a real root of this equation may not be the root desired. However, we will see that this approach does give an interesting, well-known result and at the same time provides a clue as to how to proceed with the complete dispersion relation. The solution of Eq. (7.160) can be obtained using a technique called reversion of series (see Abramowitz, and Stegun, 1965, p. 16). For this purpose, we rewrite the equation to obtain

$$\frac{k^2}{k_d^2} = \frac{1}{\xi_r^2} + \frac{2\langle u \rangle}{\xi_r^3} + \frac{3\langle u^2 \rangle}{\xi_r^4}. \quad (7.161)$$

Assuming that $k > 0$, we can write an expansion of k/k_d given by

$$\frac{k}{k_d} = \frac{1}{\xi_r^2} + \frac{\langle u \rangle}{\xi_r^3} + \frac{3\langle u^2 \rangle - \langle u \rangle^2}{2\xi_r^3}, \qquad (7.162)$$

which can be solved to give

$$\frac{1}{\xi_r} = \frac{k}{k_d} - \langle u \rangle \frac{k^2}{k_d^2} + \left(\frac{5}{2} \langle u \rangle^2 - \frac{3}{2} \langle u^2 \rangle \right) \frac{k^3}{k_d^2} + O\left(\frac{k^4}{k_d^4} \right). \qquad (7.163)$$

Equation (7.163) is written in terms of the dimensionless quantities defined in Eq. (7.82). Returning to the dimensional quantities, we leave it as an exercise for the reader to show that

$$\frac{\omega}{\omega_p} = 1 + \frac{k \langle v_{//} \rangle}{\omega_p} + \left(-2\langle v_{//} \rangle^2 + \frac{3}{2} \langle v_{//}^2 \rangle \right) \frac{k^2}{\omega_p^2} + O(k^3). \qquad (7.164)$$

For any distribution function having a zero mean velocity $\langle v_{//} \rangle = 0$, Eq. (7.164) becomes

$$\frac{\omega}{\omega_p} = 1 + \frac{3}{2} \frac{k^2 \langle v_{//}^2 \rangle}{\omega_p^2} + O(k^4), \qquad (7.165)$$

which can be written as

$$\omega^2 = \omega_p^2 + 3k^2 \langle v_{//}^2 \rangle + O(k^4), \qquad (7.166)$$

the familiar Bohm and Gross dispersion relation for long-wavelength, electron plasma waves. Equation (7.166) shows that propagation can exist only if $\omega > \omega_p$, there being a cutoff at the plasma frequency, consistent with our hypothesis that ω does not go to zero for $k \to 0$ (long-wavelength approximation).

Up to now ξ_r has been just the solution of Eq. (7.160). The only possibility that ξ_r could be a root of the entire dispersion relation, Eq. (7.159), is if a root of the entire dispersion relation was real. Thus, we must return to the complete dispersion relation to see if it is possible to find a root ξ_0 close to the real axis of

$$\mathcal{D}^+(\xi_0, k) = \varepsilon_r^+(\xi_0, k) + i\varepsilon_i^+(\xi_0, k) = 0. \qquad (7.167)$$

We now make a Taylor expansion of $\varepsilon^+(\xi_0, k)$ around the real value ξ_r already obtained, since we expect ξ_0 to be close to the real axis when $k \to 0$ and

$\xi_0 \rightarrow \xi_r$. Thus, for any given small value of k, we can write

$$0 = \varepsilon_r^+(\xi_r,k) + \frac{\partial \varepsilon_r^+}{\partial \xi}\bigg|_{\xi=\xi_r} \Delta\xi + i\left[\varepsilon_i^+(\xi_r,k) + \frac{\partial \varepsilon_i^+}{\partial \xi}\bigg|_{\xi=\xi_r} \Delta\xi\right], \quad (7.168)$$

where $\Delta\xi = \xi_0 - \xi_r$ and [see Eq. (7.159)]

$$\varepsilon_r^+(\xi_r,k) = 1 - \frac{k_d^2}{k^2}\left[\frac{1}{\xi_r^2} + \frac{2\langle u\rangle}{\xi_r^3} + \frac{3\langle u^2\rangle}{\xi_r^4}\right], \quad (7.169)$$

$$\varepsilon_i^+(\xi_r,k) = -\pi\varphi_0'(\xi_r)\frac{k_d^2}{k^2}. \quad (7.170)$$

In Eq. (7.168), the Taylor expansion of Eq. (7.167) around ξ_r, it should be noted that we have kept only the first four terms of the expansion [the first term of which is zero, in accordance with Eq. (7.160)]. Using Eqs. (7.169) and (7.170), the first-order-derivative terms in Eq. (7.168) become

$$\frac{\partial \varepsilon_r^+}{\partial \xi}\bigg|_{\xi=\xi_r} = +\frac{k_d^2}{k^2}\frac{2}{\xi_r^3}, \quad (7.171)$$

and

$$\frac{\partial \varepsilon_i^+}{\partial \xi}\bigg|_{\xi=\xi_r} = -\frac{k_d^2}{k^2}\pi\varphi_0''(\xi_r). \quad (7.172)$$

From Eqs. (7.171) and (7.172), we see that for distribution functions $\varphi_0(\xi_r)$ which are decreasing faster than $1/\xi_r$, $\partial\varepsilon_i^+/\partial\xi$ is negligible with respect to $\partial\varepsilon_r^+/\partial\xi$. But, it is generally true that most realistic distribution functions do decrease faster than $1/\xi_r$, since the total density

$$n = \int_{-\infty}^{+\infty} f_0(v_{//})\, dv_{//} = an_0 \int_{-\infty}^{+\infty} \varphi(u)\, du, \quad (7.173)$$

must be finite, so that $\varphi(u)$ must decrease faster than $1/u$ for large values of u. Thus, from Eq. (7.168), we can write

$$\Delta\xi \simeq -i\frac{\varepsilon_i^+(\xi_r,k)}{\partial\varepsilon_r^+/\partial\xi|_{\xi=\xi_r}}, \quad (7.174)$$

which is a purely imaginary quantity. Therefore, within the accuracy that we can neglect $\partial\varepsilon_i^+/\partial\xi$ at $\xi=\xi_r$, with respect to $\partial\varepsilon_r^+/\partial\xi$, at $\xi=\xi_r$, ξ_r is equal to the real part of the root ξ_0.

We now want to find an expression for the denominator of Eq. (7.174). To do this, we write the total derivative of $\varepsilon_r^+(\xi_r,k)$ with respect to k and ξ_r

$$0 = \frac{\partial \varepsilon_r^+}{\partial \xi_r} d\xi_r + \frac{\partial \varepsilon_r^+}{\partial k} dk. \tag{7.175}$$

From Eq. (7.160), we see that

$$\frac{\partial \varepsilon_r^+}{\partial k} = +\frac{2k_d^2}{k^3}\left[\frac{1}{\xi_r^2} + \frac{2\langle u \rangle}{\xi_r^3} + \frac{3\langle u^2 \rangle}{\xi_r^4}\right], \tag{7.176}$$

which can be expressed in terms of $\varepsilon_r^+(\xi_r,k)$ as

$$\frac{\partial \varepsilon_r^+}{\partial k} = +\frac{2}{k}\left[(1 - \varepsilon_r^+(\xi,k))\right]. \tag{7.177}$$

But $\varepsilon_r^+(\xi_r,k) = 0$—see Eq. (7.160)—so that Eq. (7.177) can be written

$$\left.\frac{\partial \varepsilon_r^+}{\partial k}\right|_{\xi = \xi_r} = \frac{2}{k}. \tag{7.178}$$

Substituting from Eq. (7.178) into Eq. (7.175) gives

$$\left.\frac{\partial \varepsilon_r^+}{\partial \xi}\right|_{\xi = \xi_r} = \frac{\partial \varepsilon_r^+}{\partial k}\frac{dk}{d\xi_r} = \frac{2}{k}\frac{dk}{d\xi_r}, \tag{7.179}$$

Using Eq. (7.179), Eq. (7.174) can now be written as

$$\Delta \xi = -i\frac{\pi}{2}\varphi_0'(\xi_r)\frac{k_d^2}{k}\frac{d\xi_r}{dk}. \tag{7.180}$$

In ω,k notation, Eq. (7.180) can be written

$$\frac{\omega_i}{ka} = \frac{\pi}{2}\frac{a}{n_0}\left.\frac{df_0(v_{//})}{dv_{//}}\right|_{v_{//} = \omega/k}\frac{\omega_p^2}{a^2}\frac{1}{k}\frac{d(\omega/ka)}{dk}, \tag{7.181}$$

or alternatively as

$$\omega_i = -\frac{\pi}{2}\frac{\omega_p^2}{ka}\left(\frac{d\omega}{dk} - \frac{\omega}{k}\right)\frac{1}{n_0}\left.\frac{df_0(v_{//})}{dv_{//}}\right|_{v_{//} = \omega/k} \tag{7.182}$$

Equation (7.182) is the well-known general expression for Landau damping in the long-wavelength approximation. This equation predicts either a wave decay or a wave growth, depending on the sign of the slope of the distribution function at the wave phase velocity. This supports our original assertion that a collisionless interaction between a propagating wave and the background plasma particles can reduce (or, as we see here, actually reverse) the free-streaming "damping" due to normal particle diffusion.

7.3.5. Maxwellian Plasma

We now consider a plasma whose electron velocity distribution is given by the Maxwellian distribution function

$$f_0(v_{//}) = \frac{1}{a\pi^{1/2}} \exp\left(-\frac{v_{//}^2}{a^2}\right), \tag{7.183}$$

where a is the thermal velocity defined by the equation

$$a = \left(\frac{2KT}{m}\right)^{1/2}, \tag{7.184}$$

T being the electron temperature, m the electron mass, and K the Boltzmann constant. Using the same dimensionless variables as used previously in Eq. (7.82)

$$u = \frac{v_{//}}{a}, \qquad \xi = \frac{\omega}{ka}, \tag{7.185}$$

and

$$\varphi(u) = \frac{f_0(v_{//})}{n_0}, \qquad k_d = \frac{\omega_p}{a}, \tag{7.186}$$

the distribution function can be written as

$$\varphi_0(u) = \pi^{-1/2} \exp(-u^2). \tag{7.187}$$

When computing the Hilbert transform, $Q^{\pm}(\xi)$ of $\varphi_0(u)$ we introduce the so-called plasma dispersion function, $Z(\xi)$, which, like all Hilbert transforms, has two different expressions, $Z^+(\xi)$ and $Z^-(\xi)$, given by

$$Z^{\pm}(\xi) = \pi^{-1/2} \int_{L^{\pm}} \frac{e^{-u^2}}{u - \xi} \, du. \tag{7.188}$$

[Actually, it is $Z^+(\xi)$ that has been tabulated by Fried and Conte (1961), which is usually called the plasma dispersion function.] Using Eq. (7.188), the plasma dielectric constant can be written as

$$\varepsilon^+(k,\omega_0) = 1 - \frac{k_d^2}{k^2} Z^{+\prime}(\xi), \qquad k > 0,$$

$$\varepsilon^-(k,\omega_0) = 1 - \frac{k_d^2}{k^2} Z^{-\prime}(\xi), \qquad k < 0, \tag{7.189}$$

where $Z^{\pm\prime} = dZ^{\pm}(\xi)/d\xi$.

In the inversion procedure of the Fourier transform, we have seen that we need to know the singularities of the integrand of the Fourier-transform integral, i.e., of $1/\varepsilon^{\pm}(\omega,k)$. Or, equivalently, we need to know the roots of the dispersion relation

$$\varepsilon^{\pm}(\omega,k) = 0. \tag{7.190}$$

A basic difference between the fluid theory of the previous chapters and the kinetic theory of the present chapter is that now $\varepsilon^{+}(\omega,k)$ is an entire function (see Chapter 1) of ξ which is not a polynomial function. According to Picard's theorem (Roos, 1969), an entire function that is not a polynomial takes all values, with one possible exception, an infinite number of times. Thus, the equation $Z^{+\prime}(\xi) = A$, where A can be any complex quantity, has an infinite number of roots, except for $A = -2$. Therefore, the dispersion relation $\varepsilon^{+}(k,\omega) = 0$, which can be written as

$$Z^{+\prime}\left(\frac{\omega}{ka}\right) = \frac{k^2}{k_d^2}, \tag{7.191}$$

will have an infinite number of roots if $\omega = 0$. In contrast with the situation found in fluid theory, as well as in the preceding cold-plasma and Lorentzian-plasma examples, in a Maxwellian plasma there is no unique or finite number of (ω,k) values that satisfy $\varepsilon^{+}(k,\omega) = 0$. There is therefore no simple algebraic dispersion relation that relates ω and k. The roots must be found numerically, and their corresponding contributions must be added, in order to describe the plasma response to a given perturbation.

For the case of long wavelengths, we showed that the problem is much simpler if one of the poles is close to the real axis, since its contribution, in this case, is much larger than the combined contributions of all the other poles. We now show easily, using an asymptotic expansion, that this situation exists for a Maxwellian plasma since as we shall see $\xi \to \infty$ close to the real axis.

The functions $Q^{+}(\xi)$ and $Q^{-}(\xi)$ can be related to the integral representation of the error function [Abramowitz and Stegun, (1965), p. 297], which can be written as

$$w(\xi) = \frac{i}{\pi}\int_{-\infty}^{+\infty}\frac{e^{-u^2}}{\xi - u}\,du = e^{-\xi^2}\,\mathrm{erfc}(-i\xi). \tag{7.192}$$

Considering only the case of $Q^{+}(\xi)$, we write

$$Q^{+}(\xi) = \frac{1}{\pi^{1/2}}\int_{L^+}\frac{e^{-u^2}}{u - \xi}\,du = i\pi^{1/2}e^{-\xi^2}[1 + \mathrm{erf}(i\xi)]. \tag{7.193}$$

For the case where $|z| \to \infty$ and $|\arg z| < 3\pi/4$, there exists an asymptotic expansion for the error function [Abramowitz and Stegun, (1965), p. 298], which can be written as

$$\pi^{1/2} z e^{z^2} \operatorname{erfc}(z) \simeq 1 + \sum_{n=1}^{+\infty} (-1)^n \frac{1 \cdot 3 \cdot \ldots \cdot (2n-1)}{(2z^2)^n} + R_n(z), \quad (7.194)$$

where $R_n(z)$ is a remainder term. For a Maxwellian plasma, for long wavelengths, $\xi \to \infty$ along the real axis. For $z = i\xi$, $z \to \infty$ with an argument close to $\pi/2$, which is $< 3\pi/4$. Thus, the above expansion can be used to write $Q^+(\xi)$ as

$$Q^+(\xi) = 2i\pi^{1/2} e^{-\xi^2}$$
$$-\frac{1}{\xi}\left[1 + \frac{1}{2\xi^2} + \frac{3}{4\xi^4} + \cdots + \frac{1 \cdot 3 \cdot \ldots \cdot (2n-1)}{2^n \xi^{2n}} + R_n(\xi) \right].$$
$$(7.195)$$

We obtain $Q^{+\prime}(\xi)$ by direct differentiation of Eq. (7.195), giving

$$Q^{+\prime}(\xi) = -4i\pi^{1/2}\xi e^{-\xi^2}$$
$$+\left[\frac{1}{\xi^2} + \frac{3}{2\xi^4} + \frac{15}{4\xi^6} + \frac{1 \cdot 3 \cdot \ldots \cdot (2n+1)}{2^n \xi^{2n+2}} + \frac{d[R_n(\xi)/\xi]}{d\xi} \right]. \quad (7.196)$$

This equation is similar to the general equation, Eq. (7.156), but we see here that due to the symmetry of the Maxwellian distribution, there are no odd-powered terms of $1/\xi$ in the brackets. That is,

$$\langle u \rangle = \langle u^3 \rangle = \cdots = \langle u^{2n+1} \rangle = 0. \quad (7.197)$$

As discussed earlier, in practice we must limit the number of terms used in the asymptotic expansion, otherwise the series would diverge. For a Maxwellian plasma, the remainder term $R_n(\xi)$ can be estimated numerically. Since the argument of ξ is still slightly negative for a damped wave [Abramowitz and Stegun, (1965), p. 299],

$$R_n(\xi) = \frac{1 \cdot 3 \cdot \ldots \cdot (2n-1)}{2^n \xi^{2n}} \theta(\xi, n), \quad (7.198)$$

where

$$\theta(\xi, n) = \int_0^{+\infty} e^{-t} \left[1 - \frac{t}{\xi^2} \right]^{-n-1/2} dt. \quad (7.199)$$

TABLE 7.1. Relative Values of the Successive
Terms of the Asymptotic Expansion of $Q^{+\prime}(\xi)$ for Various ξ[a]

	$\xi = 1$	$\xi = 10$	$\xi = 10^2$	$\xi = 10^3$
u_0	1	10^{-2}	10^{-4}	10^{-6}
u_1	1.5	1.5×10^{-4}	1.5×10^{-8}	1.5×10^{-12}
u_2	3.75	3.75×10^{-6}	3.75×10^{-12}	
u_3	1.31×10^1	1.31×10^{-7}	1.31×10^{-15}	
u_4	5.9×10^1	5.9×10^{-9}		
u_5	3.24×10^2	3.24×10^{-10}		
u_6	2.11×10^3	2.11×10^{-11}		
u_7	1.58×10^4	1.58×10^{-12}		
u_8	1.35×10^5	1.35×10^{-13}		
u_9	1.28×10^6	1.28×10^{-14}		
u_{10}	1.34×10^7	1.34×10^{-15}		

[a] Values lower than 10^{-15} are omitted. For values higher than a few units, the asymptotic expansion of $Q^{+\prime}(\xi)$ can be limited to a very small number of terms.

In general, for large values of ξ, the calculation of $R_n(\xi)$ is not needed, since the successive terms in the brackets decrease very quickly in magnitude. For example, Table 7.1 shows the values of the successive terms in the brackets in the expansion of $Q^{+\prime}(\xi)$ for n between 0 and 10 and for values of ξ between 1 and 10^3. In general, only the first few terms are needed to give a very accurate value. In Table 7.1, terms of value less than 10^{-15} are omitted. Keeping only the first two terms of the expansion gives

$$Q^{+\prime}(\xi) = -4i\pi\xi e^{-\xi^2} + \frac{1}{\xi^2} + \frac{3}{2\xi^4} = \frac{k^2}{k_d^2}. \tag{7.200}$$

Using the same methods as used in the preceding section on long wavelengths, the reader can easily show that the dispersion relation is given by

$$\omega_r^2 = \omega_p^2 + 3k^2\langle u^2\rangle = \omega_p^2 + \frac{3}{2}k^2a^2 \tag{7.201}$$

and the Landau damping rate by

$$\text{Im}(\omega) = -\pi^{1/2}\left(\frac{\omega_p^2}{a^3k}\right)\frac{\omega_p^2}{k^2}\exp\left(-\frac{\omega^2}{a^2k^2}\right). \tag{7.202}$$

7.4. ION-ACOUSTIC WAVES

In the preceding section we assumed that the electrons could move, but that the ions were motionless. Such an assumption is valid for high-frequency waves, since due to inertia the ions cannot follow the electron oscillations in a high-frequency field. For low-frequency oscillations, however, this assumption is no longer valid, and we must take into account both the electron and ion motion.

We define a multi-component plasma as a plasma that consists of electrons and several species of positive and negative ions. The kinetic theoretical model of such a plasma consists of a number of Vlasov equations (one for the electrons and one for each ion species) [see Eq. (7.15)] plus Poisson's equation, [see Eq. (7.16)]. The resulting system of equations, which are coupled by the electric field $E(r,t)$, can be written as

$$\frac{\partial f_m}{\partial t} + v\frac{\partial f}{\partial r} + \frac{q_m}{m_m} E \frac{\partial f_m}{\partial v} = 0, \qquad m = 0, \ldots, n,$$

$$\nabla \cdot E = \frac{1}{\varepsilon_0} \sum_m q_m \int f_m \, dv, \tag{7.203}$$

where $m = 0$ corresponds to the electrons, $m = 1, \ldots, n$ corresponds to the n ion species, the $f_m(r,v,t)$ are the corresponding distribution functions, and ρ_{ex} [see Eq. (7.16)] is assumed to be zero.

Following the procedure used in Section 7.3 for the electrons, we linearize, Fourier-and-Laplace transform, and one-dimensionalize the $n + 1$ Vlasov equations to obtain

$$f_{1m}(k,v_{//},\omega) = -i\frac{(q_m/m_m)\, df_{0m}(v_{//})/dv_{//}}{\omega - kv_{//}} E(k,\omega) - i\frac{f_{1m}(k, v_{//}, t = 0)}{\omega - kv_{//}}. \tag{7.204}$$

Linearizing, Fourier-and-Laplace transforming, and one-dimensionalizing Poisson's equation gives

$$ikE(k,\omega) = \frac{1}{\varepsilon_0} \sum_m q_m \int f_{1m}(k,v_{//},\omega) \, dv_{//} \tag{7.205}$$

which, with the preceding expression for $f_{1m}(k,v_{//},\omega)$ [see Eq. (7.204)], can be written as [see Eqs. (7.35) and (7.40)]

$$E^+(k,\omega) = \frac{N(k,\omega)}{\varepsilon^{\pm}(k,\omega)} \tag{7.206}$$

where the dielectric constant $\varepsilon(k,\omega)$ is given by [see Eq. (7.37)]

$$\varepsilon^{\pm}(k,\omega) = 1 + \frac{1}{\varepsilon_0 k^2} \int_{L^{\pm}} \frac{\sum_m (q_m^2/m_m)[df_{0m}(v_{//})/dv_{//}]}{\omega/k - v_{//}} \, dv_{//}. \tag{7.207}$$

It is important to note that for the low-frequency oscillations now being considered we must assign an equilibrium distribution function for each *ion* specie and that the *ions* now contribute to the dielectric function in the same way as do the electrons.

If we introduce an *equivalent distribution function* $F_0(v_{//})$ defined by

$$F_0(v_{//}) = \frac{1}{\omega_{p0}^2} \sum_m \omega_{pm}^2 f_{0m}(v_{//}), \tag{7.208}$$

where

$$\omega_{pm}^2 = \frac{n_m q_m^2}{\varepsilon_0 m_m} \quad \text{and} \quad \omega_{p0}^2 = \sum_m \omega_{pm}^2 \tag{7.209}$$

then Eq. (7.207) can be written as

$$\varepsilon^{\pm}(k,\omega) = 1 + \frac{\omega_{p0}^2}{k^2} \int_{L^{\pm}} \frac{dF_0(v_{//})/dv_{//}}{\omega/k - v_{//}} \, dv_{//}. \tag{7.210}$$

But Eq. (7.210) is of the same form as Eq. (7.37). Thus, taking into account the particular shape of the equivalent distribution function $F_0(v_{//})$, all the low-frequency wave behavior to be described in this section can be treated by using the same techniques used in Section 7.3.

7.4.1. Two-Component Maxwellian Plasma

We consider a plasma composed of electrons and one specie of singly charged, positive ions, with both the ions and electrons being characterized by Maxwellian distribution functions. We leave it as a straight-forward exercise for the reader to show, starting with Eq. (7.210) and using the techniques of the preceding section, that the dielectric function for this plasma can be written as [see Eqs. (7.91) and (7.96)]

$$\varepsilon(k,\omega) = 1 - \frac{k_{de}^2}{k^2} Z^{+\prime}\left(\frac{\omega}{ka_e}\right) - \frac{k_{di}^2}{k^2} Z^{+\prime}\left(\frac{\omega}{ka_i}\right), \tag{7.211}$$

where $a_e = (2KT_e/m_e)^{1/2}$, $a_i = (2KT_i/m_i)^{1/2}$, $k_{de} = \omega_{pe}/a_e$, $k_{di} = \omega_{pi}/a_i$, and $Z^{+\prime}[\omega/(ka_i)]$ and $Z^{+\prime}[\omega/(ka_e)]$ are, respectively, the derivatives of the plasma dispersion functions of the ions and electrons (see Section 7.3). Since $k_{di}^2 = k_{de}^2 T_e/T_i$, Eq. (7.211) can be written as

$$\varepsilon(k,\omega) = 1 - \frac{k_{de}^2}{k^2} \left[Z^{+\prime}\left(\frac{\omega}{ka_e}\right) + \frac{T_e}{T_i} Z^{+\prime}\left(\frac{\omega}{ka_i}\right) \right]. \tag{7.212}$$

When an ion-acoustic wave exists in the present plasma we will find that its phase velocity ω/k is much smaller than the electron thermal velocity a_e. That is, $\xi_e = \omega/k_{de} \ll 1$. Thus, in Eq. (7.212) we can replace $Z^{+\prime}(\xi_e)$ by the first few terms of its expansion in ξ_e. Such an expansion can be easily found, since $Z^+(\xi)$ [and $Z^{+\prime}(\xi)$] is related to the error function by Eq. (7.193). Thus, we can write

$$Z^+(\xi) = i\pi^{1/2}e^{-\xi^2}[1 + \text{erf}(i\xi)]. \tag{7.213}$$

For $\xi \geqslant 0$, $z = i\xi \geqslant 0$, and we can use the series expansion of the error function (Abramowitz and Stegan, (1965), p. 297),

$$\text{erf } z = \frac{2}{\pi^{1/2}}\, e^{-z^2} \sum_{n=0}^{+\infty} \frac{2^n}{1 \cdot 3 \cdot \ldots \cdot (2n-1)}\, z^{2n+1} \tag{7.214}$$

Thus, the plasma dispersion function can be written as

$$Z^+(\xi) = i\pi^{1/2}e^{-\xi^2} - 2\left[\xi - \frac{2}{3}\xi^3 + \frac{(-1)^n 2^n \xi^{2n+1}}{1 \cdot 3 \cdot \ldots \cdot (2n+1)} + \cdots\right]. \tag{7.215}$$

By direct differentiation of Eq. (7.215) we obtain the corresponding series expansion for $Z^{+\prime}(\xi)$ given by

$$Z^{+\prime}(\xi) = -2i\pi^{1/2}\xi e^{-\xi^2} - 2\left[1 - 2\xi^2 + \frac{(-1)^n 2^n \xi^{2n}}{1 \cdot 3 \cdot \ldots \cdot 2n} + \cdots\right]. \tag{7.216}$$

For $\xi \to 0$, we see that $Z^{+\prime}(\xi_e) \simeq -2$. Using this approximation Eq. (7.212) becomes

$$\varepsilon(k,\omega) \simeq 1 + 2\frac{k_{de}^2}{k^2} - \frac{k_{de}^2}{k^2}\frac{T_e}{T_i}Z^{+\prime}\left(\frac{\omega}{ka_i}\right). \tag{7.217}$$

The roots of the equation $\varepsilon(k,\omega) = 0$, where ε is given by Eq. (7.217), have been computed by Sessler and Pearson (1967), for several values of the parameter T_i/T_e. Their results are shown in Fig. 7.7. Observation of this figure shows that the predictions of the theory for the isothermal case ($T_e \simeq T_i$) are very different from those for the case where the electrons are much hotter than the ions ($T_e \gg T_i$). We now want to look a little more closely at these two separate cases.

7.4.2. Isothermal Plasma

Fried and Gould (1961) have shown that the roots of Eq. (7.212) can be described in the complex-ξ plane, as shown in Fig. 7.8. For $T_e \simeq T_i$ this equation has no zeros. The most important contribution comes from the first pole which, in a $H^+ - e^-$ plasma, gives a normalized phase velocity of $\xi_r \simeq 0.035$ and a relative damping rate of $\text{Im}(\xi) = 0.014$. A wave propagation,

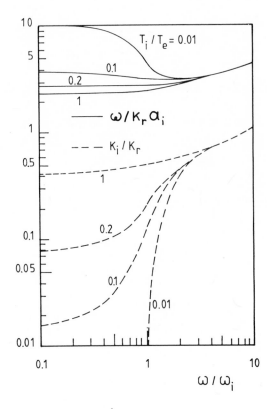

Fig. 7.7. Variation of the roots of the equation $\varepsilon(k,\omega) = 0$, as given by Eq. (7.217), with frequency. Solid line: the phase velocity $\omega/k_r a_i$. Dashed line: the relative damping rate is k_i/k_r. (After Sessler and Pearson, 1967)

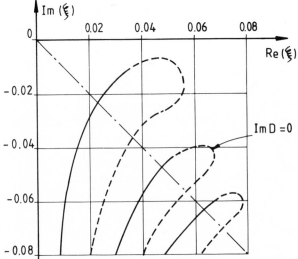

Fig. 7.8. The roots of the dispersion relation $\varepsilon = 0$ for Eq. (7.212) in an isothermal plasma. Solid line: the roots of $\text{Re}(\varepsilon) = 0$. Dotted line: the roots of $\text{Im}(\varepsilon) = 0$. (After Fried and Gould, 1961)

attributed to this pole, was first observed experimentally by Wong, D'Angelo, and Motley (1964) in the cesium plasma of a Q-machine. The details of this and other experiments on ion-acoustic waves are discussed in an excellent book on Q-machines by R. W. Motley (1975) (see, also, Chapter 10 of this book).

7.4.3. $T_e \gg T_i$

When $T_e \gg T_i$, the first pole (see Fig. 7.9) is near the real axis and the damping is small. The dispersion relation for this case can be easily found

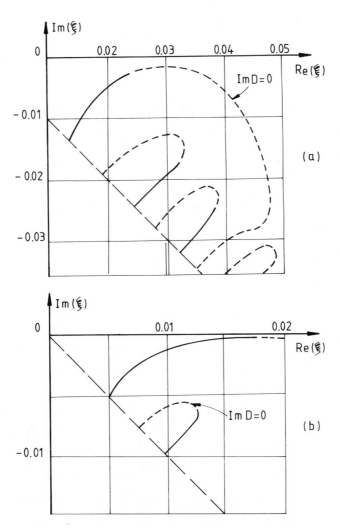

Fig. 7.9. The roots of Eq. 7.212: (a) for $T_e/T_i \simeq 4$; (b) for $T_e/T_i \simeq 25$. (After Fried and Gould, 1961)

since the wave-phase velocity, which is still much smaller than the electron thermal speed a_e, is much larger than the ion thermal speed a_i. Thus, in Eq. (7.212), since $\xi_e \ll 1$, we can make a series expansion for $Z^{+\prime}(\xi_e)$, and since $|\xi_i| \gg 1$, we can make an asymptotic expansion for $Z^{+\prime}(\xi_i)$. Doing this and keeping only the first terms of these two expansions, we leave it as an exercise for the reader to show that

$$Z^{+\prime}(\xi_e) \simeq -2,$$

$$Z^{+\prime}(\xi_i) \simeq \frac{1}{\xi_i^2} + \frac{3}{2\xi_i^4}. \tag{7.218}$$

Substituting these expressions into Eq. 7.212 gives

$$\varepsilon(k,\omega) = 1 + 2\frac{k_{de}^2}{k^2} - \frac{m_e}{m_i}\left(1 + \frac{3KT_i}{m_i}\frac{k^2}{\omega^2}\right). \tag{7.219}$$

Then, remembering that $\xi \ll 1$, the first term on the right-hand side of Eq. (7.219) can be neglected, so that the dispersion relation $\varepsilon = 0$ becomes

$$\frac{\omega^2}{k^2} \simeq \frac{KT_e + 3KT_i}{m_i}. \tag{7.220}$$

Equation (7.220) is the classical ion-acoustic wave velocity found in Chapter 2, using the fluid theory.

For the case of a multi-component plasma, where $T_2 \gg T_i$, the plasma dispersion relation is found still to have the same form

$$\varepsilon(k,\omega) = 1 + 2\frac{k_{de}^2}{k^2} - \sum_{m=1}^{n} \frac{k_{dm}^2}{k^2} Z^{+\prime}(\xi_m) = 0. \tag{7.221}$$

When the wave phase velocity is higher than the thermal velocities of all the ion species (so that negligible wave–particle interactions occur), the dispersion relation given by Eq. (7.221) becomes identical to the corresponding dispersion relation given by the fluid theory. For cold ions, for example, $Z^{+\prime}(\xi_m) \simeq 1/\xi_m^2$, and the dispersion relation given by Eq. (7.221) becomes

$$\varepsilon(k,\omega) = 1 + \frac{2k_{de}^2}{k^2} - \sum_{m=1}^{n} \frac{\omega_{pm}^2}{\omega^2} = 0. \tag{7.222}$$

KINETIC THEORY OF FORCED
OSCILLATIONS IN A
ONE-DIMENSIONAL WARM PLASMA

8.1. INTRODUCTION

In the present chapter we present the classical theory of forced oscillations of an electron gas. This theory, first given by Landau (1946), corresponds to wave excitation by a transparent grid immersed in a uniform plasma and biased at the plasma potential. As in Chapter 7, we will again use the concepts of free-streaming and collective perturbations in the plasma. We will show that, as first pointed out by Hirschfield and Jacobs (1968), the collisionless damping of macroscopic quantities such as electron density, electric field, and potential, is not always due to an energy exchange between waves and particles. In order to demonstrate this point we will compute the kinetic energy deposited in the plasma for the particular case of dipolar excitation.

From the experimental point of view the difference between collisionless damping due to collective effects (usually called Landau damping) and collisionless damping associated with free-streaming effects is important. For example, it is not possible to study nonlinear collective effects such as wave–wave coupling or particle trapping in situations where the collective effects are much smaller than free-streaming effects. Or, to put it another way, when free-streaming effects are large, experimental studies of nonlinear effects will be limited to phenomena such as echoes, for example, which do not include strong collective effects.

8.2. MICROSCOPIC THEORY OF FORCED OSCILLATIONS

Let us consider a stable, collisionless, uniform, one-dimensional, non-magnetically supported plasma composed of warm electrons and a neutralizing background of cold, motionless ions. Let $f_0(v)$ be the uperturbed electron distribution function, where v is the velocity of an electron in the x direction. We want to calculate the electrostatic response (i.e., forced oscillations) of such a plasma to an externally imposed electric field in the x direction. Let us suppose that the electric field is produced by an oscillating electric charge on an antenna immersed in the plasma and that the oscillating charge can be represented by the equation

$$\rho(x,t) = \rho(x) \exp(-i\omega_0 t) \, \Upsilon(t), \tag{8.1}$$

where ω_0 is a real, positive number and $\Upsilon(t)$ is the Heaviside function

$$\Upsilon(t) = \begin{vmatrix} 0, & \text{for } t < 0; \\ 1, & \text{for } t > 0. \end{vmatrix} \tag{8.2}$$

Thus, we want to calculate the electron distribution function perturbation $f_1(x,v,t)$ and the associated electron-current and density perturbations, respectively, $j_1(x,t)$ and $n_1(x,t)$, produced in the plasma by such an electric field.

8.2.1. The Trajectory Method

As shown in Section 8.2.2 this problem can be done using the quite general technique of solving the coupled linearized Vlasov and Poisson equations using Fourier and Laplace transforms. In this section, however, we will use the trajectory method, which gives a somewhat more physical approach. This technique was first used by Landau (1946) in the second part of his now-famous paper.

The electrical external charge ρ and the induced charge in the plasma, associated with the motion of charged particles, produce an electric field $E(x,t)$. We now want to compute the perturbation f_1 produced in the electron distribution function by this electric field. To do this we first calculate the perturbation $\Delta v(x,v,t)$ in the velocity v of an electron in the x direction. This can be easily done if we assume that the perturbations are small enough that we can neglect the production of harmonics and that any transient signals associated with the "turn on" of the external charge $\rho(x,t)$ can be neglected after a time t_0. This allows us to write the electric field produced in the plasma after a time $t > t_0$ as the monochromatic function

$$E(x, t > t_0) = E(x,\omega_0)e^{-i\omega_0 t}. \tag{8.3}$$

Let $v(x(t))$ be the velocity of an electron along its trajectory $x(t)$. We define Δv

$$\Delta v(x,v,t) = v[x(t)] - v[x(-\infty)]. \tag{8.4}$$

Noting that the force on the electron is given by $qE(x,t)$ and integrating with respect to time to obtain the change in momentum of the electron allows us to write the change in velocity $\Delta v(x,v,t)$ as

$$\Delta v(x',v,t) = \frac{q}{m} \int_{-\infty}^{t} E(x',t') \, dt', \tag{8.5}$$

where q and m are, respectively, the charge and the mass of the electron. In fact, Eq. (8.5) also includes the transient motion of the electrons. By restricting our analysis to times $t \gg t_0$, however, we can assume that Eq. (8.5) gives an approximate expression for the monochromatic Δv, i.e., for the Δv whose time behavior is given by ω_0. Thus, we can write

$$\Delta v(x',v,t) \sim \frac{q}{m} \int_{-\infty}^{t} E(x') e^{-i\omega_0 t'} dt'. \tag{8.6}$$

We now make the assumption that the perturbation is small enough to allow the integration of Eq. (8.6) to be made along the unperturbed trajectory. That is, we assume that

$$x' \simeq x + v(t' - t). \tag{8.7}$$

Equation (8.6) then becomes

$$\Delta v(x,v,t) = \delta v(x,v,\omega_0) e^{-i\omega_0 t}, \tag{8.8}$$

where

$$\delta v(x,v,\omega_0) = -\frac{q}{mv} \int_{-\infty}^{x} E(x') e^{i\omega_0(x-x')/v} dx'. \qquad \text{for } v > 0.$$

and

$$\tag{8.9}$$

$$\delta v(x,v,\omega_0) = -\frac{q}{mv} \int_{x}^{\infty} E(x') e^{i\omega_0(x-x')/v} dx'. \qquad \text{for } v < 0.$$

If we now introduce the distribution $B(x,v,\omega_0)$ defined as

$$B(x,v,\omega_0) = \tfrac{1}{2}[\text{sgn}(x) + \text{sgn}(v)] \exp\left(\frac{i\omega_0 x}{v}\right), \tag{8.10}$$

equation (8.9) can be written as the convolution product in space

$$\delta v(x,v,\omega_0) = \frac{q}{mv} E(x) * B(x,v,\omega_0), \tag{8.11}$$

where, in general, the convolution product is defined by [see Eq. (1.31)]

$$h(x) = g(x) * f(x) = \int_{-\infty}^{+\infty} g(x') f(x - x') dx'. \tag{8.12}$$

The perturbation $f_1(x,v,t)$ of the electron distribution function can be computed if we remember that as a consequence of Liouville's theorem the function is a constant along the trajectory. Thus

$$f(x,v,t) = f_0(v - \Delta v(x,v,t)) = f_0(v) + f_1(x,v,t). \tag{8.13}$$

Since Δv is a small quantity, we can make a Taylor expansion for $f_0(v - \Delta v(x,v,t))$, assuming that the distribution function is well behaved (i.e., continuous, all derivatives exist, etc.). Doing this gives

$$f_1(x,v,t) = -\frac{df_0}{dv} \, \delta v(x,v,\omega_0) e^{-i\omega_0 t}. \tag{8.14}$$

Now, from Eq. (8.8) we see that Δv is an oscillating function of t at the frequency ω_0. Thus, the perturbation $f_1(x,v,t)$, is also an oscillating function of t at the same frequency. That is, we can write

$$f_1(x,v,t) = F_1(x,v,\omega_0) \exp(-i\omega_0 t), \tag{8.15}$$

so that, comparing Eqs. (8.14) and (8.15) we can write

$$F_1(x,v,\omega_0) = -\frac{q}{mv} \frac{df_0}{dv} E(x) * B(x,v,\omega_0). \tag{8.16}$$

Knowing the perturbation of the linearized electron distribution function we can compute the linear macroscopic quantities, such as electron density perturbation, current density perturbation, etc., as a function of time t and the dipole frequency ω_0. Thus, we can write

$$n_1(x,t) = n_1(x,\omega_0) \exp(-i\omega_0 t). \tag{8.17}$$

and

$$j_1(x,t) = j_1(x,\omega_0) \exp(-i\omega_0 t), \tag{8.18}$$

where

$$n_1(x,\omega_0) = -\frac{q}{m} \int_{-\infty}^{+\infty} \frac{1}{v} \frac{df_0}{dv} E(x) * B(x,v,\omega_0) \, dv \tag{8.19}$$

and

$$j_1(x,\omega_0) = -\frac{q^2}{m} \int_{-\infty}^{+\infty} \frac{df_0}{dv} E(x) * B(x,v,\omega_0) \, dv. \tag{8.20}$$

To compute the electric field of the forced oscillation in the plasma we can use either Poisson's equation or Maxwell's second equation given, respectively, by

$$\frac{\partial E(x,\omega_0)}{\partial x} = \frac{q}{\varepsilon_0} n_1(x,\omega_0) + \frac{\rho(x)}{\varepsilon_0} \qquad (8.21)$$

and

$$i\omega_0 E(x,\omega_0) = j_0(x,\omega_0) + j_1(x,\omega_0), \qquad (8.22)$$

where $j_0(x,\omega_0)$ is the displacement current of the dipole antenna. Choosing Poisson's equation and using Eq. (8.19) we obtain

$$\frac{\partial}{\partial x} E(x,\omega_0) = -\frac{q^2}{m\varepsilon_0} \int_{-\infty}^{+\infty} \frac{1}{v} \frac{df_0}{dv} E(x) * B(x,v,\omega_0)\, dv + \frac{\rho(x)}{\varepsilon_0}. \qquad (8.23)$$

This integro-differential equation can be easily solved by the use of a Fourier transform in space, the general form of which (see Chapter 1) is given by

$$F(k) = \int_{-\infty}^{+\infty} F(x)e^{-ikx}\, dx. \qquad (8.24)$$

Using the well-known property of the convolution product

$$\int_{-\infty}^{+\infty} [g(x) * f(x)]e^{-ikx}\, dx = g(k)f(k), \qquad (8.25)$$

the transformation of Poisson's equation gives

$$ikE(k,\omega_0) = -\frac{q^2}{m\varepsilon_0} E(k,\omega_0) \int_{-\infty}^{+\infty} \frac{1}{v} \frac{df_0}{dv} B(k,v,\omega_0)\, dv + \frac{\rho(k)}{\varepsilon_0}. \qquad (8.26)$$

But, $B(k,v,\omega_0)$ can be written as

$$B(k,v,\omega_0) = -i\left\{ P\left[\frac{1}{k - \omega_0/v} \right] + i\pi\, \mathrm{sgn}(v)\, \delta\left(k - \frac{\omega_0}{v} \right) \right\}, \qquad (8.27)$$

where P is the principal part distribution. Noting that Eq. (8.27) can be written as

$$B(k,v,\omega_0) = \lim_{v \to 0+} \left[\frac{-i}{k - (\omega_0 + iv)/v} \right], \qquad (8.28)$$

and using this expression in Eq. (8.26) we obtain

$$E(k,\omega_0) = \frac{\rho(k)}{ik\varepsilon_0\varepsilon(k,\omega_0)} = \frac{E_0(k,\omega_0)}{\varepsilon(k,\omega_0)}, \qquad (8.29)$$

where $E_0(k,\omega_0)$ represents the Fourier-transformed excited electric field that would be generated in vacuum, and ε is the relative dielectric constant defined by

$$\varepsilon(k,\omega_0) = 1 + \chi(k,\omega_0), \qquad (8.30)$$

where the susceptibility $\chi(k,\omega_0)$ is given by

$$\chi(k,\omega_0) = -\frac{q^2}{m\varepsilon_0 k^2} \lim_{v\to 0+} \int_{-\infty}^{+\infty} \frac{df_0/dv}{v - (\omega_0 + iv)/k} \, dv. \qquad (8.31)$$

It is of interest to note that

$$\lim_{v\to 0+} \int_{-\infty}^{+\infty} \frac{df_0/dv}{v - (\omega_0 + iv)/k} \, dv = \int_L \frac{df_0/dv}{v - \omega_0/k} \, dv, \qquad (8.32)$$

where L represents the Landau path of integration. This path for real values of ω_0 and k is not used here for the purpose of an analytical continuation but, rather, for the explicit solution of Eq. (8.11), which obeys the causality principle

$$\begin{aligned}
\delta v(x = -\infty, v > 0, \omega_0) &= 0, \\
\delta v(x = +\infty, v < 0, \omega_0) &= 0.
\end{aligned} \qquad (8.33)$$

We can now use the expression of the Fourier transform of the electric field given by Eq. (8.29) in the expressions of the Fourier transforms of the electron distribution function, the electron density, and the electron current density. Doing this we obtain

$$F_1(k,v,\omega_0) = \lim_{v\to 0+} i\left[\frac{q}{mv} \frac{df_0}{dv} \frac{E_0(k,\omega_0)}{[k - (\omega_0 + iv)/v]\varepsilon(k,\omega_0)} \right. \qquad (8.34)$$

$$n_1(k,v,\omega_0) = i\frac{q}{mk} \frac{E_0(k,\omega_0)}{\varepsilon(k,\omega_0)} \int_L \frac{df_0/dv}{v - \omega_0/k} \, dv; \qquad (8.35)$$

$$j_1(k,v,\omega_0) = i\frac{q^2}{mk^2} \omega_0 \frac{E_0(k,\omega_0)}{\varepsilon(k,\omega_0)} \int_L \frac{df_0/dv}{v - \omega_0/k} \, dv. \qquad (8.36)$$

The expression given by Eq. (8.36) for j_1 was found by observing that

$$\int_L \frac{v(df_0/dv)}{v - \omega_0/k} \, dv = \int_L \frac{v(df_0/dv)}{v - \omega_0/k} \, dv - \int_{-\infty}^{+\infty} df_0/dv \, dv$$

$$= \frac{\omega_0}{k} \int_L \frac{df_0/dv}{v - \omega_0/k} \, dv. \qquad (8.37)$$

Equations (8.34)–(8.36) are the sought-after equations, giving the plasma response to an antenna immersed in the plasma and neglecting the transient signals after a time $t > t_0$.

8.2.2. The Fourier–Laplace Method

The preceding problem can be solved directly by making Fourier and Laplace transforms on the set of coupled linearized Vlasov and Poisson equations given by

$$\frac{\partial f_1}{\partial t} + v \frac{\partial f_1}{\partial x} + \frac{q}{m} E \frac{df_0}{dv} = 0, \tag{8.38}$$

$$\frac{\partial E}{\partial x} = \frac{q}{\varepsilon_0} \int_{-\infty}^{+\infty} f_1 \, dv + \frac{\rho(x)}{\varepsilon_0} e^{-i\omega_0 t} \Upsilon(t). \tag{8.39}$$

The Fourier-transform recipe is given by Eq. (8.24) and the Laplace transform recipe (see Chapter 1) is given by

$$F(\omega) = \int_0^\infty F(t) e^{i\omega t} \, dt, \qquad \text{Im}(\omega) > a. \tag{8.40}$$

Assuming that the initial perturbation $f_1(x, v, t = 0)$ is zero (see Chapter 1) we obtain

$$f_1(k, v, \omega) = -\frac{iq}{m} \frac{df_0}{dv} \frac{E(k, \omega)}{\omega - kv}, \tag{8.41}$$

$$E(k, \omega) = \frac{i\mathbf{E}(k, \omega)}{\omega - \omega_0}, \tag{8.42}$$

where

$$\mathbf{E}(k, \omega) = \frac{\rho(k)}{ik\varepsilon_0 \varepsilon(\omega, k)}, \tag{8.43}$$

and,

$$\varepsilon(\omega, k) = 1 - \frac{q^2}{m\varepsilon_0 k^2} \int_L \frac{df_0/dv}{v - \omega/k} \, dv. \tag{8.44}$$

In Eqs. (8.41) and (8.42), the inversion of the Laplace transforms can be made, using the theorem of residues and Jordan's lemma when $f_0(v)$ is either an entire function or a meromorphic function of v. The only contributions to the forced oscillations for large values of time are due to the two real poles at $\omega = \omega_0$ and $\omega = kv$. In fact, the singularities of $E(k,\omega)$ are the roots of $\varepsilon(\omega,k)$. All such roots must have a negative real part for the case of stable plasmas. These poles have residues that are damped solutions with respect to time, so that

$$f_1(k, v, t \to +\infty) = -i\frac{q}{m}\frac{df_0}{dv}\left[\frac{E(k,\omega_0)}{\omega_0 - kv} + \lim_{t \to +\infty}\frac{E(k,kv)}{(kv - \omega_0)}e^{-i(kv - \omega_0)t}\right]e^{-i\omega_0 t}.$$
(8.45)

Noting that

$$\lim_{t \to +\infty}\frac{e^{-i(kv - \omega_0)t}}{kv - \omega_0} = -i\pi\delta(kv - \omega_0),$$
(8.46)

we obtain

$$f_1(k, v, t \to +\infty) = F_1(k,v,\omega_0)e^{-i\omega_0 t},$$
(8.47)

where

$$F_1(k,v,\omega_0)e^{-i\omega_0 t} = -i\frac{q}{m}\frac{df_0}{dv}\left[P\left(\frac{1}{\omega_0 - kv}\right) - \pi\delta(kv - \omega_0)\right]E(k,\omega_0)$$

$$= -\frac{q}{mv}\frac{df_0}{dv}E(k,\omega_0)B(k,v,\omega_0).$$
(8.48)

Finally, we obtain the perturbation of the distribution function

$$F_1(k,v,\omega_0) = \lim_{v \to 0+} i\frac{q}{mv}\frac{df_0}{dv}\frac{E(k,\omega_0)}{k - (\omega_0 + iv)/v},$$
(8.49)

The electric field is given by

$$E(k, t > +\infty) = E(k,\omega_0)\exp(-i\omega_0 t),$$
(8.50)

where

$$E(k,\omega_0) = \frac{E_0(k,\omega_0)}{\varepsilon(k,\omega_0)},$$
(8.51)

Upon comparing Eqs. (8.49) and (8.51) with Eqs. (8.34) and (8.29), we see that the trajectory method and the Fourier–Laplace method give the same results.

8.3. DIFFICULTIES ENCOUNTERED IN THE FORCED-OSCILLATIONS PROBLEM

We must now inverse the Fourier transforms of the perturbations using the inversion recipe (see Chapter 1)

$$F(x) = \frac{1}{2\pi} \int_{-\infty}^{+\infty} F(k)e^{ikx}\, dk. \tag{8.52}$$

This equation, which looks like the inversion equation used for the Laplace transform, suggests an evaluation of the integral by use of the theory of residues. This requires an analytic continuation of $F(k)$ because the Fourier transforms that we have found [see Eqs. (8.29) and (8.34)–(8.36)] are *a priori* defined only for real values of ω_0 and k. We must, thus, study the analytic properties of $F(k)$. For this purpose, we look first at the properties of the function $H(\omega_0/k)$ given by

$$H\left(\frac{\omega_0}{k}\right) = \int \frac{df_0/dv}{v - \omega_0/k}\, dv. \tag{8.53}$$

If $f_0(v)$ is an entire or meromorphic function of v, the above equation corresponds to two different entire or meromorphic functions of (ω_0/k), depending on the sign of the real part of k, given by

$$H\left(\frac{\omega_0}{k}\right) = H^-\left(\frac{\omega_0}{k}\right)\Upsilon(-k) + H^+\left(\frac{\omega_0}{k}\right)\Upsilon(k), \tag{8.54}$$

where

$$H^\pm\left(\frac{\omega_0}{k}\right) = \int_{L^\pm} \frac{df_0/dv}{v - \omega_0/k}\, dv, \tag{8.55}$$

and $L^\pm$ represent the Landau paths for positive and negative values of k (see Fig. 8.1). Therefore Eq. (8.52) can be written as

$$F(x) = \frac{1}{2\pi} \int_{-\infty}^{0} F^-(k)e^{ikx}\, dk + \frac{1}{2\pi} \int_{0}^{\infty} F^+(k)e^{ikx}\, dk. \tag{8.56}$$

The functions $F^\pm(k)$ can be analytically continued in the complex-k plane, since the functions $H^\pm(\omega_0/k)$ are entire or meromorphic functions. We find here, however, that the displacement of the path of integration in Eq. (8.56) introduces a "branch cut" in the complex-k plane. Thus, the calculation of the integral in Eq. (8.52) cannot be made by making a simple sum of the

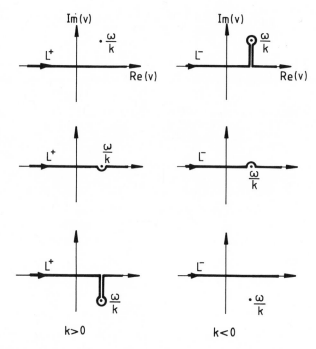

Fig. 8.1. Landau integration paths $L^{\pm}$ as a function of $\mathrm{Im}(\omega/k)$. L^{+} is used for the analytic continuation of Hilbert transforms for $k > 0$, while L^{-} is used, correspondingly, for $k < 0$.

residues. This causes the problem of forced oscillations to be a much more difficult problem to solve than the initial-value problem solved in the previous chapter without inversion of the Fourier transforms. We will return to this difficulty in Chapter 9. For the moment we want to discuss the physical interpretation of Landau damping by comparing the collective and free-streaming aspects of the charge perturbations.

8.4. FREE-STREAMING AND COLLECTIVE EFFECTS

One of the important properties of Maxwell's equations is that they are *linear* equations. The total electric field is, thus, the sum of the exciting field $E_0(x,\omega_0) \exp(-i\omega_0 t)$ and the induced field $E_1(x,\omega_0 t) \exp(-i\omega_0 t)$ produced by the electric charges in the plasma. The exciting field is related to the external charges by the equation [see Eq. (8.29)]

$$E_0(k,\omega_0) = \frac{\rho(k)}{ik\varepsilon_0}. \tag{8.57}$$

Moreover, since we have a uniform plasma we have been able to make Fourier transforms of the perturbations. We will now show that $E_1(k,\omega_0)$ is proportional to $E_0(k,\omega_0)$. For this purpose we first note that

$$\frac{1}{\varepsilon(k,\omega_0)} = 1 - \frac{\chi(k,\omega_0)}{\varepsilon(k,\omega_0)}, \tag{8.58}$$

and from Eq. (8.29)

$$E(k,\omega_0) = E_0(k,\omega_0) + E_1(k,\omega_0), \tag{8.59}$$

where

$$E_1(k,\omega_0) = -E_0(k,\omega_0) \frac{\chi(k,\omega_0)}{\varepsilon(k,\omega_0)}. \tag{8.60}$$

Using Eq. (8.59) to replace the total field in terms of the source field and the induced field, Eqs. (8.34)–(8.36) can be written as

$$F_1 = F_{1L} + F_{1C}, \tag{8.61}$$

$$n_1 = n_{1L} + n_{1C}, \tag{8.62}$$

$$j_1 = j_{1L} + j_{1C}, \tag{8.63}$$

where

$$F_{1L} = \lim_{v \to 0+} i \frac{q}{mv} \frac{df_0}{dv} \frac{E_0(k,\omega_0)}{k - (\omega_0 + iv)/v}; \tag{8.64}$$

$$F_{1C} = \lim_{v \to 0+} i \frac{q}{mv} \frac{df_0}{dv} \frac{E_1(k,\omega_0)}{k - (\omega_0 + iv)/v}; \tag{8.65}$$

$$n_{1L} = i \frac{q}{mk} E_0(k,\omega_0) \int_L \frac{df_0/dv}{v - \omega_0/k} dv; \tag{8.66}$$

$$n_{1C} = i \frac{q}{mk} E_1(k,\omega_0) \int_L \frac{df_0/dv}{v - \omega_0/k} dv; \tag{8.67}$$

$$j_{1L} = -i\omega_0\chi(k,\omega_0)E_0(k,\omega_0); \tag{8.68}$$

$$j_{1C} = -i\omega_0\chi(k,\omega_0)E_1(k,\omega_0). \tag{8.69}$$

In Eqs. (8.61)–(8.69), F_{1L}, n_{1L}, and j_{1L} represent the free-streaming terms of the perturbations produced by the source while F_{1C}, n_{1C}, and j_{1C} represent the collective effects of the plasma. If the induced electric field is negligible, then the perturbations reduce to the free-streaming terms.

The presentation of each of the perturbations as the sum of separate collective and free-streaming terms is purely arbitrary since, obviously, we can indefinitively apply the result of Eq. (8.55) to obtain

$$\frac{1}{\varepsilon(k,\omega_0)} = 1 - \chi(k,\omega_0) + \chi^2(k,\omega_0) + \cdots \tag{8.70}$$

It is interesting to note from Eqs. (8.31), (8.60), and (8.66) that the Fourier transform of the source field can be given as a function of the Fourier transform of the induced density perturbation by

$$E_1(k,\omega_0) = \frac{qn_{1L}}{ik\varepsilon_0\varepsilon(k,\omega_0)} \tag{8.71}$$

Thus, the induced field can be computed using the free-streaming perturbation of the distribution function produced by the external charges as a source term.

8.4.1. Damping Associated with Free-Streaming—Pseudowaves

Let us consider for simplicity a source term $\rho(x)$ that represents an electric dipole whose electric field is described by the equation

$$E_0(x,\omega_0) = V_0\delta(x). \tag{8.72}$$

From Eq. (8.64), after doing a Fourier transform on Eq. (8.72), we find

$$F_{1L} = \lim_{v\to 0+} i\frac{qV_0}{mv}\frac{df_0}{dv}\frac{1}{k - (\omega_0 + iv)/v}. \tag{8.73}$$

Then, using Eq. (8.28) and (8.10), the inversion of the Fourier transform of Eq. (8.73) gives

$$F_{1L}(x,v,\omega_0) = -\frac{qV_0}{mv}\frac{df_0}{dv}\frac{1}{2}[\text{sgn}(x) + \text{sgn}(v)]e^{i\omega_0 x/v}. \tag{8.74}$$

If we consider positive distances $x > 0$, we then have

$$n_{1L}(x > 0, \omega_0) = -\frac{qV_0}{m}\int_0^\infty \frac{1}{v}\frac{df_0}{dv}e^{i\omega_0 x/v}\,dv. \tag{8.75}$$

For the case of a Maxwellian electron distribution function, Eq. (8.75) gives the *free-streaming* density perturbation as

$$n_{1L}(x > 0, \omega_0) = n_0 \frac{qV_0}{KT} I(z),$$ (8.76)

where T and K are, respectively, the electron temperature and Boltzmann's constant, and

$$I(z) = \frac{1}{\pi^{1/2}} \int_0^\infty e^{-u^2} e^{iz/u} \, du$$ (8.77)

and

$$u = \frac{v}{a}; \qquad a^2 = \frac{2KT}{m}; \qquad z = \frac{\omega_0 x}{a}$$ (8.78)

In Eq. (8.77) the function $\exp(iz/u)$ is oscillatory when $z \to \infty$, so that $I(z) \to 0$. Thus, even if the distribution perturbation $F_{1L}(x,v,\omega_0)$ does not disappear, the macroscopic quantity, namely, the electron density perturbation, is damped and vanishes. Such a damping is not the result of some exchange of energy between the wave and the particles. This damping is the result of phase mixing (see Sections 7.1, 7.2).

The integral $I(z)$ is a well-known function that appears frequently in the theory of thermal-neutron absorption (Laporte, 1937; Zahu, 1937; Torrey, 1941; Kruse and Ramsey, 1951). In the original paper by Landau (1946) a somewhat similar integral was used, which we will discuss in the following chapter.

In general, the integral $I(z)$ is evaluated from its asymptotic expansion obtained by the so-called saddle-point technique (Morse and Feshbach, 1953; Budden, 1966). For the integral

$$I_m(z) = \frac{1}{\pi^{1/2}} \int_0^\infty u^m e^{-u^2 + iz/u} \, du, \qquad m = 0, 1, \ldots$$ (8.79)

this gives (Abramowitz and Stegun, 1965)

$$I_m(z) \sim \frac{\zeta_0^m}{3^{1/2}} e^{-3\xi_0^2} \sum_{m=1}^\infty \frac{a_n}{(3\xi_0^2)^n},$$ (8.80)

where

$$\xi_0 = \left(\frac{z}{2}\right)^{1/3} \exp\left(\frac{-i\pi}{6}\right),$$ (8.81)

$$a_0 = 1, \qquad a_1 = \left(\frac{1}{12}\right)(3m^2 + 3m - 1),$$ (8.82)

and in general we have a recurrence relation given by

$$12(n + 2)a_{n+2} = -(12n^2 + 36n - 3m^2 - 3m + 25)a_{n+1}$$
$$+ \cdots \tfrac{1}{2}(m - 2n)(2n + 3 - m)(2n + 3 + 2m)a_n. \quad (8.83)$$

The numerical evaluation of the integral in Eq. (8.77) shows that for $z > 5$ the asymptotic expansion, when limited to only the first term, is a very good approximation. Thus, we can write

$$I(z > 5) \approx \frac{e^{-3\xi_0^2}}{\sqrt{3}} = \frac{1}{\sqrt{3}} \exp\left[-\frac{3}{2}\left(\frac{z}{2}\right)^{2/3}(1 - i\sqrt{3}) \right]. \quad (8.84)$$

Equation (8.84) shows that $I(z)$ can be *locally* represented by a *pseudo-wave* having the mathematical form $\exp[ik(x)x]$, where the real and imaginary parts of the wave number are given by

$$k_r(x) = \frac{\partial\varphi}{\partial x}; \qquad k_i(x) = -\frac{1}{A}\frac{\partial A}{\partial x}, \quad (8.85)$$

with A and φ being real and being defined by the equation

$$I(z) = A(z)e^{i\varphi(z)}. \quad (8.86)$$

Applying Eqs. (8.85) and (8.86) to Eq. (8.84), we find

$$k_r(x) = \sqrt{3}\left(\frac{\omega_0}{2a}\right)^{2/3}x^{-1/3}; \quad (8.87)$$

$$k_i(x) = \left(\frac{\omega_0}{2a}\right)^{2/3}x^{-1/3}. \quad (8.88)$$

From Eqs. (8.87) and (8.88) we can compute the *phase velocity* and *damping rate* of the pseudowaves as

$$v_\varphi(x) = \frac{\omega_0}{k_r(x)} = \frac{1}{\sqrt{3}}(2a)^{2/3}(\omega_0 x)^{1/3} = \frac{2^{2/3}}{\sqrt{3}}a\left(\frac{\omega_0 x}{a}\right)^{1/3}; \quad (8.89)$$

$$\frac{k_i(x)}{k_r(x)} = \frac{1}{\sqrt{3}}. \quad (8.90)$$

From Eq. (8.89) we can see that the phase velocity is a slowly varying function of *distance* for $\omega_0 x/a \gg 1$. Thus, the local definitions of the wave number given above have some meaning. A condition can be given for the *wave picture* to be useful, corresponding to the adiabaticity condition

$$\left| \frac{1}{k_r} \frac{\partial k_r}{\partial x} \right| \ll \frac{k_r}{2\pi}. \tag{8.91}$$

Thus, when collective effects are negligible, the free-streaming perturbation produced by the source produces a macroscopic signal that can be locally represented as a damped wave. This phenomenon was given the name *linear pseudowave* by Doucet and Gresillon (1970), who observed a linear, weakly-collective perturbation resembling the theoretical description given by Sessler and Pearson (1967). It is to be noted that this phenomenon is different from the nonlinear pseudowave introduced by Alexeff *et al.* (1968), which has a phase velocity that varies with the amplitude of the voltage signal applied to the excitation grid.

The free-streaming phenomenon described above is frequently described in the literature using the adjective "ballistic." However, by some language abuse the same terminology has been used to categorize the asymptotic macroscopic perturbations that still exist at large distances, even when collective effects cannot be neglected. For collective effects, we will show that we can introduce a wave number locally, as we did for free-streaming effects. Quite understandably, the lack of agreement on the exact terminology to be used in describing free-streaming effects has caused some confusion in the literature.

8.4.2. Damping Associated with Collective Effects—Asymptotic Perturbations

Using the same dipolar excitation as used above we will now study the density perturbations produced at *large* distances from the source, but in this treatment we will introduce the collective effects.

Using the trajectory method we can write [see Eq. (8.9)]

$$\delta v(x \to +\infty, v > 0, \omega_0) \sim \frac{q}{mv} \int_{-\infty}^{+\infty} E(x',\omega_0) e^{i\omega_0(x-x')/v} dx', \tag{8.92}$$

where $\delta v(x \to +\infty, v < 0, \omega_0) \sim 0$. But

$$\int_{-\infty}^{+\infty} E(x',\omega_0) e^{-i\omega_0 x'/v} dx' = E(k = \omega_0/v, \omega_0). \tag{8.93}$$

Thus, using Eqs. (8.14), (8.29), (8.72), and (8.93)

$$f_1(x \to +\infty, v, \omega_0) = -\frac{qV_0}{mv} \frac{df_0}{dv} \frac{e^{i\omega_0 x/v} \Upsilon(v)}{\varepsilon(k = \omega_0/v, \omega_0)}. \tag{8.94}$$

The asymptotic density perturbation is, therefore, given by

$$n_1(x \to +\infty, \omega_0) = -\frac{qV_0}{m} \int_0^\infty \frac{1}{v} \frac{df_0}{dv} \frac{e^{i\omega_0 x/v} \, dv}{\varepsilon(k = \omega_0/v, \omega_0)}. \tag{8.95}$$

Equation (8.95) shows that if we neglect the collective effects by assuming that the plasma susceptibility is zero, i.e., that $\chi(k,\omega_0) = 0$ or $\varepsilon(k,\omega_0) = 1$, the asymptotic density perturbation is again the free-streaming one [see Eq. (8.75)].

Let us consider the case of a Maxwellian distribution. Using the reduced variables given in Eq. (8.78) we can write Eq. (8.95) as

$$n_1(z \gg 1, \omega_0) = \frac{n_0 q V_0}{KT} \frac{1}{\pi^{1/2}} \int_0^\infty \frac{e^{-u^2 + iz/u}}{\varepsilon(\omega_0, u)} \, du. \tag{8.96}$$

The integral appearing in Eq. (8.96) can be calculated—again using the saddle-point technique—to give

$$n_1(z \gg 1, \omega_0) \sim n_0 \frac{qV_0}{KT} \frac{1}{\sqrt{3}} \frac{e^{-3\xi_0^2}}{\varepsilon(\omega_0, \xi_0)}, \tag{8.97}$$

where

$$\xi_0 = \left(\frac{z}{2}\right)^{1/3} \exp\left(\frac{-i\pi}{6}\right). \tag{8.98}$$

The expression given by Eq. (8.97) is correct only when the variations of $\varepsilon(\omega_0, u)$ are small around the saddle point. This condition is fulfilled for $z \gg 1$, for which case

$$\varepsilon(\omega_0, \xi_0) \sim 1 - \frac{\omega_p^2}{\omega_0^2} = \varepsilon_p. \tag{8.99}$$

We obtain then the asymptotic expression

$$n_1(z \gg 1, \omega_0) \simeq n_0 \frac{qV_0}{KT} \frac{\exp(-3\xi_0^2)}{\sqrt{3}\,\varepsilon_p} \simeq n_{1L}(z \gg 1, \omega_0) \left[1 + \frac{\omega_p^2/\omega_0^2}{1 - \omega_p^2/\omega_0^2}\right]. \tag{8.100}$$

Landau (1946) was the first to do a treatment of this problem similar to that given above. Many other researchers, such as Feix (1964), Gould (1964), Sessler and Pearson (1967), Hirshfield and Jacobs (1968), and so on, have treated this problem for slightly different physical situations.

A very interesting aspect of Eq. (8.100) is that no mode corresponding to any root of the dielectric constant appears. From Eq. (8.93) it follows that the expression given in Eq. (8.100) represents the result of the free-streaming perturbation of the distribution function, such perturbation being produced not only by the source excitation field $E_0(x,\omega_0)$, but also by the induced field (collective effect) $E_1(x,\omega_0)$. Thus, the asymptotic perturbation can be understood to be the result of the superposition of all the Landau modes, which modes are infinite in number for the case of a Maxwellian plasma.

The mathematical complexity of the dielectric constant in kinetic theory arises from two contradictory aspects of the plasma behavior: (1) The collective and fluid aspect (described by fluid theory), which is included in the first Landau pole contribution of the kinetic theory; and (2) The "neutral gas" aspect, corresponding to the free-streaming phenomenon. This latter aspect is difficult to represent by a set of discrete modes, even though the Landau technique, in principle, allows this to be done. In fact, it can be seen from Eq. (8.100) that for $\omega_0 \gg \omega_p$, the plasma behavior is similar to that of a neutral gas, so that the Landau poles no longer have any individual meaning and no longer represent the collective effects.

8.5. PHYSICAL MEANING OF LANDAU DAMPING

The collisionless damping of macroscopic perturbations is usually presented as an unusual phenomenon resulting from an energy exchange between a wave and the charged particles of the plasma in which the wave is propagating. Such a presentation is not correct because it suggests that the suppression of energy exchange between the wave and the particles would eliminate the damping. We have seen, however, that this is not true, since the free-streaming phenomenon just described was found to be strongly damped without any wave-particle energy exchange being considered. What is more surprising, perhaps, is the collective effect itself: A real wave, with coherent-particle motion, arises from the free-streaming, the free-streaming being an incoherent motion of the particles resulting in phase mixing. We will now study how the random kinetic energy of the particles can be transformed into coherent motion, the total energy being, alternatively, either kinetic or electrical. In this transformation, the damping is related to the energy exchange between the wave and the particles and is, quite generally, *smaller* than the free-streaming damping. It is possible, even, to have an instability if the amount of work done by the electric field of the wave on the particles is negative.

The wave arising from the free-streaming is quite a complicated mechanical problem, particularly in the case of forced oscillations. We will not try to describe it but, rather, will limit our effort to correctly describing the energy balance. In this calculation, the plasma acts like a spring that is first made to oscillate before exhibiting an increase in temperature.

8.5.1. Kinetic Energy in a Homogeneous One-Dimensional Plasma

The variation of $W_k(x,t)$, the kinetic energy of the particles at the position x between the time t and the time $t = -\infty$ is given by

$$\delta W_k(x,t) = \frac{m}{2}\left[\int_{-\infty}^{+\infty} f(w,x,t)w^2 dw - \int_{-\infty}^{+\infty} f_0(v)v^2 \, dv\right], \quad (8.101)$$

where

$$w = v + \Delta v(x,v,t), \quad (8.102)$$

with v being the initial velocity of an electron, $f_0(v)$ the unperturbed electron distribution function, and Δv the velocity increase of the electron at position x and time t with the new velocity w. From Liouville's theorem, $f(x,w,t) = f_0(v)$, so that

$$\delta W_k(x,t) = \frac{m}{2}\left[\int_{-\infty}^{+\infty} f_0(v)(v + \Delta v)^2\left(1 + \frac{\partial}{\partial v}\Delta v\right)dv - \int_{-\infty}^{+\infty} f_0(v)v^2 \, dv\right].$$
$$(8.103)$$

Doing an integration by parts we obtain

$$\delta W_k(x,t) = -\frac{m}{2}\int_{-\infty}^{+\infty}\frac{df_0}{dv}\left[v^2\,\Delta v + v\,\Delta v^2 + \frac{\Delta v^3}{3}\right]dv \quad (8.104)$$

We now consider a forced-oscillations problem, such that

$$\Delta v = \lambda\delta v_1\cos(\omega_0 t + \phi_1) + \lambda^2\delta v_2\cos(2\omega_0 t + \phi_2) + \cdots, \quad (8.105)$$

and try to calculate the average variation of the kinetic energy of the particles over one complete cycle of oscillation at the dipole frequency ω_0, given by

$$\langle\delta W_k(x)\rangle = \frac{\omega_0}{2\pi}\int_0^{2\pi/\omega_0}\varepsilon W_k(x,t)\,dt. \quad (8.106)$$

With an error of only the fourth order in λ, the energy perturbation can be expressed in terms of only the linear-velocity perturbation as

$$\langle\delta W_k(x)\rangle = -\frac{m}{2}\int_{-\infty}^{+\infty}\frac{df_0}{dv}(\delta v_1)^2 v\,dv, \quad (8.107)$$

where

$$\delta v_1^2 = \delta v(x,\omega_0) \; \delta v^*(x,\omega_0), \tag{8.108}$$

and δv^* is the complex conjugate of δv the velocity perturbation given by Eq. (8.11).

A discussion of stability is outside the scope of this book, but we can see directly from Eq. (8.107) that any distribution function having a single maximum (often called a bell-shaped distribution function) is stable. To do that one needs only to compute $\langle \delta W_k(x) \rangle$ in a frame moving with velocity v_0 such that $(df/dv) = 0$ for $v = v_0$, in order to show that $\langle \delta W_k(x) \rangle > 0$. The kinetic energy being positive is sufficient for energy conservation to imply stability.

8.5.2. Energy Density Deposited in a Plasma Excited by a Dipole

The integral in Eq. (8.107) is difficult to evaluate for arbitrary values of x. However, for the case $x \to \infty$ the problem is not diffcult. As an example, we will compute $\langle \delta W_k(x) \rangle$ for the case of dipolar excitation for $\omega_0 > \omega_{pe}$.

Starting from Eq. (8.107) and using Eqs. (8.9) and (8.11) we have

$$\delta v(x \to \infty, v, t) = \frac{qV_0}{mv} \frac{e^{i\omega_0 x/v}}{\varepsilon^+(\omega_0/v,\omega_0)}, \tag{8.109}$$

and for the energy density

$$\langle \delta W_k(x \to \infty) \rangle = -\frac{n_{0e}qV_0}{4} \frac{qV_0}{m} \frac{\omega_0^2}{\omega_{pe}^2} \frac{1}{\pi} \int_0^\infty \frac{1}{v^3} \mathrm{Im}\left[\frac{1}{\varepsilon^+(\omega_0/v,\omega_0)}\right] dv, \tag{8.110}$$

which, for a Maxwellian electron distribution becomes

$$\langle \delta W_k(x \to \infty) \rangle = +\frac{n_{0e}qV_0}{8} \frac{qV_0}{KT_e} \frac{1}{\pi^{1/2}} \int_0^\infty \left[\frac{e^{-u^2}}{|\varepsilon^+(f,u)|^2}\right] du, \tag{8.111}$$

where $f = \omega_0/\omega_{pe}$, the normalized frequency.

In a way similar to the technique introduced in Section 9.2.4, we can write a pseudo-causal relationship (see Chapter 9)

$$\frac{1}{\pi} \int_0^\infty \mathrm{Im}[u^3|\varepsilon^+(f,u)|]^{-1} dv = -\frac{1}{f^2} - \frac{C_0}{u_0^2} \tag{8.112}$$

where C_0 is given by Eq. (9.42) in which $\xi_0 = u_0$, the root of $\varepsilon^+(f,u)$ in the upper half-plane $\mathrm{Im}(u) > 0$. Introducing the expression given by Eq. (8.112) into Eq. (8.110) we get

$$\langle \delta W_k(x \to \infty) \rangle = + \frac{n_{0e}qV_0}{8} \frac{qV_0}{KT_e} \left(1 + \frac{C_0 f^2}{u_0^2} \right). \qquad (8.113)$$

Then, $\langle \delta W_k \rangle$ for the case of $\omega_0 > \omega_{pe}$ and $x \to \infty$ can be expressed as

$$\langle \delta W_k(x \to \infty) \rangle = + \frac{n_{0e}qV_0}{8} \frac{qV_0}{KT_e}. \qquad (8.114)$$

   This energy density for dipolar excitation is the same as the energy density obtained from Eq. (8.111) for free-streaming with $\varepsilon^+ = 1$. Thus, for $\omega_0 > \omega_{pe}$ we write

$$\langle \delta W_k(x \to +\infty) \rangle = \langle \delta W_k^*(x > 0) \rangle, \qquad (8.115)$$

where

$$\langle \delta W_k^* \rangle = \frac{n_0 qV_0}{4} \frac{qV_0}{KT_e} \qquad (8.116)$$

is the energy density deposited in the plasma when there is no collective effect.

   Thus, for frequencies higher than the plasma frequency, at distances large enough to neglect the induced electric field the energy density deposited in the plasma is independent of collective effects. The coherent plasma motion associated with the collective effect just delays the time required for the energy deposited to appear permanently as kinetic energy: As long as the induced field is not yet damped, the energy alternates between being kinetic energy and electrical energy. For frequencies below the plasma frequency, however, it can be shown that the situation is different, because of a plasma-shielding effect (Debye shielding) which appears and limits the energy deposition (Buzzi, 1974).

COMPUTING TECHNIQUES FOR ELECTROSTATIC PERTURBATIONS

9.1. INTRODUCTION

In Chapter 8 we showed that the plasma response to forced oscillations is given by a Fourier transform which, in the general case, can be evaluated only by numerical methods. In this chapter we discuss three numerical techniques for calculating electrostatic perturbations in a plasma.

The first method is the Gould technique, which starts from the integral form of the plasma response given by Landau and solves the problem numerically for the case of ion-acoustic waves. The second technique is the Derfler–Simonen technique, which, in principle, is more basic than the Gould technique, but which, in practice, presents some difficulties. We show that these two methods are, in fact, complementary. Finally, we present a hybrid technique by Buzzi which is a combination of the other two methods that avoids the difficulties of those methods.

9.2. DIELECTRIC CONSTANT OF A MAXWELLIAN ELECTRON CLOUD

Before discussing the numerical methods, we make a short study of the properties of the dielectric constant to complement the study already found in Chapter 8. We consider only the case of a boundary-value problem (complex k, real ω) in a "Maxwellian electron cloud," i.e., a plasma consisting of cold ions and Maxwellian electrons. Most of these properties are well known except, perhaps, the *pseudo-causality* relationships and the relationships existing between the excitation coefficients presented in Section 9.2.4.

9.2.1. Dielectric Constant

Since we have already discussed in some detail the derivation of the dispersion relation and the dielectric constant for a Maxwellian plasma (see Section 7.3.5) the first part of the following discussion is repeated here for the convenience of the reader.

We use the reduced variables of Chapter 7 [see Eq. (7.82)]

$$u = \frac{v}{a}; \qquad \xi = \frac{\omega_0}{ka};$$

$$k_d^2 = \frac{\omega_p^2}{a^2}; \qquad \varphi_0 = \frac{f_0(v)}{n_0},$$

(9.1)

where $\omega_p^2 = n_0 q^2 / \varepsilon_0 m$, a is the electron thermal velocity, and the reduced electron distribution function φ_0 is normalized so that

$$\int_{-\infty}^{+\infty} \varphi_0(u)\, du = 1.$$

(9.2)

Letting $a = (2KT/m)^{1/2}$, the reduced Maxwellian electron distribution function can be written as

$$\varphi_0(u) = \pi^{-1/2} \exp(-u^2).$$

(9.3)

The relative dielectric constant $\varepsilon(k,\omega_0)$ can then be shown to be given by [see Eq. (7.189)]

$$\varepsilon(k > 0, \omega_0) = \varepsilon^+(k,\omega_0) = 1 - \frac{k_d^2}{k^2} Z^{+\prime}(\xi),$$

$$\varepsilon(k < 0, \omega_0) = \varepsilon^-(k,\omega_0) = 1 - \frac{k_d^2}{k^2} Z^{-\prime}(\xi),$$

(9.4)

where

$$Z^{\pm\prime}(\xi) = \frac{d}{d\xi} Z^\pm(\xi); \qquad Z^\pm(\xi) = \frac{1}{\pi^{1/2}} \int_{L^\pm} \frac{e^{-u^2}}{u - \xi}\, du.$$

(9.5)

The function $Z^+(\xi)$ is the plasma dispersion and has been numerically calculated by Fried and Conte (1961). Using the properties of Hilbert transforms, the plasma dispersion function can be represented by the error function with a complex argument as [see Eq. (7.193)]

$$Z^+(\xi) = i\pi^{1/2} e^{-\xi^2}[1 + \text{erf}(i\xi)],$$

(9.6)

where (see Abramowitz and Stegun, 1965, p. 297)

$$\text{erf}(v) = \frac{2}{\pi^{1/2}} \int_0^v e^{-t^2}\, dt.$$

(9.7)

Introducing further reduced quantities defined by

$$f = \frac{\omega_0}{\omega_p} \qquad \eta = \frac{ka}{\omega_0} = 1/\xi,$$

(9.8)

we can write

$$\varepsilon^{\pm}(f,\xi) = 1 - \frac{\xi^2}{f^2} Z^{\pm\prime}(\xi) = 1 - (f^2\eta^2)^{-1} Z^{\pm\prime}\left(\frac{1}{\eta}\right). \tag{9.9}$$

From this equation, $\varepsilon^{\pm}(f,\xi)$ is seen to be an entire function of ξ having only one singular point, at infinity, which is an essential singularity [coming from the term $\exp(-\xi^2)$]. In the k (or η) plane, $\varepsilon^{\pm}$ is thus an analytic function everywhere, except at the point $k = 0$ ($\eta = 0$), which is an essential singularity. Further, one can verify that for $|k| \to 0$ (for which case $|\xi| \to \infty$)

$$\varepsilon^+ \to 1 - \frac{\omega_p^2}{\omega_0^2}, \qquad \text{for } -\frac{\pi}{4} < \arg(\xi) < \frac{5\pi}{4},$$

$$\tag{9.10}$$

$$\varepsilon^- \to 1 - \frac{\omega_p^2}{\omega_0^2}, \qquad \text{for } -\frac{5\pi}{4} < \arg(\xi) < \frac{\pi}{4},$$

and for $|k| \to \infty$ (for which case, $|\xi| \to 0$)

$$\varepsilon^{\pm} \to 1. \tag{9.11}$$

Additionally, using the Hilbert-transform property $Q^{\pm}(\xi^*) = [Q^{\mp}(\xi)]^*$, we can deduce that

$$\varepsilon^{\pm}(f,\xi^*) = \varepsilon^{\mp}(f,\xi)^*, \qquad \text{for real } f, \tag{9.12}$$

and since the distribution function is an even function, using the Hilbert-transform property $Q^{\pm\prime}(-\xi^*) = -[Q^{\pm\prime}(\xi)]^*$,

$$\varepsilon^{\pm}(f,-\xi^*) = \varepsilon^{\pm}(f,\xi)^*, \qquad \text{for real } f. \tag{9.13}$$

9.2.2. Roots of $\varepsilon^+(f,\xi)$

When we want to use the inversion technique of a Fourier transform, we need to know the singularities of the integrand, as discussed in Chapter 7. Since the integrand includes the function $1/\varepsilon^{\pm}(\omega_0,k)$, we are interested to know what are the roots of $\varepsilon^{\pm}(\omega_0,k)$ in the complex-k plane for a given real ω_0. From Eq. (9.12) it follows that the roots of $\varepsilon^-(\omega_0,k)$ are known when we know the roots of $\varepsilon^+(\omega_0,k)$.

A very important difference between the dielectric constant obtained from the Landau approach and the dielectric constant given by any fluid-type or macroscopic equation is that, in the first case $\varepsilon^+(\omega_0,\xi)$ is an entire function of ξ which is not a polynomial. It follows from Picard's theorem (Roos, 1969) that every entire function that is not a polynomial of the variable ξ can take every arbitrary complex value a, except one, for an infinite number of values of the variable ξ. Thus, the equation

$$Z^{+\prime}(\xi) = a, \tag{9.14}$$

where a is any complex number, is satisfied for an infinite set of values of ξ, except for $a = -2$, which is the exception in Picard's theorem. For $a = -2$, the only possible value for ξ is $\xi = 0$. For $\varepsilon^+(\omega_0,k) = 0$, Eq. (9.4) gives the dispersion relation as

$$Z^{+\prime}\left(\frac{\omega_0}{ka}\right) = \frac{k^2}{k_d^2}. \tag{9.15}$$

Thus, we see that the dispersion relation will have an infinite number of roots in the complex-k plane, except for $\omega_0 = 0$. If $\omega_0 = 0$, we have only two roots, $\pm i2^{1/2}k_d$. And, since the distribution function is an even function, we can also predict that if k_n is a root, $-k_n^*$ is also a root of the dispersion relation.

Hunting the roots in the complex-k plane is a game that is discussed in Chapter 11. Here, following the results given by Simonen (1966), we restrict ourselves to showing (see Fig. 9.1) the "ballet" of the roots in the complex-k plane as a function of ω_0. We briefly describe the various classes of roots that exist:

(1) $k = k_{de}$ is always on the positive half of the imaginary axis, $\mathrm{Im}(k) > 0$. For $\omega_0 = 0$, $k_{de} = i2^{1/2}k_d$, and $|k_{de}|$ is an increasing function of ω_0. Letting ξ_{de} be the corresponding root in the complex-ξ plane, we can show that for $f \to 0$

$$\xi_{de} = \frac{\omega_0}{k_{de}a} = -i\frac{f}{2^{1/2}}\left(1 - \frac{\pi^{1/2}}{2^{3/2}}f\right). \tag{9.16}$$

(2) $k = k_0$ is always on the negative half of the imaginary axis, $\mathrm{Im}(k) < 0$. This root exists only for $0 \leqslant \omega_0 < \omega_p$. For $\omega_0 = 0$, $k_0 = -k_{de}$, and for $\omega_0 \to \omega_p$, $k_0 \to 0$. Letting ξ_0 be the corresponding root in the complex-ξ plane, we can show that for $f \to 0$

$$\xi_0 = i\frac{f}{2^{1/2}}\left(1 + \frac{\pi^{1/2}}{2^{3/2}}f\right), \tag{9.17}$$

and for $f \to 1$, but $f < 1$, that

$$\xi_0 = if\left[\frac{3}{2(1 - f^2)}\right]^{1/2}. \tag{9.18}$$

The root k_0 (also called the Debye pole) corresponds to the evanescent electronic wave below the plasma frequency.

(3) $k = k_{\pm 1}$ both exist only for $\omega_0 > \omega_{pe}$. When $\omega_0 \to \omega_{pe}$, we find that $k_{-1} = -k_1^*$ and $k_1 \to 0$. These two roots, called first Landau poles, correspond to the macroscopic electronic (Bohm and Gross) wave occurring at frequencies just above the plasma frequency (see Chapter 7). Letting $\xi_{\pm 1}$ be the corresponding roots in the complex-ξ plane, we find the well-known result

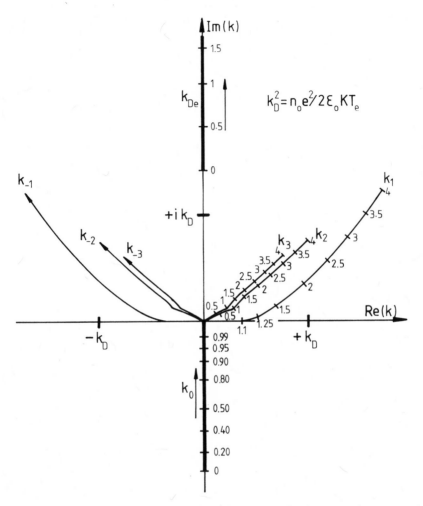

Fig. 9.1. Landau poles of the dielectric constant $\varepsilon^+(\omega_0/k)$ in the complex-k plane for $\omega_0 > 0$ and real. The numbers shown on the curves give numerical values of the ratio ω_0/ω_p.

$$v = \mathrm{Re}(\xi_1) = f\left[\frac{3}{2(f^2 - 1)}\right]^{-1/2};$$ (9.19)

$$\mathrm{Im}(\xi_1) = -i\tfrac{2}{3}\pi^{1/2}v^6 e^{-v^2};$$ (9.20)

$$\xi_{-1} = -\xi_1^*.$$ (9.21)

(4) $k = k_{\pm n}$, where $n \geqslant 2$ are the Landau poles of higher order. They all move toward the point $k = 0$ when $\omega_0 \to 0$. As the frequency increases they separate, as shown in Fig. 9.1.

9.2.3. *Expansion of* $1/\varepsilon^+$ *in Partial Fractions*

The dispersion relation $\varepsilon^{\pm}(\omega_0,k) = 0$ has only first-order roots in the k plane; thus the functions $1/\varepsilon^+(f,\xi)$ are meromorphic functions of first order (i.e., having only poles of order 1) in the ξ plane. Moreover, they are bounded by $|\xi| \to \infty$ along any path going between the poles. We can therefore represent these functions using a partial-fraction expansion (Morse and Feshbach, 1953) as

$$F(\xi) = F(0) + \sum_m \left[\frac{A_m}{\xi - \xi_m} + \frac{A_m}{\xi_m} \right], \tag{9.22}$$

where the sum $\sum_m$ is over all of the first-order poles ξ_m, and the A_m are the corresponding residues. Such a representation of $1/\varepsilon^+$ was first used by Derfler and Simonen (1966). Following them, we can apply the expansion given by Eq. (9.22) to $1/\varepsilon^{\pm}$ to give

$$\frac{1}{\varepsilon^{\pm}(f,\xi)} = 1 + \sum_m \frac{a_m^{\pm} \xi}{\xi_m^{\pm}(\xi - \xi_m^{\pm})}, \tag{9.23}$$

where

$$a_m^{\pm} = \frac{1}{\partial \varepsilon^{\pm}/\partial \xi \big|_{\xi = \xi_m}} = -\left[\frac{2}{\xi_m^{\pm}} + \xi_m^{\pm 2} Q^{\pm \prime \prime}(\xi_m^{\pm}) \right]^{-1}, \tag{9.24}$$

and $Q^{\pm \prime \prime}(\xi)$ is the second-order derivative with respect to ξ of the Hilbert transform of the distribution function $\varphi_0(u)$. Since $\xi_m^- = (\xi_m^+)^*$ and $Q^{\pm}(\varepsilon_m) = [Q^{\pm}(\varepsilon_m)]^*$, then $Q^{\pm \prime \prime}(\varepsilon_m^*) = [Q^{\pm \prime \prime}(\varepsilon_m)]^*$, so that

$$a_m^- = (a_m^+)^*. \tag{9.25}$$

Moreover, if $\varphi_0(u)$ is an even function, the roots can be selected by couples $\xi_m^{\pm}$, such that $\xi_{-m}^{\pm} = -(\xi_m^{\pm})^*$. Then, from the following property of Hilbert transforms,

$$Q^{\pm \prime \prime}(-\xi_m^{\pm *}) = -[Q^{\pm \prime \prime}(\xi_m^{\pm})]^* = Q^{\pm \prime \prime}(\xi_{-m}^{\pm}), \tag{9.26}$$

we see that for an even distribution function

$$a_{-m}^{\pm} = -(a_m^{\pm})^* \qquad \text{for } \xi_{-m}^{\pm} = -(\xi_m^{\pm})^*. \tag{9.27}$$

The function $1/\varepsilon^{\pm}$ can also be expanded with respect to the variable $\eta = 1/\xi$ to obtain

$$\frac{1}{\varepsilon^{\pm}(f,\eta)} = 1 + \sum_m \frac{b_m^{\pm}}{\eta - \eta_m^{\pm}}, \tag{9.28}$$

where

$$b_m^{\pm} = -\frac{a_m^{\pm}}{\xi_m^{\pm 2}}. \tag{9.29}$$

Using Eqs. (9.25) and (9.27) we can deduce some relationships between these coefficients, depending on the parity of the distribution function. For *all* $\varphi_0(u)$ functions we have

$$\eta_m^- = (\eta_m^+)^* \quad \text{and} \quad b_m^- = (b_m^+)^*. \tag{9.30}$$

For an even function $\varphi_0(u)$ we have

$$b_{-m}^\pm = -(b_m^\pm)^* \qquad \text{for } \eta_{-m}^\pm = -(\eta_m^\pm)^*. \tag{9.31}$$

We now consider Maxwellian distribution functions. Let the successive roots be as defined previously,

$$\zeta_m^+ = \zeta_0, \zeta_{de}, \zeta_{\pm 1}, \ldots, \zeta_{\pm n} \qquad n = 1, 2, \ldots. \tag{9.32}$$

Also, let $b_m^+ = B_m$. Then, taking into account Eq. (9.30) we can write Eq. (9.28) as

$$\frac{1}{\varepsilon^+(f,\xi)} = 1 + \frac{B_{de}}{\eta - \eta_{de}} + \frac{B_0}{\eta - \eta_0} + \sum_{n=1}^{\infty} \left(\frac{B_n}{\eta - \eta_n} - \frac{B_n^*}{\eta + \eta_n^*} \right),$$

$$\frac{1}{\varepsilon^-(f,\xi)} = 1 + \frac{B_{de}^*}{\eta - \eta_{de}^*} + \frac{B_0^*}{\eta - \eta_0^*} + \sum_{n=1}^{\infty} \left(\frac{B_n^*}{\eta - \eta_n^*} - \frac{B_n}{\eta - \eta_n} \right). \tag{9.33}$$

Using Eq. (9.24) the relationship $Z^{+\prime}(\zeta_m) = f^2/\zeta_m^2$ and the well-known relationships between Z^+, $Z^{+\prime}$, and $Z^{+\prime\prime}$, we find B_m to be given by

$$B_m = \sigma_m \frac{f^2}{\zeta_m[3f^2 + 2\zeta_m^2(1 - f^2)]}, \tag{9.34}$$

where

$$\sigma_0 = \begin{cases} 1, & \text{for } 0 \leqslant \omega_0 < \omega_p; \\ 0, & \text{for } \omega_0 > \omega_p; \end{cases} \qquad \begin{aligned} \sigma_1 &= 1 - \sigma_0; \\ \sigma_m &= 1, \quad \text{for } m > 1. \end{aligned} \tag{9.35}$$

9.2.4. Additional Properties of the Dielectric Constant

We give here a few other properties of the dielectric constant which are not frequently found in the literature, but which can be useful. These properties can be called *pseudo-causality* relationships, since they are quite similar to the well-known Kramers–Kronig relationships (see Chapter 1).

Applying Cauchy's theorem in the positive half-plane, $\text{Im}(\xi') > 0$, we have

$$\int_{-\infty}^{+\infty} \frac{d\xi'}{(\xi' - \xi)\varepsilon^+(f,\xi')} + \lim_{R \to \infty} \int_C \frac{d\xi'}{(\xi' - \xi)\varepsilon^+(f,\xi')}$$

$$= 2\pi i \left[\frac{1}{\varepsilon^+(f,\xi)} + \frac{\sigma_0}{(\xi_0^+ - \xi)(\partial\varepsilon^+/\partial\xi)_{\xi=\xi_0}} \right], \tag{9.36}$$

where as before C is the part of the circle, $\xi = R \exp(i\theta)$, located in the positive half-plane $\mathrm{Im}(\xi') > 0$, and ξ_0^+ is the root corresponding to the Debye pole. For $\mathrm{Im}(\xi) \to 0+$, we have after some algebra,

$$\mathrm{Re}\left[\frac{1}{\varepsilon^\pm(f,\xi)} - \frac{1}{\varepsilon_p}\right] = \pm\frac{1}{\pi}\int_{-\infty}^{+\infty}\frac{d\xi'}{\xi'-\xi}\,\mathrm{Im}\left[\frac{1}{\varepsilon^\pm(f,\xi')}\right]$$
$$- \mathrm{Im}\left[\frac{2i\sigma_0}{(\xi_0^\pm - \xi)(\partial\varepsilon^\pm/\partial\xi)_{\xi_0^+}}\right]; \qquad (9.37)$$

$$\mathrm{Im}\left[\frac{1}{\varepsilon^\pm(f,\xi)}\right] = \pm\frac{1}{\pi}\int_{-\infty}^{+\infty}\frac{d\xi'}{\xi'-\xi}\,\mathrm{Re}\left[\frac{1}{\varepsilon^\pm(f,\xi')}\right]$$
$$+ \mathrm{Re}\left[\frac{2i\sigma_0}{(\xi_0^\pm - \xi)(\partial\varepsilon^\pm/\partial\xi)_{\xi_0^+}}\right], \qquad (9.38)$$

where ξ is real, and $\varepsilon_p = 1 - \omega_0^2/\omega_p^2$. Equations (9.37) and (9.38) are similar to the Kramers–Kronig relationships, except for the contribution of the Debye poles ξ_0^+ and ξ_0^-. For $\xi = 0$, Eqs. (9.37) and (9.38) become

$$\frac{1}{\pi}P\int_{-\infty}^{+\infty}\frac{d\xi}{\xi'}\,\mathrm{Im}\left[\frac{1}{\varepsilon^\pm(f,\xi')}\right] = \pm 2\left\{\frac{1}{2}\left(1 - \frac{1}{\varepsilon_p}\right)\right.$$
$$\left. + \mathrm{Im}\,\frac{i\sigma_0}{[\xi(\partial\varepsilon^\pm/\partial\xi)]_{\xi_0^\pm}}\right\}; \qquad (9.39)$$

$$\frac{1}{\pi}P\int_{-\infty}^{+\infty}\frac{d\xi}{\xi'}\,\mathrm{Re}\left[\frac{1}{\varepsilon^\pm(f,\xi')}\right] = \pm 2\left\{\mathrm{Re}\left(\frac{i\sigma_0}{[\varepsilon(\partial\varepsilon^\pm/\partial\xi)]_{\xi_0^\pm}}\right)\right.$$
$$\left. - \frac{1}{2}\mathrm{Im}\left[\frac{1}{\varepsilon^\pm(f,0)}\right]\right\}. \qquad (9.40)$$

The relations given by Eqs. (9.39) and (9.40) are particularly interesting for the case of an even distribution function. Then

$$\frac{1}{\pi}\int_0^{+\infty}\frac{d\xi'}{\xi'}\,\mathrm{Im}\left[\frac{1}{\varepsilon^\pm(f,\xi')}\right] = \pm\frac{1}{2}\left(1 - \frac{1}{\varepsilon_p}\right) \pm \frac{\sigma_0}{[\xi(\partial\varepsilon^\pm/\partial\xi)]_{\xi_0^\pm}}. \qquad (9.41)$$

For a Maxwellian plasma we can write

$$\frac{\sigma_0}{\xi_0^\pm(\partial\varepsilon^\pm/\partial\xi)_{\varepsilon_0^\pm}} = -B_0\xi_0 = -C_0 = -\frac{f^2\sigma_0}{3f^2 + 2\xi_0^2(1-f^2)}. \qquad (9.42)$$

On the other hand, the expansion of $1/\varepsilon^\pm$ in partial fractions allows us to obtain some interesting relationships involving the coefficients b_m, which appear in Eq. (9.28). For this purpose we use a MacLaurin expansion of $1/\varepsilon^{\pm}(f,\xi)$

$$\frac{1}{\varepsilon^{\pm}(f,\xi)} \simeq 1 - \left(\frac{\partial\varepsilon^{\pm}}{\partial\xi}\right)_{\xi=0}\xi - \frac{1}{2}\left[\frac{\partial^2\varepsilon^{\pm}}{\partial\xi^2} - 2\left(\frac{\partial\xi^{\pm}}{\partial\xi}\right)^2\right]_{\xi=0}\xi^2. \quad (9.43)$$

For $\xi = 0$, the derivatives in Eq. (9.43) have the values

$$\left(\frac{\partial\varepsilon^{\pm}}{\partial\xi}\right)_{\xi=0} = 0; \qquad \left(\frac{\partial^2\varepsilon^{\pm}}{\partial\xi^2}\right)_{\xi=0} = -\frac{2}{f^2}Q^{\pm\prime}(0). \quad (9.44)$$

Thus, Eq. (9.43) becomes

$$\frac{1}{\varepsilon^{\pm}(f,\xi)} \simeq 1 + \frac{Q^{\pm\prime}(0)}{f^2}\xi^2. \quad (9.45)$$

But from Eq. (9.23), in the vicinity of $\xi = 0$ we have

$$\frac{1}{\varepsilon^{\pm}(f,\xi)} = 1 + \xi\sum_m B_m + \xi^2\sum_m b_m/\xi_m. \quad (9.46)$$

Comparing Eqs. (9.45) and (9.46) we discover that

$$\sum_m b_m^{\pm} = 0; \qquad \sum_m \frac{b_m^{\pm}}{\xi_m^{\pm}} = \frac{Q^{\pm\prime}(0)}{f^2}. \quad (9.47)$$

When the distribution function is an even function we have

$$\sum_m b_m^{\pm} = \sum_m \text{Im}(b_m^{\pm}) = 0; \quad (9.48)$$

and

$$\sum_m \frac{b_m^{\pm}}{\xi_m^{\pm}} = \sum_m \text{Re}(b_m^{\pm}) = \frac{Q^{\pm\prime}(0)}{f^2}. \quad (9.49)$$

For Maxwellian electrons, Eq. (9.49) gives

$$\sum_m \frac{b_m^{\pm}}{\xi_m^{\pm}} = \sum_m \text{Re}(b_m^{\pm}) = -\frac{2}{f^2}. \quad (9.50)$$

If we apply the value $\eta = 0$ in Eq. (9.28) we find that

$$\sum_m b_m^{\pm}\xi_m^{\pm} = 1 - \frac{1}{\varepsilon_p} = \frac{1}{1-f^2}. \quad (9.51)$$

More generally, for any even distribution function we find that

$$\sum_m b_m^{\pm} \xi_m^{\pm} = \sum_m \mathrm{Re}(b_m \xi_m) = 1 - \frac{1}{\varepsilon_p}. \tag{9.52}$$

Thus, except for the additional contribution of the Debye pole that we have included here, it is seen that the above calculations are similar to the Kramers–Kronig relations discussed in Chapter 1.

9.3. THE GOULD TECHNIQUE

The original article by Gould (1964) dealt with the dipole excitation of ion-acoustic waves. As a variation on the theme of that paper we apply the same technique to electron plasma waves.

9.3.1. Principle

We consider an external dipole charge given by

$$\rho(x) = \varepsilon_0 V_0 \delta'(x). \tag{9.53}$$

Substituting from Eq. (8.51) we can write the Fourier–Laplace transform of the potential $V(k,\omega_0)$ as

$$V(k,\omega_0) = \frac{iE(k,\omega_0)}{k} = \frac{iV_0}{k\varepsilon(k,\omega_0)}, \tag{9.54}$$

from which, doing the inverse transform, we find that

$$V(x,\omega_0) = -\frac{V_0}{2} \left[-\frac{i}{\pi} P \int_{-\infty}^{+\infty} \frac{e^{ikx}}{k\varepsilon(k,\omega_0)} \, dk \right]. \tag{9.55}$$

When $k \to 0$, then $\varepsilon^{\pm}(k,\omega_0) \to \varepsilon_p = 1 - \omega_p^2/\omega_0^2$, and we can write the integral in Eq. (9.55) as

$$P \int_{-\infty}^{+\infty} \frac{e^{ikx}}{k\varepsilon} \, dk = P \int_{-\infty}^{+\infty} \frac{e^{ikx}}{k} \left(\frac{1}{\varepsilon} - \frac{1}{\varepsilon_p} \right) dk + \frac{1}{\varepsilon_p} P \int_{-\infty}^{+\infty} \frac{e^{ikx}}{k} \, dk. \tag{9.56}$$

Using the reduced variables $\eta = ka/\omega_0 = 1/\xi$, $z = \omega_0 x/a$ and the well-known relation

$$\int_{-\infty}^{-\infty} \frac{e^{ikx}}{k} \, dk = i\pi \, \mathrm{sgn}(x), \tag{9.57}$$

Eq. (9.55) can be written as

$$V(x,\omega_0) = -\frac{V_0}{2}\left[\frac{1}{\varepsilon_p}\,\text{sgn}(x) + \Phi(z)\right],$$ (9.58)

where

$$\Phi(z) = -\frac{i}{\pi}\int_{-\infty}^{+\infty}\frac{e^{i\eta z}}{\eta}\left(\frac{1}{\varepsilon} - \frac{1}{\varepsilon_p}\right)d\eta.$$ (9.59)

Introducing the functions $F^+(\eta)$ and $F^-(\eta)$ defined by

$$F^\pm(\eta) = \frac{1}{\eta}\left[\frac{1}{\varepsilon^\pm(f,\eta)} - \frac{1}{\varepsilon_p}\right],$$ (9.60)

$\Phi(z)$ can be written as

$$\Phi(z) = -\frac{i}{\pi}\int_{-\infty}^{0}F^-(\eta)e^{i\eta z}\,d\eta - \frac{i}{\pi}\int_{0}^{\infty}F^+(\eta)e^{i\eta z}\,d\eta.$$ (9.61)

Without any loss of generality, we can choose $\omega_0 \geq 0$. In this case, z has the same sign as x. Using Jordan's lemma and the fact that $F^-(\eta)$ is an analytical function in the whole positive half-plane $\text{Im}(\eta) \geq 0$, we have for $z > 0$

$$\frac{i}{\pi}\int_{-\infty}^{+\infty}F^-(\eta)e^{i\eta z}\,d\eta = -C,$$ (9.62)

where

$$C = 2\,\text{Res}[e^{i\eta z}F^-(\eta)], \qquad \text{for }\text{Im}(\eta) \geq 0.$$ (9.63)

For $\text{Im}(\eta) \geq 0$, the only singular point for $\omega_0 < \omega_p$ is the Debye pole $\eta^- = 1/\xi_0^-$. Thus, we find that

$$C = 2C_0^* e^{iz/\xi_0^*},$$ (9.64)

where

$$C_0 = \frac{-\sigma_0}{[\xi(\partial\varepsilon^+/\partial\xi)]_{\xi=\xi_0}}.$$ (9.65)

From Eq. (9.61) we can write for $z > 0$

$$-\frac{i}{\pi}\int_{-\infty}^{0}F^-(\eta)e^{i\eta z}\,d\eta = \frac{i}{\eta}\int_{0}^{\infty}F^-(\eta)e^{i\eta z}\,d\eta + 2\sigma C_0^* e^{iz/\xi_0^*}.$$ (9.66)

Similarly, we find for $z < 0$

$$-\frac{i}{\pi} \int_0^\infty F^+(\eta) e^{i\eta z} \, d\eta = \frac{i}{\pi} \int_{-\infty}^0 F^+(\eta) e^{i\eta z} \, d\eta - 2C_0 e^{iz/\xi_0}. \qquad (9.67)$$

Since [see Eqs. (9.12) and (9.60)] $F^-(\eta) = [F^+(\eta)]^*$, we can use the result of Eq. (9.67) in Eq. (9.61) to obtain

$$\Phi(z > 0) = I^+(z) + 2C_0^* e^{iz/\xi_0^*},$$
$$\Phi(z < 0) = I^-(z) - 2C_0 e^{iz/\xi_0}, \qquad (9.68)$$

where

$$I^+(z) = \frac{2}{\pi} \int_0^\infty \text{Im}\left(\frac{1}{\eta \varepsilon^+}\right) e^{i\eta z} \, d\eta, \qquad (9.69)$$

$$I^-(z) = -\frac{2}{\pi} \int_{-\infty}^0 \text{Im}\left(\frac{1}{\eta \varepsilon^+}\right) e^{i\eta z} \, d\eta. \qquad (9.70)$$

The potential given by Eq. (9.58) can then be written as

$$V(x > 0, \omega_0 > 0) = -\frac{V_0}{2}\left[\frac{1}{\varepsilon_p} + 2C_0^* e^{iz/\xi_0^*} + I^+(z)\right], \qquad (9.71)$$

$$V(x < 0, \omega_0 > 0) = \frac{V_0}{2}\left[\frac{1}{\varepsilon_p} + 2C_0 e^{iz/\xi_0} - I^-(z)\right]. \qquad (9.72)$$

Thus, Gould's technique consists of the numerical evaluation of the Laplace transforms $I^\pm(z)$. We note that this method of calculation of the potential is valid for any distribution for which the Hilbert transform $Q^{\pm\prime}(\xi)$, is an entire function or a meromorphic function of ξ, and for any external charge that does not introduce any new singularities in $F^\pm(\eta)$. We will now investigate some of the properties of the Laplace transforms $I^\pm(z)$ before evaluating them.

9.3.2. Some Properties of $I^\pm(z)$

A one-dimensional ideal electric dipole can be closely approximated by two closely spaced, parallel grids which are nearly transparent with respect to the particles. The impedance of this system will be independant of the nature of the dielectric medium between the grids. We can express the potential difference between the two grids as

$$\Delta V = V(x = 0-, \omega_0) - V(x = 0+, \omega_0) = V_0, \qquad (9.73)$$

where V_0 is the applied voltage between the grids in a vacuum. Substituting from Eqs. (9.71) and (9.72) into Eq. (9.73) gives

$$\Delta V = \frac{V_0}{2} \left[\frac{2}{\varepsilon_p} + 2(C_0 + C_0^*) - I^-(0) + I^+(0) \right]. \qquad (9.74)$$

But

$$I^+(0) - I^-(0) = \frac{2}{\pi} \int_{-\infty}^{+\infty} \text{Im}\left(\frac{1}{\eta \varepsilon^+} \right) d\eta = \frac{2}{\pi} \int_{-\infty}^{+\infty} \text{Im}\left(\frac{1}{\xi \varepsilon^+} \right) d\xi. \qquad (9.75)$$

Using the pseudo-causality relation given by Eq. (9.41) we have

$$I^+(0) - I^-(0) = 2 - \frac{2}{\varepsilon_p} + 4\,\text{Im}(-iC_0) = 2 - \frac{2}{\varepsilon_p} - 2(C_0 + C_0^*). \qquad (9.76)$$

We can see that if we use this relation in Eq. (9.74), the potential obeys Eq. (9.73), as we have assumed in our model.

If the electron distribution is an even function, we would have $V(-x) = -V(x)$ for dipole excitation, so that

$$V(x = 0^+, \omega_0) = -\frac{V_0}{2}. \qquad (9.77)$$

To check this result, we start from the expressions of the potential given in Eqs. (9.71) and (9.72)

$$V(x = 0^+, \omega_0) = -\frac{V_0}{2} \left[\frac{1}{\varepsilon_p} + 2C_0^* + I^+(0) \right]. \qquad (9.78)$$

The value $I^+(0)$ of the Laplace transform $I^+(z)$ can be found from Eq. (9.41), giving

$$I^+(0) = 1 - \frac{1}{\varepsilon_p} - 2C_0. \qquad (9.79)$$

Thus

$$V(x = 0^+, \omega_0) = -\frac{V_0}{2} [1 + 2(C_0^* - C_0)]. \qquad (9.80)$$

For an even distribution function, C_0 is a real number, and Eq. (9.77) is justified.

For the case of odd distribution functions, we can find a close approximation of $I^+(0)$. For $\omega_0 > \omega_p$, this is very simple because from Eq. (9.79)

$$I^+(0) = \frac{1}{1 - f^2}. \tag{9.81}$$

For low frequencies, and for a Maxwellian plasma, we can evaluate C_0 from Eq. (9.42) using the approximation given by Eq. (9.17) for ξ_0. Doing this gives an approximate expression for $I^+(0)$ in Eq. (9.79)

$$I^+(0) \sim -\frac{\pi^{1/2}}{2^{3/2}} f. \tag{9.82}$$

The result given by Eq. (9.82) can be obtained using a more general technique, even for a non-Maxwellian plasma. For this purpose, we start again with the integral expression of $I^+(z)$

$$\operatorname{Im}\left(\frac{1}{\eta \varepsilon^+}\right) = \frac{\pi f^2 \eta \varphi_0'(1/\eta)}{\{f^2 \eta^2 - \operatorname{Re}[Q^{+\prime}(1/\eta)]\}^2 + \pi^2 \varphi_0'^2(1/\eta)}. \tag{9.83}$$

But for $\eta > \eta_0$ we have

$$Q^{+\prime} \frac{1}{\eta} \simeq Q^{+\prime}(0) + \frac{Q^{+\prime\prime}(0)}{\eta}. \tag{9.84}$$

If the distribution function is again an even function, then

$$\operatorname{Re}\left[Q^{+\prime}\left(\frac{1}{\eta}\right)\right] \simeq Q^{+\prime}(0) = -A^2 < 0, \qquad A > 0 \tag{9.85}$$

and

$$\varphi'\left(\frac{1}{\eta}\right) \sim \frac{\varphi_0''(0)}{\eta}. \tag{9.86}$$

Therefore, if $f^2 \eta_0^2 \ll A^2$, then $I^+(0)$ is approximately given by

$$I^+(0) \sim f^2 \mathfrak{J}_1 + f^2 \mathfrak{J}_2 \tag{9.87}$$

where $\mathfrak{J}_1$ is given by

$$\mathfrak{J}_1 = 2 \int_0^{\eta_0} \frac{\eta \varphi_0''(1/\eta) \, d\eta}{\{\operatorname{Re}[Q^{+\prime}(1/\eta)]\}^2 + \pi^2 \varphi_0'^2(1/\eta)} = C^{te}, \tag{9.88}$$

and $\mathcal{J}_2$ is given by

$$\mathcal{J}_2 = 2 \int_{\eta_0}^{\infty} \frac{\varphi_0''(0)}{(f^2\eta^2 + A^2)^2} \, d\eta \simeq \varphi_0''(0) \int_{-\infty}^{+\infty} \frac{d\eta}{(f^2\eta^2 + A^2)^2} \simeq \frac{\pi}{2f} \frac{\varphi_0''(0)}{A^3}.$$
(9.89)

It is noted that $\mathcal{J}_2$ is a function of f, whereas $\mathcal{J}_1$ is not. For $f \ll 1$, Eq. (9.87) becomes

$$I^+(0) \sim \frac{\frac{1}{2} f \varphi_0''(0)}{A^3}.$$
(9.90)

As a check on the validity of Eq. (9.90) we can let φ_0 be a Maxwellian distribution. For that case, $\varphi_0''(0) = -2\pi^{1/2}$ and $A = 2^{1/2}$, so that we obtain again the result given in Eq. (9.82).

9.3.3. Some Comments on the Calculation of the Laplace Transforms, $I^{\pm}(z)$

The numerical evaluation of the Laplace transforms $I^{\pm}(z)$ is a problem that presents some difficulties first discussed by Gould (1974) for the case of ion-acoustic waves. The main difficulty is that we have to numerically evaluate an integral at a large number of points. The integration step must not be too large because of Shannon's theorem (see for instance, Middleton, 1960). On the other hand, because the integral must be done over a large interval, we cannot use steps that are too small, otherwise the number of steps becomes too large.

Of considerable help in the evaluation of $I^{\pm}(z)$ is the fact that one can learn valuable information concerning the presence and qualitative character of wave modes simply by looking at the integrand $\text{Im}(1/\eta\varepsilon^+)$ as a function of η. For example, when the integrand presents a sharp peak around a wave number value k_1, we can predict the existence of a wave with a wave number k, the real part of which is close to k_1, and the imaginary part of which is proportional to the width of the peak (Gould, 1964). The existence of such a peak in the integrand of $I^{\pm}(z)$ is always indicative of strong collective wave behavior (Hirshfield et al., 1971).

We terminate the discussion of the Gould technique by noting some of its advantages and disadvantages. Among the advantages is the fact that we do not need to find the roots of the dispersion relation. A disadvantage appears when there is a strong collective effect. In this case a sharp peak appears in the integrand of $I^{\pm}(z)$. This sharp peak introduces a singularity that makes a direct calculation of the integral impossible. For example, this technique cannot be used for electron plasma waves for $\omega_p < \omega_0 \leqslant 1.1\omega_p$, when $\text{Im}(k) < 0.01 \, \text{Re}(k)$.

An interesting aspect of Gould's method is that it can be generalized to the calculation of other quantities besides the potential. Also, it can be applied to the case of non-Maxwellian plasmas having stable distribution functions. However, these functions must be meromorphic or entire functions with respect to the velocity variable.

9.4. THE DERFLER–SIMONEN TECHNIQUE

We consider again the calculation of the potential $V(x,\omega_0)$ for the case of dipole excitation. From Eq. (9.58) we have

$$V(x,\omega_0) = -\frac{V_0}{2}\left[\frac{1}{\varepsilon_p}\,\text{sgn}(x) + \Phi(z)\right], \tag{9.91}$$

where

$$\Phi(z) = -\frac{i}{\pi}\int_{-\infty}^{0}F^{-}(\eta)e^{i\eta z}\,d\eta - \frac{i}{\pi}\int_{0}^{\infty}F^{+}(\eta)e^{i\eta z}\,d\eta, \tag{9.92}$$

and [see Eq. (9.60)]

$$F^{\pm}(\eta) = \frac{1}{\varepsilon^{\pm}(f,\eta)} - \frac{1}{\varepsilon_p}. \tag{9.93}$$

Now, however, we want to use the technique used first by Derfler and Simonen (1966).

9.4.1. Principle

This technique uses an expansion of $1/\varepsilon^{\pm}$ in partial fractions for the calculation of $\Phi(z)$.

Using Eq. (9.28) we can write

$$\frac{1}{\eta}\left[\frac{1}{\varepsilon^{\pm}(f,\eta)} - \frac{1}{\varepsilon_p}\right] = \frac{1 - 1/\varepsilon_p}{\eta} + \sum_{m}\frac{b_m^{\pm}}{\eta(\eta - \eta_m^{\pm})},$$

$$= (1 - 1/\varepsilon_p - \sum_{m}b_m^{\pm}/\eta_m^{\pm})/\eta + \sum_{m}\frac{b_m^{\pm}}{\eta_m^{\pm}(\eta - \eta_m^{\pm})}. \tag{9.94}$$

Taking Eq. (9.52) into account we can write

$$F^{\pm}(\eta) = \sum_{m}\frac{C_m^{\pm}}{\eta - \eta_m^{\pm}}, \tag{9.95}$$

where

$$C_m^{\pm} = -a_m^{\pm}/\zeta_m^{\pm}. \tag{9.96}$$

Using this expression of $F^\pm$ in $\Phi(z)$ [see Eq. (9.92)] we obtain

$$\Phi(z) = -\frac{i}{\pi}\left(\sum_m C_m^* \int_{-\infty}^0 \frac{e^{i\eta z}}{\eta - \eta_m^*}\, d\eta + \sum_m C_m \int_0^\infty \frac{e^{i\eta z}}{\eta - \eta_m}\, d\eta\right), \quad (9.97)$$

where we have used the facts that $\eta_m^- = \eta_m^*$ and $C_m^- = (C_m^+)^* = (C_m)^*$. There-fore, $\Phi(z)$ is represented by the sum of the contributions of each pole, as in the case of the initial-value problem discussed in Chapter 8. But here the function appearing in each contribution is no longer an exponential function. Rather, it is the plasma wave function introduced by Derfler and Simonen (1966). We show in Section 9.7 that this plasma wave function can be repre-sented by the so-called exponential integral function. Using this result we can write Eq. (9.97) as

$$\Phi(z) = \sum_m \Phi_m(z), \quad (9.98)$$

where

$$\Phi_m(z > 0) = -\frac{i}{\pi}\left\{ C_m[\Psi^+(i\eta_m z) + 2\pi i \sigma(\eta_m)e^{i\eta_m z}] + \cdots \right.$$

$$\left. \cdots + C_m^*[-\Psi^+(i\eta_m^* z) + 2\pi i \beta(\eta_m^*)e^{i\eta_m^* z}]\right\}; \quad (9.99)$$

$$\Phi_m(z < 0) = -\frac{i}{\pi}\left\{ C_m[\Psi^-(i\eta_m z) - 2\pi i \sigma(\eta_m^*)e^{i\eta_m z}] + \cdots \right.$$

$$\left. \cdots + C_m^*[-\Psi^-(i\eta_m^* z) - 2\pi i \beta(\eta_m)e^{i\eta_m^* z}]\right\}. \quad (9.100)$$

In Eqs. (9.99) and (9.100) we have used the definitions

$$\Psi^\pm(v) = e^v E_1(v \pm i0+), \quad (9.101)$$

where E_1 is the integral differential function (see, for example, Abramowitz and Stegun, 1965)

$$E_1(z) = \int_z^\infty \frac{e^{-t}}{t}\, dt, \quad (|\arg(z)| < \pi), \quad (9.102)$$

and (see Section 9.7)

$$\sigma(a) = \begin{cases} 1, & \text{for } 0 < \arg(a) \leqslant \pi/2, \\ 0, & \text{everywhere else}; \end{cases}$$

$$\beta(a) = \begin{cases} 1, & \text{for } \pi/2 < \arg(a) < \pi, \\ 0, & \text{everywhere else}. \end{cases} \quad (9.103)$$

When the distribution function is an even function, then $\eta_{-m} = -\eta_m^*$ and $C_{-m} = C_m^*$, for $\arg(\eta_m) \neq \pm\pi/2$. Then, using the fact that $\Psi^+(v^*) = [\Psi^+(v)]^*$

$$\Phi_m(z > 0) + \Phi_{-m}(z > 0) = 2\left\{C_m e^{i\eta_m z} - \frac{i}{\pi} \operatorname{Re}[C_m \varphi_m(z)]\right\}, \quad (9.104)$$

where

$$\varphi_m(z) = \Psi^+(i\eta_m z) - \Psi^+(-i\eta_m z). \quad (9.105)$$

For $\eta_m = i|\eta_m|$, we have

$$\Phi_m(z > 0) = C_m e^{-|\eta_m|z} + \frac{i}{\pi} C_m P_m(z), \quad (9.106)$$

and for $\eta_m = -i|\eta_m|$

$$\Phi_m(z > 0) = C_m e^{-|\eta_m|z} - \frac{i}{\pi} C_m P_m(z), \quad (9.107)$$

where

$$P_m(z) = e^{-|\eta_m|z} E_i(|\eta_m|z) + \Psi^+(|\eta_m|z), \quad (9.108)$$

and

$$E_i(v) = -P \int_{-v}^{\infty} \frac{e^{-t}}{t} \, dt. \quad (9.109)$$

Using these results, we can now write $\Phi(z)$ in the form

$$\Phi(z > 0) = C_0 \left[e^{-|\eta_0|z} - \frac{i}{\pi} P_0(z)\right] + C_{de} \left[e^{-|\eta_{de}|z} + \frac{i}{\pi} P_{de}(z)\right] + \cdots$$

$$\cdots + \sum_{m=1}^{\infty} 2\left\{C_m e^{i\eta_m z} - \frac{i}{\pi} \operatorname{Re}[C_m \varphi_m(z)]\right\}. \quad (9.110)$$

The preceding expression is the one used by Derfler and Simonen (1966) for the calculation of $\Phi(z)$.

We note that for any value of z it is always possible to choose $z \ll 1$ such that $|\eta_m|z \ll 1$. Using the fact that

$$\Psi^+(v \to 0) \simeq -\ln(v) - \gamma, \quad (9.111)$$

where γ is the Euler constant, we obtain

$$\lim_{z \to 0^+} \varphi_n(z) = -i\pi. \quad (9.112)$$

Similarly, for $P_m(z)$ we obtain

$$\lim_{z \to 0^+} P_m(z) = 0. \tag{9.113}$$

Therefore, Eq. (9.110) can be written as

$$\Phi(z = 0+) = C_0 + C_{de} + 2 \sum_{m=1}^{\infty} \text{Re}(C_m) = \sum_m b_m \xi_m. \tag{9.114}$$

Using Eq. (9.52) we can also write

$$\Phi(z = 0+) = 1 - \frac{1}{\varepsilon_p}. \tag{9.115}$$

Finally, substituting from Eq. (9.115) into Eq. (9.91) we find again [see Eq. (9.77)] that

$$V(x = 0+, \omega_0) = -\frac{V_0}{2}. \tag{9.116}$$

Unfortunately, when $z \gg 1$ this technique does not lead to a simple expression, because the inequality $|\eta_m|z \gg 1$ can be satisfied for $m \ll N$, but not for $m \gg N$. Thus, one is not able to make an asymptotic expansion or a Taylor expansion that is everywhere valid.

9.4.2. Discussion of the Method

The representation of the plasma response by a sum of wave functions is of interest, in practice, only when the sum converges rapidly. This problem was first studied by Simonen (1966). The main features can be summarized as follows:

(1) Since

$$C_0 + C_{de} + 2 \sum_{m=1}^{\infty} \text{Re}(C_m) = 1 - \frac{1}{\varepsilon_p}, \tag{9.117}$$

a given pole can make a dominant contribution at some distance from the source only if its excitation coefficient is approximately equal to $(1 - 1/\varepsilon_p)$. It can be shown that the only poles that satisfy this condition are

$$
\begin{aligned}
k &= k_0, k_{de} &&\text{for } \omega_0 \to 0, \\
k &= k_0 &&\text{for } \omega_0 \to \omega_p, \omega_0 < \omega_p, \\
k &= k_{\pm 1} &&\text{for } \omega_0 \to \omega_p, \omega_0 > \omega_p.
\end{aligned} \tag{9.118}
$$

(2) We have seen that every pole gives rise to a wave function that is the sum of an exponential function and another function which we will call the *remainder* term [see Eqs. (9.104)–(9.107)]. Experience gives the following empirical rule: A given dominant pole, satisfying condition (1), will make a dominant contribution at distances where the exponential term exceeds the remainder term. In connection with this rule, we can note that at large distances the remainder term varies like $1/z$, i.e., $\Psi^+(v \to \infty) \sim 1/v$. Thus, at large distances, the remainder term always dominates the exponential term, since for $z \to \infty$ this term is damped like $\exp(-cz)$, which is faster than $1/z$. Thus, at large distances the perturbation can never be represented by the contribution of a single dominant pole.

(3) If no pole satisfies condition (1), then the convergence of the series, $\sum_m \Phi_m(z)$, is extremely slow (Simonen, 1966).

In conclusion, we can say that when collective effects (Debye sheath, electron plasma waves around ω_p, etc.) are not the dominant effects, the use of the Derfler–Simonen technique is not appropriate.

9.5. THE HYBRID TECHNIQUE

The Derfler–Simonen and Gould techniques appear to be complementary in the sense that, when one is appropriate the other is not, and conversely. Combining these two approaches, however, gives a hybrid technique that can be used for both the weakly collective and the strongly collective cases.

9.5.1. Principle

The basic idea of the hybrid technique is to use the wave function technique of Derfler and Simonen (1966) to calculate the contribution of a dominant pole, whereas the contributions of other poles are calculated by the evaluation of an integral using Gould's technique (1964).

For the purpose of illustration we look again at the problem of the potential calculation presented in Sections 9.2 and 9.3. From Eq. (9.68) we have

$$\Phi(z > 0) = 2C_0^* e^{iz/\xi_0^*} + I^+(z), \tag{9.119}$$

where

$$I^+(z) = \frac{2}{\pi} \int_0^\infty \mathrm{Im}[F^+(\eta)] e^{i\eta z}\, d\eta. \tag{9.120}$$

Derfler and Simonen's technique consists in writing

$$\mathrm{Im}[F^+(\eta)] = \sum_m \left[\frac{C_m}{\eta - \eta_m} - \frac{C_m^*}{\eta - \eta_m^*} \right]. \tag{9.121}$$

Instead of considering the sum $\sum_m$ on all the poles, we limit ourselves to n poles by writing

$$\text{Im}[F^+(\eta)] = \frac{1}{2} \sum_{m=1}^{n} \left[\frac{C_m}{\eta - \eta_m} - \frac{C_m^*}{\eta - \eta_m^*} \right] + R(\eta), \qquad (9.122)$$

where

$$R(\eta) = \text{Im} \left[\frac{1/\varepsilon^+ - 1/\varepsilon_p}{\eta} - \frac{C_m}{\eta - \eta_m} \right]. \qquad (9.123)$$

If we define $f_m(z)$ as

$$f_m(z) = \frac{2}{\pi} \int_0^\infty \text{Im} \left(\frac{C_m}{\eta - \eta_m} \right) e^{i\eta z} \, d\eta, \qquad (9.124)$$

then Eq. (9.120) becomes

$$I^+(z) = \sum_{m=1}^{n} f_m(z) + \frac{2}{\pi} \int_0^\infty R(\eta) e^{i\eta z} \, d\eta. \qquad (9.125)$$

Thus, we have only to compute the contributions of a finite sum of n poles and one integral. Obviously, we will choose the poles to be those that make a dominant contribution. Looking at the integral we note that since we have eliminated the dominant poles, $R(\eta)$ no longer contains the sharp peaks that were responsible for the difficulty in the numerical integration in Gould's technique. The integrand is always a "gentle" function. Therefore, we can use the hybrid method to treat both the weakly collective and the strongly collective cases.

9.5.2. Discussion of the Method

We have seen that the hybrid method seems to combine the advantages of the Derfler–Simonen and Gould techniques without having inherited any of their disadvantages. Nevertheless, perhaps as a consequence of Alexeff's *law of conservation of wretchedness* (1963), we find that new difficulties can appear. One such difficulty is the calculation of the integral

$$\mathcal{J}(z) = \frac{2}{\pi} \int_0^\infty R(\eta) e^{i\eta z} \, d\eta. \qquad (9.126)$$

As was the case in Gould's technique, the evaluation of $\mathcal{J}(z)$ is difficult for small distances since the integrand slowly decreases to zero as η goes to infinity. The integral can be divergent, even for $z \to 0$, if the series $\sum_m f_m(z)$ is divergent, a difficulty that can happen for some distribution functions that are odd with respect to the velocity variable v.

Another possible difficulty rises when $z \to \infty$, since the sum of the n poles can be approximated by $\sum_m f_m(z) \sim \alpha_n/z$, so that $J(z) \sim -\alpha_n/z + I^+(z)$ for large values of z, with $|\Phi^+(z)| \ll |\alpha_n|/z$. In other words, for large values of z, $I^+(z)$ is obtained from the difference of two approximately equal terms, which from a numerical point of view can cause problems.

9.6. CONCLUSIONS

In the calculation of macroscopic and microscopic perturbations, we can use the hybrid method which combines, in a simple way, the Derfler–Simonen and Gould techniques. The choice of the contributions to be represented separately by wave functions must be based on a knowledge of the collective character of the plasma and on the distance of the perturbation from the source of excitation. Once a proper choice is made, the hybrid method can be used to solve both weakly collective and strongly collective problems for distances up to values at which asymptotic expressions become valid.

One of the main advantages of the hybrid method, when compared with more sophisticated techniques such as the saddle-point method [see, for example, Massel (1967)], is that it is an easy technique to use for a large class of distribution functions. For example, one can easily write a computer program that will numerically calculate the electrostatic perturbations for any entire or meromorphic distribution function (Buzzi, 1974).

9.7. APPENDIX: PLASMA WAVE FUNCTIONS

We consider integrals of the type

$$I(x) = \int_0^\infty \frac{e^{ikx}}{k - a} \, dk, \tag{9.A.1}$$

where x is real and $\arg(a) \neq k\pi$. For $x > 0$, we can apply Cauchy's theorem and Jordan's lemma in the first quadrant of the complex-k plane (i.e., for $0 \leqslant \arg(k) \leqslant \pi/2$). This gives,

$$I(x) = \int_0^{i\infty} \frac{e^{ikx}}{k - a} \, dk + 2i\pi\sigma(a)e^{iax}, \tag{9.A.2}$$

where

$$\sigma(a) = \begin{cases} 1, & \text{for } 0 < \arg(a) < \pi/2, \\ 0, & \text{everywhere else.} \end{cases} \tag{9.A.3}$$

We note that the choice of $\sigma(a)$ determines, without any ambiguity, the path of integration for $\arg(a) = \pi/2$

$$\int_0^{i\infty} \frac{e^{ikx}}{k-a}\, dk = P \int_0^{i\infty} \frac{e^{ikx}}{k-a}\, dk - i\pi e^{iax}. \qquad (9.A.4)$$

Making the change in variables

$$t = -i(k-a)x, \qquad (9.A.5)$$

Eq. (9.A.2) can be written as

$$I(x) = e^{iax} \int_{iax}^{\infty} \frac{e^{-t}}{t}\, dt + 2i\pi\sigma(a)e^{iax}. \qquad (9.A.6)$$

However,

$$\int_z^{\infty} \frac{e^{-t}}{t}\, dt = E_1(z), \qquad \text{for } |\arg(z)| < \pi, \qquad (9.A.7)$$

where $E_1(z)$ is the integral exponential function. For $\arg(a) = \pi/2$ we have

$$I(x) = e^{iax}E_1(iax + i0+) + 2i\pi\sigma(a)e^{iax}. \qquad (9.A.8)$$

We define

$$\Psi^+(v) = e^v E_1(v + i0+). \qquad (9.A.9)$$

Then, Eq. (9.A.2) can be written as

$$\int_0^{\infty} \frac{e^{ikx}}{k-a}\, dk = \Psi^+(iax) + 2i\pi\sigma(a)e^{iax}, \qquad x > 0. \qquad (9.A.10)$$

Similarly,

$$\int_0^{\infty} \frac{e^{ikx}}{k-a}\, dk = \Psi^-(iax) - 2i\pi\sigma(a^*)e^{iax}, \qquad x < 0, \qquad (9.A.11)$$

where

$$\Psi^-(v) = e^v E_1(v - i0+). \qquad (9.A.12)$$

Similarly, we find that

$$\int_{-\infty}^0 \frac{e^{ikx}}{k-a}\, dk = -\Psi^+(iax) + 2i\pi\beta(a)e^{iax}, \qquad x > 0, \qquad (9.A.13)$$

where

$$\beta(a) = \begin{cases} 1, & \text{for } \pi/2 < \arg(a) < \pi, \\ 0, & \text{everywhere else,} \end{cases} \tag{9.A.14}$$

and

$$\int_{-\infty}^{0} \frac{e^{ikx}}{k-a} \, dk = -\Psi^-(iax) - 2i\pi\beta(a^*)e^{iax}, \qquad x < 0. \tag{9.A.15}$$

We also note that $E_1(z)$ can be represented using cosine and sine integral functions, and that $\Psi^\pm(v)$ is a confluent hypergeometric function (Abramowitz and Stegun, 1965).

Furthermore, one often uses the properties

$$E_1(z) = -\gamma - \ln(z) - \sum_{n=1}^{\infty} \frac{(-1)^n z^n}{nn!}, \qquad |\arg(z)| < n, \tag{9.A.16}$$

$$e^z E_1(z) \simeq \frac{1}{z}\left(1 - \frac{1}{2} + \frac{1\cdot 2}{z^2} - \frac{1\cdot 2\cdot 3}{z^3} + \cdots\right), \qquad |\arg(z)| < 3\pi/2; \tag{9.A.17}$$

and

$$E_1(z^*) = [E_1(z)]^*, \tag{9.A.18}$$

where γ is Euler's constant.

ION-ACOUSTIC WAVES IN MAXWELLIAN PLASMAS
A BOUNDARY-VALUE PROBLEM

10.1. INTRODUCTION

The history of the study of Landau damping of ion-acoustic waves is an interesting and at times colorful one. The initial concept of so-called Landau damping was introduced by L. D. Landau (1946) who demonstrated theoretically that if the speed of an acoustic wave in a plasma is slightly larger than the average thermal speed of one of the charge species (either ions or electrons) in the plasma, a wave–particle interaction should occur which would lead to a net transfer of energy from the wave to the slower moving particles, thereby causing the wave to be damped.

The first experimental effort to demonstrate Landau damping was made by Wong, Motley, and D'Angelo (1964). The experiments were performed in alkaline plasmas (cesium and potassium) produced in a Q-machine having two emitters. Since these experiments are described in detail in an excellent book on Q-machines by Motley (1975), we do not describe them here, other than to say that the plasmas used were isothermal ($T_e = T_i$) plasmas, and that the wave propagation was studied along the direction of the magnetic field.

As described in detail in Section 9.3, Gould (1965) made a theoretical study of wave excitation and propagation in simple, nonmagnetically supported plasmas as a function of T_e/T_i. For the case of an isothermal plasma ($T_e/T_i = 1$) corresponding to Wong *et al.*'s study of ion-acoustic wave propagation along magnetic field lines, Gould's calculation showed that: (1) In a complex k plane, the first Landau pole coresponding to the ion-acoustic wave does not have the smallest imaginary part. However, at distances not too close and not too far from the excitation source, this first Landau pole gives the dominant contribution, even though $T_e = T_i$ corresponds to the case where the contribution is the most strongly damped. (2) For distances larger than some threshold distance λ_e the potential perturbation is dominated by the electronic contribution (i.e., by the contributions of electrons for frequencies below the plasma frequency, as discussed by Gould, 1964). Although Wong *el al.* observed a signal that they interpreted as the ion-acoustic wave contribution predicted by Gould, this interpretation is questionable, as later shown experimentally by Jahns and Van Hoven (1972).

An additional complexity and uncertainty in the interpretation of Wong *et al.*'s experiments was introduced by Hirshfield and Jacob (1968) who pointed out, as we have seen in Chapter 7, that free-streaming effects can produce a perturbation that can easily be mistaken for a damped wave. In particular, they showed that for an isothermal plasma both the damping rate and the speed of such a perturbation are very nearly equal to those predicted by Gould's theory for ion-acoustic waves. To make the situation more interesting, Doucet and Gresillon (1970) presented evidence for low-frequency experiments in isothermal plasmas showing the noncollective perturbations proposed by Hirshfield and Jacob. Then Estabrook and Alexeff (1972) incorporated nonlinear effects in computer simulation experiments to try to show that the low-frequency perturbations seen by Doucet and Gresillon could perhaps be explained as weakly collective effects. Finally, Buzzi (1972, 1974) showed that the nonlinear effects found by Estabrook and Alexeff could be obtained from a linear theory that incorporated both free-streaming and collective effects. In Section 10.3, as an example of a nontrivial application of kinetic theory, we present Buzzi's theoretical treatment (1974) of forced oscillations in an isothermal plasma, showing under what conditions we expect to observe collective and free-streaming effects in such a plasma.

Perhaps the most clearcut and convincing demonstration of Landau damping of ion-acoustic waves was made by Alexeff *et al.* (1967) using a nonisothermal ($T_e \gg T_i$), nonmagnetically supported, discharge-type plasma. As described in detail in Section 10.4, *contaminant* ions, whose thermal speed closely matched the wave speed, could be controllably introduced into the plasma. In that experiment, the presence and degree of Landau damping of the ion-acoustic waves could be produced and controlled at will by controlling the number of contaminant ions.

10.2. DISPERSION RELATION FOR ION-ACOUSTIC WAVES

Since ion-acoustic wave motion involves the motion of both ions and electrons, we must modify the kinetic theory treatment of the preceding chapters, which required that the ions be motionless. We first generalize to a multicomponent plasma and then restrict the discussion to a two-component plasma of positive and negative species. Then, following the pattern used in Chapter 9 we find the plasma dielectric constant, discuss the numerical method of hunting the roots of the dispersion relation, and expand the dielectric function in terms of partial fractions in order to find additional properties of this quantity at low frequencies.

10.2.1. The Exact Dielectric Constant and the Boltzmann Approximation

The kinetic theory of forced oscillations presented in previous chapters can be generalized to a multicomponent plasma. Letting s represent a specie, the dielectric constant $\varepsilon(k,\omega)$ can be written as

$$\varepsilon(k,\omega_0) = 1 - \sum_s \frac{\xi_s^2}{f_s^2} Q_s'(\xi_s), \tag{10.1}$$

where (see Chapter 9)

$$\omega_{ps}^2 = \frac{n_{0s}q_s^2}{\varepsilon_0 m_s}; \qquad f_s = \frac{\omega_0}{\omega_{ps}}; \qquad \xi_s = \frac{\omega_0}{ka_s},$$

and a_s is an arbitrary velocity value such that

$$f_{0s}(v) = \frac{n_{0s}}{a_s} \varphi_{0s}(u), \tag{10.2}$$

where $u = v/a_s$, and

$$\int_{-\infty}^{+\infty} \varphi_0(u)\,du = 1.$$

Similar to what was done in Chapter 1, the Q's can be written as

$$Q_s'(\xi_s) = Q^{-\prime}(\xi_s)\Upsilon(-k) + Q^{+\prime}(\xi_s)\Upsilon(k), \tag{10.3}$$

where

$$Q^{\pm\prime}(\xi_s) = \int_{L^\pm} \frac{d\varphi_{0s}/du}{u - \xi_s}\,du. \tag{10.4}$$

Throughout the present chapter, we consider ion waves in a two-component Maxwellian plasma, letting $s = i$ for the ions and $s = e$ for the electrons. Using the plasma dispersion function, the dielectric constant can then be written as

$$\varepsilon^\pm(\omega_0,k) = 1 - \frac{\xi_i^2}{f_i^2} Z^{\pm\prime}(\xi_i) - \frac{\xi_e^2}{f_e^2} Z^{\pm\prime}(\xi_e) \tag{10.5}$$

where

$$\xi_{i,e} = \frac{\omega_0}{ka_{i,e}}; \qquad a_{i,e}^2 = \frac{2KT_{i,e}}{m_{i,e}}. \tag{10.6}$$

Using the following definitions

$$\xi_i = \xi, \qquad q_i = Z_c e; \qquad f_i = f; \qquad \frac{T_i}{T_e} = \Theta^2; \qquad \frac{m_e}{m_i} = m^2, \qquad (10.7)$$

where $-e$ is the electron charge, Eq. (10.5) can be written as

$$\varepsilon^{\pm}(f,\xi) = 1 - \frac{\xi^2}{f^2} \left[Z^{\pm\prime}(\xi) + \frac{\Theta^2}{Z_c} Z^{\pm\prime}(m\Theta\xi) \right]. \qquad (10.8)$$

When ξ is small enough that $m\Theta\xi \ll 1$, we can use the approximation

$$Z^{\pm\prime}(m\Theta\xi) \simeq Z^{\pm\prime}(0) = -2. \qquad (10.9)$$

The dielectric constant is then given by

$$\varepsilon^{\pm}(f,\xi) = 1 - \frac{\xi^2}{f^2} \left[Z^{\pm\prime}(\xi) - \frac{2\Theta^2}{Z_c} \right], \qquad (10.10)$$

where $m\Theta\xi \ll 1$. The approximation given by Eq (10.10) can be obtained directly by assuming that the electrons obey the linearized Boltzmann equation

$$n_e \simeq n_{0e} \left(1 - \frac{eV}{KT_e} \right), \qquad (10.11)$$

and that the ion motion is described by the Vlasov equation. We have to remember that the Boltzmann expression of the dielectric constant given by Eq. (10.10) is not correct for large ξ values, i.e., for small wave numbers. This is important because the asymptotic behavior along the x axis is determined by the behavior of the Fourier transform of the perturbations for small k values. Thus, we can expect that the dielectric constant cannot be described by the Boltzmann expression for distances far from the source. The following simple reasoning can be used to determine when the Boltzmann equation is valid. First, we can say that the Boltzmann equation is valid as long as the potential "seen" by the particle along its trajectory is time-independent. But this will be approximately true if

$$x \ll L = \frac{a_e}{\omega_0}, \qquad (10.12)$$

where L is the mean distance of the electron motion during one period of the electric field oscillation. For distances at which Eq. (10.12) is not satisfied, the expression given by Eq. (10.8) must be used in the calculation of electric perturbations.

10.2.2. The Roots of the Dispersion Relation

We now want to make a few remarks concerning the positions of the roots of $\varepsilon^+(f,\xi)$ in the complex-ξ plane for real values of $f = \omega_0/\omega_{pi}$. First, using Eqs. (9.12) and (9.13) (see Chapter 9), we can deduce that: (1) If ξ_m is a root of the dispersion relation $\varepsilon^+(f,\xi) = 0$, then ξ_m^* is a root of $\varepsilon^-(f,\xi) = 0$; and (2) if ξ_m is a root of the dispersion relation $\varepsilon^+(f,\xi) = 0$, then $-\xi_m^*$ is also a root of the same equation.

If we look at the general inventory of roots of ε^+, we find that it is similar to that of the electron waves, except that we now have two different sets, one corresponding to the ions and one to the electrons. The precise determination of the roots, whose values can depend strongly on the mass ratio of the two-particle species, is discussed in Chapter 11. For the case of a plasma produced by the photodissociation of thalium iodide (Davidovits *et al.*, 1969), for example, the mass ratio of Tl^+ and I^- is approximately equal to 1 and, as shown by Buzzi and Henry (1972), the two sets of poles are close together and follow each other in their motion in the complex plane when the frequency variations are in the range that characterizes the two ion species.

In general, we can describe the roots as follows:

(1) $\xi = \xi_d$. This root is always on the negative half of the imaginary axis, $\text{Im}(\xi) < 0$. For low frequencies ($f = \omega_0/\omega_{pi} \ll 1$), it can be shown from the finite development (Taylor's expansion, with remainder) of $Z^{\pm\prime}$ around zero that

$$\xi_d \simeq -i \frac{f}{[2(1 + \Theta^2/Z_c)]^{1/2}} \left[1 - \frac{\pi^{1/2}}{2^{3/2}} \frac{1 + m\Theta^3/Z_c}{(1 + \Theta^2/Z_c)^{3/2}} f \right]. \tag{10.13}$$

In the complex-k plane, for $f \to 0$, this root is given by

$$k_d \simeq i \frac{(1 + Z_c/\Theta^2)^{1/2}}{\lambda_d} \left[1 + \frac{\pi^{1/2}}{2^{3/2}} \frac{1 + m\Theta^3/Z_c}{(1 + \Theta^2/Z_c)^{3/2}} f \right], \tag{10.14}$$

where λ_d is the classical Debye length,

$$\lambda_d = \left[\frac{\varepsilon_0 K T_e}{n_{0e} e^2} \right]^{1/2}. \tag{10.15}$$

The reader can show that a relationship similar to Eq. (9.16) (see Chapter 9) exists when $m_i/m_{e-} \to \infty$, $\omega_0/\omega_{pi} \gg 1$ and $\omega_0/\omega_{pe} \ll 1$.

(2) $\xi = \xi_0$. This root is always on the positive half of the imaginary axis, $\text{Im}(\xi) > 0$. For $f = \omega_0/\omega_{pi} \ll 1$

$$\xi_0 = i \frac{f}{2(1 + \Theta^2/Z_c)} \left[1 + \frac{\pi^{1/2}}{2^{3/2}} \frac{1 + m\Theta^3/Z_c}{(1 + \Theta^2/Z_c)^{3/2}} f \right]. \tag{10.16}$$

Above the plasma frequency $\omega_0 > \omega_{p0} = (\omega_{pe}^2 + \omega_{pi}^2)^{1/2}$, this root disappears and the electron wave occurs. For the purpose of ion wave studies, we can assume that ξ_0 exists since we usually have only low frequencies $(\omega_0 \ll \omega_{p0})$.

(3) $\xi = \xi_{\pm 1}$. These roots correspond to the classical ion-acoustic wave. They are the least damped roots of the set of roots associated with the ions. At the limit of low frequencies $(f = 0)$, they are solutions of the equation

$$Z^{+\prime}(\xi) + \frac{\Theta^2}{\mathcal{Z}_c} Z^{+\prime}(m\Theta\xi) = 0. \tag{10.17}$$

Thus, at very low frequencies, these roots are independent of the frequency f. An analytical expression can be found for the case where the phase velocity lies between the electron and ion thermal velocities $(a_e \gg |\xi a_i| \gg a_i)$. For that case, except for the damping, we find the results of the adiabatic theory

$$\xi_r^2 = [\text{Re}(\xi_1)]^2 \simeq \left[\frac{\mathcal{Z}_c}{4\Theta^2}\right]\left\{(1 - f^2) + \left[(1 - f^2) + \frac{12\Theta^2}{\mathcal{Z}_c}\right]^{1/2}\right\}; \tag{10.18}$$

$$\xi_i = \text{Im}(\xi_1) \simeq -\frac{2\pi^{1/2}\xi_r^5(e^{-\xi_r^2} + m\Theta^3/\mathcal{Z}_c)}{3 + 4\Theta^2\xi_r^4/\mathcal{Z}_c}; \tag{10.19}$$

$$\xi_{-1} = -\xi_1^*. \tag{10.20}$$

It should be emphasized that the preceding expressions require that $a_e \gg |\xi a_i| \gg a_i$, a condition that is never satisfied if $T_e \simeq T_i$ or $T_e < T_i$, or if $f \gg 1$. For the case where Eq. (10.18) is valid, Eq. (10.19) shows that the ion wave damping is due to the ions if $\exp(-\xi_r^2) \gg m\Theta^3/\mathcal{Z}_c$, but is due to the electrons if $\exp(-\xi_r^2) \ll m\Theta^3/\mathcal{Z}_c$. It should be noted that when the damping is due to the ions, the accuracy of the computation of $\text{Im}(\xi_1) = \xi_i$ is generally poor. This is due to the fact that even though the use of the approximation involved in Eq. (10.18) gives a small error in ξ_r, this results in a large error in ξ_i due to the exponential term $\exp(-\xi_r^2)$ in Eq. (10.19). A similar situation arises, incidentally, for the case of the electrons [see Eqs. (9.19)–(9.21) in Chapter 9].

(4) $\xi = \xi_{\pm ni}$, where $(n \geq 2)$. These are the higher-order roots associated with the ions.

(5) $\xi = \xi_{\pm ne}$, where $(n \geq 2)$. These correspond to the set of roots associated with the electrons. We have remarked already that for $f \to 0$, the roots, $\xi_{\pm ni}(n > 1)$ and $\xi_{\pm ne}(n \geq 2)$, are given by Eq. (10.17), and that their position in the complex plane varies very little as the frequency is varied.

We conclude this description of the roots by stating that the calculation of roots is a routine numerical calculation now, and that one no longer need construct new tables. Nevertheless, for the case of the ion waves, one needs to initialize the program for hunting the roots. This initialization can be done, however, by using the values given by Sessler and Pearson (1967) or by Gresillon (1971).

10.2.3. Expansion of $\varepsilon^{\pm}(f,\eta)$ in Partial Fractions

In a way similar to what was done in Section 9.2.2 we can show that

$$\frac{1}{\varepsilon^{+}(f,\eta)} = 1 + \sum_{m} \frac{B_m}{\eta - \eta_m};$$

$$\frac{1}{\varepsilon^{-}(f,\eta)} = 1 + \sum_{m} \frac{B_m^{*}}{\eta - \eta_m}, \tag{10.21}$$

where $\eta = 1/\xi$ and $\eta_m = 1/\xi_m$, the ξ_m being the roots in the complex-ξ plane of $f^2 \varepsilon^{+}(f,\xi) = 0$. As in the case of ion waves, we always consider only frequency values for which the root ξ_0 exists. Taking into account the possible roots described in the preceding section, $1/\xi^{+}$ can be written in the form

$$\frac{1}{\varepsilon^{+}(f,\eta)} = 1 + \frac{B_d}{\eta - \eta_d} + \frac{B_0}{\eta - \eta_0} + \sum_{n=1}^{\infty} \left[\frac{B_n}{\eta - \eta_n} - \frac{B_n^{*}}{\eta + \eta_n} + \frac{B_n^{e}}{\eta - \eta_n^{e}} - \frac{B_n^{e*}}{\eta + \eta_n^{e*}} \right). \tag{10.22}$$

As long as the electrons can be considered to be Boltzmannian, this expression can be written as

$$\frac{1}{\varepsilon^{+}(f,\eta)} = 1 + \frac{B_d}{\eta - \eta_d} + \frac{B_0}{\eta - \eta_0} + \sum_{n=1}^{\infty} \left(\frac{B_n}{\eta - \eta_n} - \frac{B_n^{*}}{\eta + \eta_n^{*}} \right), \tag{10.23}$$

where the coefficients of excitation B_m are given by

$$B_m = \frac{f^2}{\xi_m} \left\{ 2f^2 - \xi_m^3 \left[Z^{+\prime\prime}(\xi_m) + \frac{m\Theta^3 Z^{+\prime\prime}(m\Theta\xi_m)}{Z_c} \right] \right\}^{-1}. \tag{10.24}$$

Making a Taylor expansion of $Z^{+\prime\prime}$ around $\xi = 0$ and using the values of ξ_d and ξ_0 given, respectively, by Eqs. (10.13) and (10.16), we obtain the low-frequency ($f \ll 1$) approximations

$$B_d \simeq \frac{1}{2\xi_d} \left[1 - \frac{\pi^{1/2}}{2^{3/2}} \frac{1 + m\Theta^3/Z_c}{(1 + \Theta^2/Z_c)^{3/2}} f \right],$$

$$B_0 \simeq \frac{1}{2\xi_0} \left[1 + \frac{\pi^{1/2}}{2^{3/2}} \frac{1 + m\Theta^3/Z_c}{(1 + \Theta^2/Z_c)^{3/2}} f \right]. \tag{10.25}$$

For the special case where $f \to 0$, B_d and B_0 become diverging quantities (like $1/f$) since for that case Eqs. (10.25) reduce to

$$B_d \simeq -B_0 \simeq \frac{i2^{1/2}(1 + \Theta^2/Z_c)^{1/2}}{f}. \tag{10.26}$$

The behavior of B_d and B_0 for low frequencies is very different from that of the other B_m for low frequencies. This can be seen from Eq. (10.24) which shows that, for $f \ll 1$,

$$B_m \simeq f^2, \tag{10.27}$$

and that ξ_m is almost constant. In particular, when the root ξ_1 is given by Eq. (10.18), we find that, for $f \ll 1$,

$$B_1 \simeq -\frac{f^2(+f^2)}{2\xi_1} \simeq -\Theta f^2 \frac{[1 - 3\Theta^2/2Z_c + 3f^2/2]}{(2Z_c)^{1/2}}. \tag{10.28}$$

Finally, following the procedure used in Section 9.2.4 we can show that

$$\frac{1}{\pi} \int_0^{+\infty} \frac{d\xi}{\xi} \, \text{Im} \left[\frac{1}{\varepsilon^\pm(f,\xi)} \right] = \pm \left(\frac{1}{2} - C_0 - \frac{1}{2\varepsilon_p} \right), \tag{10.29}$$

where

$$\frac{1}{\varepsilon_p} = 1 - \frac{\omega_{pe}^2 + \omega_{pi}^2}{\omega_0^2} \simeq -\frac{m_e}{m_i} Z_c \left[1 + (1 - f^2) \frac{m_e}{m_i} Z_c \right],$$

$$C_0 = B_0\xi_0 = \frac{f^2}{2f^2 - \xi_0^3[Z^{+\prime\prime}(\xi_0) + (m\Theta^3 Z^{+\prime\prime}(m\Theta\xi)/Z_c)]} \tag{10.30}$$

Substituting from Eqs. (10.25), we then have for low frequencies

$$\frac{1}{\pi} \int_0^\infty \frac{d\xi}{\xi} \, \text{Im} \left[\frac{1}{\varepsilon^\pm(f,\xi)} \right] \simeq \pm \left[\frac{m_e}{2m_i} Z_c + \frac{\pi^{1/2}}{2^{5/2}} \frac{1 + m\Theta^3/Z_c}{(1 + \Theta^2/Z_c)^{3/2}} f \right]. \tag{10.31}$$

Similarly, we can demonstrate the following properties

$$\sum B_m = \sum \text{Im}(B_m) = 0,$$

$$\sum \left(\frac{B_m}{\xi_m} \right) = \sum \text{Re} \left(\frac{B_m}{\xi_m} \right) = -\frac{2(1 + \Theta^2/Z_c)}{f^2}, \tag{10.32}$$

$$\sum C_m = \sum \text{Re}(C_m) = 1 - \frac{1}{\varepsilon_p},$$

where $C_m = B_m\xi_m$.

10.3. ION-ACOUSTIC WAVES IN AN ISOTHERMAL PLASMA

In this section we numerically evaluate the results of Gould's theory for the case of an isothermal plasma in order to find the potential perturbation produced in such a plasma by a dipole. We then compute this potential perturbation by computing the separate contributions to the perturbation by the ion-acoustic waves, the ballistic ions, and the electron plasma waves, all of which are generated by the dipole. A similar calculation is done to find the corresponding *density* perturbations produced not only by a dipole but also by a monopole. The results of all these computations are then used to make a critical evaluation of the Wong *et al.* experiment. Finally, we discuss the roles played by collective and free-streaming effects in both isothermal and non-isothermal plasmas, and suggest how these sometimes-competing effects can be distinguished from each other in low-frequency experiments.

10.3.1. Interpretation of Gould's Numerical Results

Gould's calculation (see Section 9.3.1) deals with the potential produced in a plasma by an electric dipole. From Eq. (9.71) for positive values of ω_0 and x, the potential is given by

$$V(x,\omega_0) = -\frac{V_0}{2}\left[\frac{1}{\varepsilon_p} + 2C_0^* e^{iz/\xi_0^*} + I^+(z)\right], \qquad (10.33)$$

where

$$\varepsilon_p = 1 - \frac{\omega_{pi}^2 + \omega_{pe}^2}{\omega_0^2};$$

$$I^+(z) = \frac{2}{\pi}\int_0^\infty \text{Im}\left(\frac{1}{\eta\varepsilon^+}\right)e^{i\eta z}\,d\eta. \qquad (10.34)$$

The equations for C_0, ξ_0, and $\varepsilon^+(f, \xi = 1/\eta)$ are given in Section 10.2.3. We first examine the behavior of $I^+(z)$ as a function of z for the case where $f = \omega_0/\omega_{pi} \ll 1$.

10.3.1.1. Behavior of $I^+(z)$ Close to the Antenna. Applying to Eq. (10.34) the procedure used in Section 9.3.2, which permitted the determination of $I^+(f \ll 1, 0)$, we find that

$$I^+(z \ll 1, f \ll 1) \simeq -\frac{4f^2}{\pi^{1/2}}\left(1 + \frac{m\Theta^3}{Z_c}\right)\int_0^\infty \frac{e^{i\eta z}\,d\eta}{[f^2\eta^2 + 2(1 + \Theta^2/Z_c)]^2}. \qquad (10.35)$$

After some laborious but straightforward calculations, one finds using the results of the Appendix in Section 9.7

$$I^+(z \ll 1, f \ll 1) = \frac{i\lambda f H(x/\lambda_0)}{\pi}, \qquad (10.36)$$

where

$$\lambda = \frac{\pi^{1/2}}{2^{3/2}} \frac{1 + m\Theta^3/\mathcal{Z}_c}{(1 + \Theta^2/\mathcal{Z}_c)^{3/2}}; \qquad \lambda_0 = \frac{\lambda_d}{(1 + \mathcal{Z}_c/\Theta^2)^{1/2}}, \qquad (10.37)$$

and

$$H\left(\frac{x}{\lambda_0}\right) = \left[\Psi^+\left(\frac{-x}{\lambda_0}\right) + 2\pi i e^{-x/\lambda_0}\right]\left(\frac{x}{\lambda_0} + 1\right) + \Psi^+\left(\frac{x}{\lambda_0}\right)\left[\frac{x}{\lambda_0} - 1\right]. \quad (10.38)$$

Continuing to assume that f and z are small quantities, we consider the two extreme cases corresponding to $x/\lambda_0 \ll 1$ and to $x/\lambda_0 \gg 1$. Making Taylor and asymptotic expansions of Ψ^+, as shown in Section 9.7, we obtain for $z \ll f \ll 1$

$$I^+\left(\frac{x}{\lambda_0} \ll 1, f \ll 1\right) \simeq -\lambda f\left(1 + \frac{ix}{\pi \lambda_0}\right)\ln\left(\frac{x}{\lambda_0}\right), \qquad (10.39)$$

and for $f \ll z \ll 1$

$$I^+\left(\frac{x}{\lambda_0} \gg 1, f \ll 1\right) \simeq \frac{-if^2}{\pi^{1/2}(1 + \Theta^2/\mathcal{Z}_c)^2 z} = -\frac{4i\lambda f}{\pi} \frac{\lambda_0}{x}. \qquad (10.40)$$

We note particularly that for short distances $(x/\lambda_0 \ll 1)$ the contribution to the potential given by Eq. (10.39) will be dominated by the oscillating Debye sheath [see Eqs. (10.14) and (10.15)] since for $f \ll 1$

$$2C_0^* e^{-iz/\xi_0^*} \simeq (1 + \lambda f) \exp\left[\frac{-(1 - \lambda f)x}{\lambda_0}\right]. \qquad (10.41)$$

The opposite situation occurs for long distances $(x/\lambda_0 \gg 1)$, where the contribution given by Eq. (10.40) is the dominant one. In this case the electric potential variations will be proportional to $1/z$ until the contributions of the other poles, in succession, finally dominate.

10.3.1.2. Contribution of the Ion-Acoustic Wave. We call the contribution of the ξ_1 root of ε^+—which varies like $\exp(i\eta_1 z)$, where $\eta_1 = 1/\xi_1$—the *ion wave contribution*. This contribution can be more explicitly seen if we express $I^+(z)$ in the form

$$I^+(z) = R^+(z) + I_i^+(z), \qquad (10.42)$$

where

$$R^+(z) = \frac{1}{i\pi} \int_c \frac{1}{\eta}\left(\frac{1}{\varepsilon^+} - \frac{1}{\varepsilon^-}\right)e^{i\eta z}\, d\eta, \qquad (10.43)$$

$$I_i^+(z) = 2C_1 e^{i\eta_1 z}, \qquad (10.44)$$

where [see Eq. (10.32)] C_1 is given by

$$C_1 = \frac{-1}{[\xi(\partial\varepsilon^+/\partial\xi)]_{\xi=\xi_i}}. \tag{10.45}$$

The contour c in Eq. (10.43) is the same as that in Fig. 10.1, and corresponds to the contour of the saddle-point method.

The contribution I_i^+ is neither dominant at short distances, as discussed above, nor is it dominant at very large distances, as we now show using Derfler and Simonen's technique (see Chapter 9). We start with the potential

$$V(x,\omega_0) = -\frac{V_0}{2}\left[\frac{\text{sgn}(x)}{\varepsilon_p} + \Phi(z)\right], \tag{10.46}$$

where $\Phi = \sum \Phi_m(z)$. Now the contribution from the two poles ξ_1 and ξ_{-1} is given by

$$\Phi_1(z>0) + \Phi_{-1}(z>0) = 2C_1 e^{+i\eta_1 z} - i\frac{\text{Re}[C_1\varphi_1(z)]}{\pi}, \tag{10.47}$$

where

$$\varphi_1(z) = \Psi^+(i\eta_1 z) - \Psi^+(-i\eta_1 z). \tag{10.48}$$

Thus, for $|\eta_1 z| \gg 1$, we have

$$\varphi_1(z) \simeq \frac{2}{i\eta_1 z}. \tag{10.49}$$

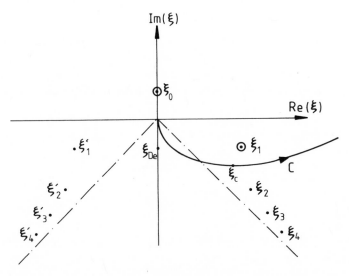

Fig. 10.1. Path of integration c [see Eq. (10.43)] in the complex-$\xi = 1/\eta$ plane. This is the integration path corresponding to the so-called saddle-point technique.

Therefore, at large distances, the term $\mathrm{Re}[C_1\varphi_1(z)]$—see Eq. (10.47)—which decreases like $1/z$, dominates over the exponentially decreasing term corresponding to $I_i^+(z)$. In a similar way, we see that the contribution of $I_i^+(z)$ does not dominate at very small distances. A numerical evaluation of Eq. (10.40) shows that, even for $\mathrm{Im}(\eta_1)/\mathrm{Re}(\eta_1) \geq 0.5$, the exponential term in Eq. (10.47) is always smaller than the term proportional to $\mathrm{Re}(C_1\varphi_1)$.

Although these considerations are useful for predicting the dominant behavior of $I_i^+(z)$ (i.e., the importance of the collective effect), a numerical calculation is required to be able to predict at what distances and frequencies the ion wave term will dominate. For example, for $T_e = T_i$ and $f \ll 1$, a numerical calculation shows that $I_i^+(z)$ is dominant for $z \geq 5$, the upper limit depending on the mass ratio m_e/m_i.

We recall, finally, that for $f \ll 1$, $\xi_1 \simeq \mathrm{const}$, so that C_1 is proportional to $f^2 = \omega_0^2/\omega_{pi}^2$ (see Section 10.2.3). Thus, the only dominant perturbation produced in a plasma by an electric dipole at low frequencies is the oscillating Debye sheath (see Section 10.3.1.1) and its wake [see Eq. (10.40)].

10.3.1.3. Asymptotic Ion Contribution. We can write $I_i^+(z)$—see Eq. (10.34)—as

$$I^+(z) = \mathcal{J}_i^+(z) + \mathcal{J}_e^+(z), \tag{10.50}$$

where

$$\mathcal{J}_i^+(z) = -\frac{4}{\pi^{1/2}} f^2 \int_0^\infty \frac{\xi^4 e^{-\xi^2}}{|f^2\varepsilon^+|^2} e^{i\eta z} \, d\eta; \tag{10.51}$$

$$\mathcal{J}_e^+(z) = -\frac{4}{\pi^{1/2}} f^2 \frac{m\Theta^3}{Z_c} \int_0^\infty \frac{\xi^4 e^{-m^2\Theta^2\xi^2}}{|f^2\varepsilon^+|^2} e^{i\eta z} \, d\eta. \tag{10.52}$$

The integrands in $\mathcal{J}_i^+(z)$ and $\mathcal{J}_e^+(z)$ are proportional, respectively, to the derivatives of the ion and the electron distribution functions. We note, also, that $\mathcal{J}_e^+(z)$ vanishes in the approximation $m = 0$ where m is the square root of the electron to the ion mass ratio (i.e., when the electrons obey Boltzmann's equation).

We now want to compute the asymptotic expression of $\mathcal{J}_i^+(z)$, i.e., when $z \gg 1$. For this purpose, we take advantage of the existence of the exponential term $\exp(-\xi^2)$ and apply the saddle-point technique (see Section 8.4.1). Doing this, we find that

$$\mathcal{J}_i^+(z \gg 1) = -\frac{4}{\pi^{1/2}} f^2 \int_0^\infty \frac{\xi^2 e^{-(\xi^2 - iz/\xi)}}{|f^2\varepsilon^+|^2} \, d\xi \simeq -\frac{4f^2}{3^{1/2}} \frac{\xi_c^2 e^{-3\xi_c^2}}{|f^2\varepsilon^+(f,\xi_c)|^2}, \tag{10.53}$$

where

$$\xi_c = \left(\frac{z}{2}\right)^{1/3} \exp\left(\frac{-i\pi}{6}\right). \tag{10.54}$$

In order for Eq. (10.53) to be valid, it is necessary that $|\xi_c| \gg 1$. Therefore, we approximate $|f^2\varepsilon^+|$ by

$$|f^2\varepsilon^+|^2 \simeq \left| f^2 - 1 + \frac{2\xi_c^2\Theta^2}{Z_c} \right|^2, \qquad (10.55)$$

where we have assumed that $m\vartheta|\xi_c| \ll 1$. Making these approximations explicitly in Eq. (10.53) gives

$$\mathcal{J}_i^+(z) \simeq -\frac{4f^2}{3^{1/2}} \frac{\xi_c^2 e^{-3\xi_c^2}}{|(f^2 - 1 + 2\xi_c^2\Theta^2/Z_c)|^2}. \qquad (10.56)$$

We call this expression the asymptotic ion contribution to the electric potential. This contribution is also sometimes called the *ballistic* contribution, since it contains the same factor, $\exp(-3\xi_c^2)$, that appears in the free-streaming contribution (see Section 8.4.1). In fact, it is only at high frequencies ($f = \omega_0/\omega_p \gg 1$) that we see explicitly the free-streaming term produced by dipolar excitation. For this case, Eq. (10.56) can be shown to be given approximately by

$$\mathcal{J}_i^+(z) \simeq -\frac{4}{3^{1/2}} \frac{\xi_c^2}{f^2} \exp(-3\xi_c^2). \qquad (10.57)$$

At low frequencies, the meaning of $\mathcal{J}_i^+$ is slightly different. We will see in Section 10.3.2, when computing the ion density, that the asymptotic ion contribution results from the free streaming of the distribution-function perturbation produced by the oscillating Debye sheath instead of by the dipolar field.

10.3.1.4. Asymptotic Electron Contribution. We now compute $\mathcal{J}_e^+(z)$—see Eq. (10.52)—for $z = \omega_0 x/a_i \gg 1$, but in the limit of $zm\Theta \ll 1$. This corresponds to distances that are large with respect to the ion scale length but which are still small with respect to the electron scale length. We then approximate $\mathcal{J}_e^+(z)$ by

$$\mathcal{J}_e^+(z) \simeq -\frac{4}{\pi^{1/2}} f^2 \frac{m\Theta^3}{Z_c} \int_0^\infty \frac{e^{i\eta z}\, d\eta}{[(f^2 - 1)\eta^2 + 2\Theta^2/Z_c]^2} \simeq -\frac{f^2 m Z_c}{\pi^{1/2}\Theta} \int_0^\infty e^{i\eta z}\, d\eta. \qquad (10.58)$$

Then, for $1 \ll z \ll (m\Theta)^{-1}\omega_0 \ll \omega_{pe}$,

$$\mathcal{J}_e^+(z) \simeq -i \frac{f^2 m Z_c}{\pi^{1/2}\Theta} \frac{1}{z}. \qquad (10.59)$$

We easily show that $\mathcal{J}_e^+(z)$ represents the contribution of the electron plasma wave for frequencies below the electron plasma frequency ω_{pe} and near (in terms of the scale length of the electrons) the source. For this purpose, we use the result (see Section 9.3.2)

$$I^+\left(\frac{\omega_0 x}{a_e} \ll 1\right) \simeq \left(\frac{\omega_0}{\omega_{pe}}\right)\left(-\frac{2}{\pi^{1/2}}\right)\int_0^\infty \frac{e^{iv\omega_0 x/a_e}\,dv}{[(\omega_0^2/\omega_p^2)v^2 + 2]^2}, \quad (10.60)$$

where $v = ka_e/\omega_0$. The integral is computed in a way similar to that used for Eq. (10.35). Upon making the integration and requiring that $x/\lambda_d \gg 1$, Eq. (10.59) results. We estimate the various contributions to the potential and compare them with the numerical calculations of Gould for the case of an isothermal $(T_e = T_i)$ cesium plasma.

In Fig. 10.2, we present the results of Gould's calculation for the case of low frequencies $(f = 0.1)$. We represent the amplitude and the phase of the

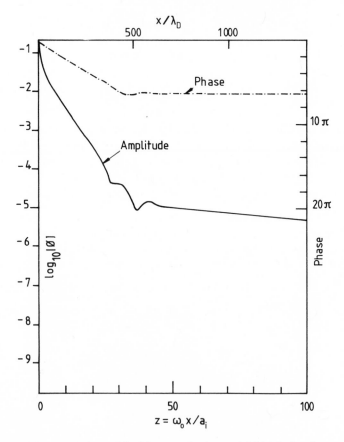

Fig. 10.2. Amplitude and phase of Gould's normalized potential $\Phi(z)$ for low-frequency dipole excitation in an isothermal cesium plasma $(m_e/m_i \simeq 500, f = \omega_0/\omega_p = 0.1, T_e = T_i)$ as a function of distance from the antenna.

quantity Φ, which is defined by

$$\Phi(z) = -\frac{1}{\pi} \int_0^\infty \mathrm{Im}\left(\frac{1}{\eta f^2 \varepsilon^+}\right) e^{i\eta z}\, d\eta. \qquad (10.61)$$

The potential, a real quantity, is given by

$$V(x) = f^2 \Phi V_0 - \frac{V_0}{2}\left[\frac{1}{\varepsilon_p} + 2C_0^* e^{iz/\xi_0^*}\right], \qquad (10.62)$$

where V_0 is the applied voltage between the grids of the dipole.

In Fig. 10.3, still at low frequencies ($f = 0.1$), we compare the amplitude of $\Phi(z)$ as given in Fig. 10.2 with the various contributions we have just calculated. Note that we do not give the interpretation of $\Phi(z)$ close to the source,

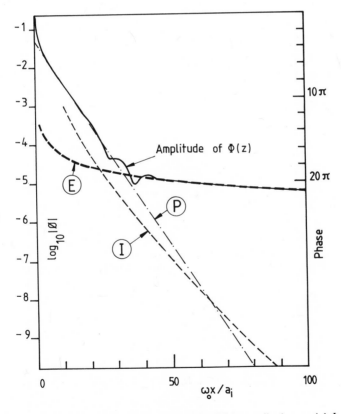

Fig. 10.3. Interpretation of the spatial variation of Gould's normalized potential Φ, given by the solid curve, for the same conditions as in Fig. 10.2. P gives the (ion-acoustic wave) contribution from the first Landau pole [Eq. (10.63)], I is the (ballistic) asymptotic ion contribution [Eq. (10.64)], and E is the electron plasma wave contribution [Eq. (10.65)].

because the corresponding contribution in Eq. (10.36) is valid only for very low frequencies. However, this limitation does not apply to the other contributions. In Fig. 10.3 the ion wave contribution is labeled P, the asymptotic ion contribution is labeled I, and the electron wave contribution is labeled E. The equations for these contributions are given, respectively, by

$$P = -\frac{C_1}{f^2} e^{i\eta_1 z};$$ (10.63)

$$I = \frac{2}{\sqrt{3}} \frac{\xi_c^2 e^{-3\xi_c^2}}{|(f^2 - 1) + 2\xi_c^2/Z_c|^2};$$ (10.64)

$$E = \frac{im Z_c}{2\pi^{1/2}\Theta z}.$$ (10.65)

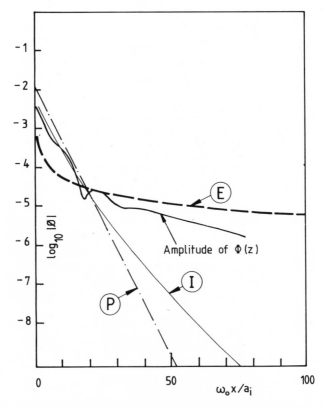

Fig. 10.4. Interpretation of the spatial variation of Gould's normalized potential Φ for the same conditions as in Fig. 10.3, except here the interpretation is for the high-frequency case ($f = \omega_0/\omega_p = 3.0$).

From observation of Fig. 10.3, it is clear that at low frequencies the potential is predominantly due to the contribution of the ion wave pole for $5 \leqslant z \leqslant 30$. For $z > 30$, the electron wave contribution easily dominates over both the ion wave contribution and the asymptotic contribution of the ions.

The situation is different at high frequencies, as seen in Fig. 10.4, which shows the case for $f = \omega_0/\omega_{pi} = 3$. Here, the asymptotic ion contribution, which can be closely identified with the free-streaming term, has an amplitude (that at low frequencies is slightly less than that of the ion wave contribution) which dominates the ion wave contribution. Thus, for distances up to $z \simeq 20$, we have a mixture of collective and ballistic effects. For distances, $z > 20$, however, the electronic contribution easily dominates both the ionic contributions.

In conclusion, we give the following succinct summary for the case of an isothermal plasma corresponding to the well-known and much-discussed experiment of Wong *et al.*: (1) At low frequencies the ion-acoustic wave contribution easily dominates for distances $5 \leqslant z \leqslant 30$, whereas for $z > 30$ the electron plasma wave contribution dominates; the asymptotic ion contribution is never important. (2) At high frequencies, for distances up to $z \simeq 20$, both the ion-acoustic wave and the asymptotic ion contributions are important, though, of the two, the first is the more important; for $z > 20$, the electron plasma wave contribution easily dominates the ionic contributions (Wong, 1965).

These results raise a question concerning the validity of the interpretation by Hirshfield and Jacob (1968)—in terms of the free-streaming effect—of the Wong, *et al.* experiment. However, before making any additional statements concerning the collective or noncollective aspects of this experiment, we have to remember that, as is usual in such plasma wave measurements, it was not the potential that was measured in their experiment but, rather, the electron-density perturbations. But as shown by Wong *et al.* in their original paper and later demonstrated by Doucet *et al.* (1968), and as discussed previously in Chapter 3, *under certain conditions* there is a direct relationship between the density perturbation and the resulting potential perturbation. As we show later, however, such a simple relationship does not always exist. Thus, in order to make a more direct comparison between experiment and theory we compute the electron and ion *density* perturbations produced in the plasma by a dipole.

10.3.2. Density Perturbations

Using Eq. (8.35) for the case of dipolar excitation in an isothermal Maxwellian plasma of ions and electrons, for $z > 0$, we find for the ions

$$n_i = n_{e0} \frac{eV_0}{2KT_i} f^2 N_i(z), \qquad (10.66)$$

where

$$\mathcal{N}_i(z) = -\frac{1}{\pi} \int_0^\infty \mathrm{Im}\left[\frac{Z^{+\prime}(1/\eta)}{\eta f^2 \varepsilon^+}\right] e^{i\eta z} \, d\eta; \tag{10.67}$$

and for the electrons

$$n_e = n_{e0} \frac{eV_0}{2KT_i} f^2 \mathcal{N}_e(z), \tag{10.68}$$

where

$$\mathcal{N}_e(z) = \frac{\Theta^2}{\pi} \int_0^\infty \mathrm{Im}\left[\frac{Z^{+\prime}(m\Theta\xi)}{\eta f^2 \varepsilon^+}\right] e^{i\eta z} \, d\eta. \tag{10.69}$$

We can use the methods of the previous section, which allowed us to calculate the three *potential* contributions P, I and E, to calculate the corresponding *density* perturbations. Doing this, we find the P ion and electron density perturbations (i.e., those resulting from ion-acoustic waves) to be given by

$$\mathcal{N}_{ip}(z) = -\frac{C_1}{f^2} Z^{+\prime}(\xi_1) e^{i\eta_1 z}; \tag{10.70}$$

$$\mathcal{N}_{ep}(z) = \frac{C_1}{f^2} \Theta^2 Z^{+\prime}(m\Theta\xi_1) e^{i\eta_1 z}. \tag{10.71}$$

Taking the ratio of these density perturbations and using Eq. (10.8) we find

$$\frac{Z_c \mathcal{N}_{ip}}{\mathcal{N}_{ep}} = 1 - \frac{f^2}{\xi^2 \Theta Z^{+\prime}(m\vartheta\xi)/Z_c} \tag{10.72}$$

We see that for low frequencies (i.e., for $f \ll 1$) the second term on the right-hand side of Eq. (10.72) is small with respect to the first term, so that we find the well-known result of quasi-neutrality for low-frequency density perturbations in a plasma. But we see clearly that when the frequency increases, the neutrality no longer exists, which then produces wave dispersion. We note, without proof, the following interesting point: The quasi-neutrality at low frequencies is not limited to the ion wave; for each root ξ_m considered, it is possible to find a relationship similar to that given by Eq. (10.72) indicating quasi neutrality for all low-frequency density perturbations in a plasma.

Proceeding as in the previous section, we find the I ion and electron density perturbations, i.e., those found for the asymptotic behavior of the ions, to be given by

$$\mathcal{N}(z)_{iI} \simeq \frac{2}{3^{1/2}} \frac{f^2 + 2\Theta^2 \xi_c^2 / Z_c}{|(f^2 - 1) + 2\Theta^2 \xi_c^2 / Z_c|^2} e^{-3\xi_c^2}; \tag{10.73}$$

$$\mathcal{N}(z)_{eI} \simeq \frac{2}{3^{1/2}} \frac{2\Theta^2 \xi_c^2}{|(f^2 - 1) + 2\Theta^2 \xi_c^2 / Z_c|^2} e^{-3\xi_c^2}. \tag{10.74}$$

For low frequencies, quasi neutrality is again easily shown since

$$\mathcal{N}_{il}(z, f \ll 1) \simeq \frac{1}{3^{1/2}} \frac{\mathcal{Z}_c}{\Theta^2} \frac{\xi_c^2}{|\xi_c^2|} e^{-3\xi_c^2} \simeq \frac{\mathcal{N}_{el}}{\mathcal{Z}_c}. \tag{10.75}$$

For high frequencies, we find that

$$\mathcal{N}_{il}(z, f \gg 1) \simeq \frac{2}{3^{1/2}} \frac{f^2 + 2\Theta^2 \xi_c^2 / \mathcal{Z}_c}{f^2 + 2\Theta^2 \xi_c^2 / \mathcal{Z}_c} e^{-3\xi_c^2}. \tag{10.76}$$

$$\mathcal{N}_{el}(z, f \gg 1) \simeq \frac{2}{3^{1/2}} \frac{2\Theta^2 \xi_c^2}{|f^2 + 2\Theta^2 \xi_c^2 / \mathcal{Z}_c|^2} e^{-3\xi_c^2}. \tag{10.77}$$

Thus, upon comparing Eqs. (10.76) and (10.77), we find that quasi-neutrality does not exist at high frequencies. As for the preceding ion-acoustic wave case this is because the ions have large inertia at high frequencies and can no longer follow the motion of the electrons.

Before proceeding to find the E density contributions, i.e., those introduced by the electron plasma wave, we wish to briefly discuss a concept that we call the *oscillating Debye sheath approximation*. In this approximation we make the assumption that the only electric field that the particles in the vicinity of the dipole see is that of the oscillating Debye sheath. At high frequencies this assumption allows us to calculate density perturbations that agree with those calculated using the actual electric field produced by the dipole itself. However, at low frequencies this approximation is not accurate. As an example of this approximation, we calculate the asymptotic ion density contributions that would be produced by such an oscillating sheath, both for $f \ll 1$ and $f \gg 1$, and then compare the results with Eqs. (10.75) and (10.76), respectively.

To calculate the asymptotic ion density contributions produced by the oscillating Debye sheath, we first use Eq. (8.9) to compute the velocity perturbations produced by the oscillating sheath. The velocity perturbations are then used to compute the corresponding distribution function perturbation f_1 and then the corresponding density perturbation n_1. For the velocity perturbation we find

$$\delta v(x \gg 1, v > 0, \omega_0) \simeq \frac{q}{mv} E(k, \omega_0) e^{i\omega_0 x / v} \tag{10.78}$$

$$\delta v(x \gg 1, v < 0, \omega_0) \simeq 0.$$

We then use the corresponding distribution function perturbation to compute the average ion density perturbation

$$\bar{n}_i(z) \simeq n_{0e} \frac{eV_0}{KT_i} \int_0^\infty \frac{e^{-u^2 + iz/u}}{f^2 \xi^+(f,u)} du. \tag{10.79}$$

For $f \ll 1$, we can approximate $f^2 \varepsilon^+$ by

$$f^2 \xi^+ \simeq f^2 + 2\xi^2(1 + \Theta^2/\mathcal{Z}_c), \tag{10.80}$$

which is equivalent to assuming that the only electric field applied to the particles is that of the oscillating Debye sheath. Equation (10.79) then gives

$$\bar{n}_i(z, f \ll 1) \simeq n_{0e} \frac{eV_0}{KT_i} \frac{f^2}{3^{1/2}} \frac{e^{-3\xi_c^2}}{2\xi_c^2(1 + \Theta^2/\mathcal{Z}_c)}. \tag{10.81}$$

For $f \gg 1$, we use the approximation

$$f^2 \xi^+ \simeq f^2 - 1 + \frac{2\xi^2\Theta^2}{\mathcal{Z}_c} \simeq f^2 + \frac{2\xi^2\Theta^2}{\mathcal{Z}_c}. \tag{10.82}$$

Equation (10.82) has roots, $k = \pm i/\lambda_d$, i.e., we have considered that the ions are motionless in the Debye-sheath calculation. Equation (10.79) then gives

$$\bar{n}_i(z, f \gg 1) \simeq n_{0e} \frac{eV_0}{KT_i} \frac{f^2}{3^{1/2}} \frac{e^{-3\xi_c^2}}{f^2 + 2\xi^2\Theta^2/\mathcal{Z}_c}. \tag{10.83}$$

Comparing Eqs. (10.81) and (10.83), respectively, with Eqs. (10.75) and (10.76), by use of Eq. (10.66), we find that

$$\left|\frac{n_i}{\bar{n}_i}\right| = \frac{1 + \mathcal{Z}_c T_e}{T_i}; \tag{10.84}$$

$$\left|\frac{n_i}{\bar{n}_i}\right| = 1. \tag{10.85}$$

Thus, for high frequencies, the oscillating-sheath approximation gives correct results. For low frequencies, however, we obtain incorrect results by assuming that the only field acting on the particles is the field of the oscillating Debye sheath, the approximation being worse for $T_e \gg T_i$.

Continuing with the calculation of the density perturbations produced by a dipole, we find the normalized-ion and electron-density perturbations due to the electron plasma wave to be given, respectively, by

$$\mathcal{N}_{iE}(z) \sim \frac{m\mathcal{Z}_c}{\pi^{1/2}\Theta} \left(-\frac{i}{z^3}\right), \tag{10.86}$$

$$\mathcal{N}_{eE}(z) \sim \frac{m\mathcal{Z}_c^2}{\pi^{1/2}\Theta} \left(-\frac{i}{z^3}\right)(1 - f^2). \tag{10.87}$$

For $f \ll 1$, we see that Eqs. (10.86) and (10.87) are identical so that quasi-neutrality is again seen at low frequencies. For $f \gg 1$, however, we have

$$\frac{\mathcal{Z}_c \mathcal{N}_{iE}}{\mathcal{N}_{eE}} \simeq -f^2, \tag{10.88}$$

so that, as expected, because of the large inertia of the ions, quasi neutrality is not preserved at high frequencies.

Observation of Eqs. (10.86) and (10.87) shows that for large distances the electronic *density* contribution is much weaker, relative to the total perturbation, than is the electronic *potential* contribution, the density perturbation decreasing as $1/z^3$ versus the potential perturbation decreasing as $1/z$ [see Eq. (10.65)]. At $z \sim 50$, for example, the electronic-density contribution relative to the total is approximately three orders of magnitude less than that of the electronic potential contribution. One may suspect, therefore, that in terms of the total signal the ion wave contribution may dominate the electron plasma wave contribution over much larger distances when density perturbations rather than potential perturbations are being considered. This suspicion

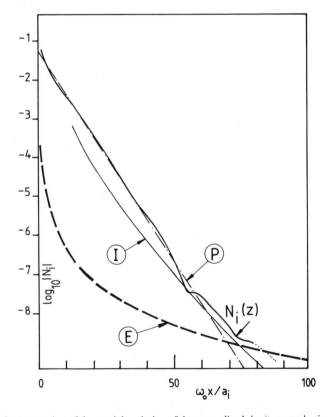

Fig. 10.5. Interpretation of the spatial variation of the normalized *density* perturbation $\mathcal{N}_i(z)$ [see Eq. (10.67)] produced in an isothermal cesium plasma by low-frequency $(\omega_0/\omega_p = 0.1)$ dipole excitation (same conditions as for Fig. 10.2). From this comparison of *density* perturbations, contrary to what was observed in Fig. 10.2 where a comparison was made of *potential* perturbations, it is seen that over a short distance the ballistic contribution (I) dominates the ion wave contribution (P) before the electron contribution (E) finally dominates.

is confirmed in Figs. 10.5 and 10.6, which show the results of numerical calculations giving both the total ion-density perturbation, $\mathcal{N}_i(z)$ [see Eq. (10.67)] and the three contributions to this total as given by Eqs. (10.70), (10.73), and (10.86). Figure 10.5 shows that $P \geqslant E$ for distances up to $z \sim 70$ (versus $z \sim 35$ for *potential* perturbations; see Fig. 10.3) for the low-frequency case. For high frequencies, Fig. 10.6 shows that $P \geqslant E$ for distances up to $z \sim 50$ (versus $z \sim 20$ for potential perturbations; see Fig. 10.4).

Figure 10.7 makes a comparison of the potential perturbation V (normalized by the driving voltage V_0 on the dipole) and the normalized ion and density perturbations $\mathcal{N}_i$ and $\mathcal{N}_e$. The comparison is made at a fixed distance of approximately 30 Debye lengths from the dipole, as a function of the normalized excitation frequency. The density perturbations were obtained by

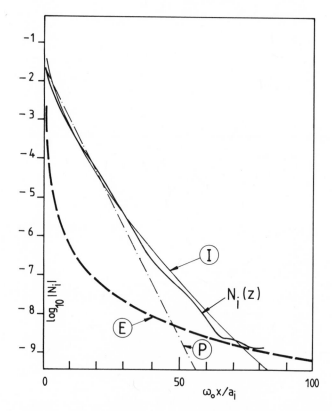

Fig. 10.6. Interpretation of the spatial variation of the normalized *density* perturbation $\mathcal{N}_i(z)$ for the same conditions as in Fig. 10.5, except that here *high*-frequency $(\omega_0/\omega_p = 3)$ excitation is being considered. For this case, it is seen that the ballistic contribution (I) dominates the ion-acoustic wave contribution (P) over an appreciable range of distance.

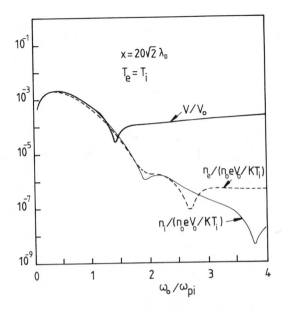

Fig. 10.7 A comparison of the variation with frequency of the amplitudes of the potential and density perturbations produced in an isothermal cesium plasma by dipole excitation at a distance of approximately 30 Debye lengths from the dipole antenna.

numerical integration of Eqs. (10.67) and (10.69). The normalized potential perturbation was obtained by numerical integration of Eq. (10.34) and subsequent substitution in Eq. (10.33).

Observation of Fig. 10.7 reveals several things of interest. First, the normalized potential curve clearly changes character at $\omega_0 \sim 1.4\, \omega_{pi}$. This reflects the fact that, as already discussed in the preceding section dealing with potential perturbations, there are *two* regimes of variation of the *potential* with frequency: one below the ion plasma frequency in which the ion-acoustic wave contribution dominates, and one above the ion plasma frequency in which the electron plasma wave dominates. Second, the normalized *density* perturbation curves show *three* regimes of variation with frequency: one for $\omega_0 \lesssim 1.8\, \omega_{pi}$ for both the ions and the electrons; one for $1.8\, \omega_{pi} \lesssim \omega_0 \lesssim 2.6\, \omega_{pi}$ for the electrons and $1.8\, \omega_{pi} \lesssim \omega_{pi} \lesssim 3.8\, \omega_{pi}$ for the ions; and one for $\omega_0 > 3\omega_{pi}$ for the electrons. Based on the discussions in this section, the first regime can be easily understood as being the frequency range in which the ion-acoustic wave contribution dominates. In this range, as expected, quasi neutrality is seen to be closely observed. The second regime can be understood as being the frequency range in which we have a mixture of ion-acoustic wave and ion-ballistic (or free-streaming) contributions. Over most of this range, quasi neutrality begins to break down and (though not indicated here) strong ion-acoustic wave dispersion begins to occur as the frequency is increased. Also in this region, the condition given by Eq. (10.12) is no longer fulfilled, so that the Boltzmann equation is no longer valid. The third region corresponds to the high-frequency range where the electron plasma wave is always the dominant contributor to the density perturbation. It can be noted also that, on comparing the qualitative behavior of the potential and density curves, there is an approximately linear relationship between the potential perturbation and the density perturbations at low frequencies, as found both by Wong *et al.* and by Doucet *et al.* (1968); see also Fig. 3.10. Above ω_{pi}, however, this linear relationship no longer exists.

Figure 10.8 shows the importance of each of the three contributors to the normalized *ion density* $N_i(z,f)$ as a function of distance and frequency. This figure shows the frequency–distance domain in which each of the contributors dominates. The three contributions were computed using Eqs. (10.70), (10.73), and (10.86), respectively, for the ion-acoustic wave, ballistic, and electron plasma wave contributions. Inside a given domain, the criterion used to indicate dominance was that the contribution of the dominant contributor in that domain be at least a factor of e^2 larger than those of the other two

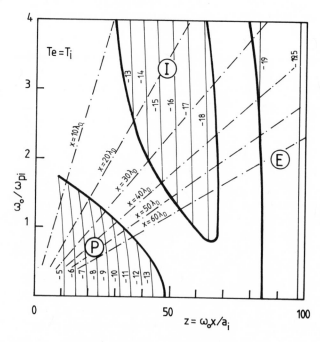

Fig. 10.8. The frequency–distance domains in which each of the contributors to the normalized density $N_i(z)$ dominate. The criterion used to determine a domain of dominance was that the dominant contribution in that domain be at least e^2 larger than that of any other contribution.

contributors. The approximately vertical lines shown in each domain provide an indication of the relative magnitude of the contribution in that domain, relative to $\ln|N_i(z,f)|$. The series of dot-dash curves radiating from the origin of the coordinate system allows a comparison at various distances from the dipole exciter of how the magnitude of the contribution of each contributor varies with frequency in its domain of dominance.

10.3.3. Remarks about Wong et al.'s Experiment

It is of interest to reconsider the possible interpretations of the experiment of Wong *et al.* We begin by briefly describing the experiment and summarizing what was done in that work. We then point out a possible incompatibility between the experiment and the theoretical model. A theoretical model that is perhaps more suitable, is then discussed, and comparison is made between the predictions of this model and the experimental measurements. Finally, we note an alternative interpretation of this experiment in terms of the propagation of electron plasma waves in a finite-geometry plasma.

First, we remind the reader that this was an experiment dealing with wave propagation parallel to the axis of a magnetically supported isothermal cesium plasma in a so-called *Q*-machine (called "*Q*" because such a plasma was found to be very *q*uiescent). Using negatively biased metallic probes, measurements were made of the electron-*density* perturbations produced along the plasma column on one side of a planar grid that was oriented perpendicular to the plasma column and on which were imposed sinusoidally varying voltages of frequency $\omega_0 \ll \omega_{pi}$. These measurements were then compared with the *potential* perturbations given by Gould's theory of dipolar excitation. The *qualitative* behavior of the amplitudes of the observed and computed perturbations with respect to frequency and spatial variation strongly suggested that the observed perturbation was a Landau-damped ion-acoustic wave, and this was the interpretation that was made. However, it was noted that the amplitude of the measured perturbation was approximately an order of magnitude larger than the computed amplitude.

Two additional factors raise some questions with respect to the ion wave interpretation of these researchers. First, a single grid was used as the wave source in the experiment. In principle, such a source would seem to be more accurately modeled by a monopole, rather than by the dipole used in Gould's theory. Second, finite-geometry effects may have been important. Although electron plasma waves, for example, would not be expected to propagate below the electron plasma frequency in an infinite plasma (see Fig. 3.1). We have seen in Chapter 5 that finite-geometry effects, even in a cold plasma (see Fig. 5.3 and Section 5.2.2.1), can allow the electron plasma wave (the *slow wave*) to propagate at lower frequencies. Thus, it is not *a priori* clear that electron plasma waves were not present in this experiment.

The theory of monopolar excitation is not greatly different from Gould's theory of dipolar excitation. Doing such a theory, one finds that

$$n_e(z) = n_{0e} \frac{E_0/k_{di}}{2KT_i} f i \frac{\Theta^2}{\pi} \int_0^\infty \mathrm{Im}\left[\frac{Z^{+\prime}(m\Theta\xi)}{f^2\eta^2\varepsilon^+}\right] e^{i\eta z} \, d\eta, \qquad (10.89)$$

where $k_{di} = \omega_{pi}/a_i$. One then finds the electron density contributions due to ion-acoustic waves and electron-acoustic waves for monopolar excitation to be given, respectively, by

$$n_{ep}(z) = n_{0e} \frac{E_0/k_{di}}{2KT_i} f\left(-\frac{2iC_1\xi_1}{f^2}\Theta^2\right) e^{i\eta_1 z}; \qquad (10.90)$$

$$n_{eE}(z) = n_{0e} \frac{E_0/k_{di}}{2KT_i} f\left[\frac{m Z_c^2}{2\pi^{1/2}\Theta}(1-f^2)\frac{i}{z}\right]. \qquad (10.91)$$

We denote the amplitude of the experimentally measured electron density perturbation as n_{em}. Since we want to make a comparison between the measured amplitude and the Φ-calculation of Gould, we have adjusted this quantity so that $n_{em}(z=0) \simeq \Phi(z=0)$. Doing this, we can say that

$$\frac{n_{em}}{(n_{eE})\mathrm{monop}} = \frac{n_{em}}{(\Phi_e)\mathrm{dip}} \frac{(\Phi_e)\mathrm{dip}}{(\Phi_P)\mathrm{dip}} \frac{(n_{ep})\mathrm{monop}}{(n_{eE})\mathrm{monop}} = \frac{n_{em}}{(\Phi_e)\mathrm{dip}}\left(-\frac{2\xi_1\Theta^2 z}{Z_c}\right). \qquad (10.92)$$

For $z \simeq 50$, one finds that $n_{em}/\Phi_e \simeq 10$. Thus, for the same z value

$$\left|\frac{n_{em}}{(n_{eE})\mathrm{monop}}\right| \simeq 1000. \qquad (10.93)$$

Thus, because of a greatly decreased excitation efficiency of monopolar excitation as compared to dipolar excitation, we see that instead of the measured signal having an amplitude that is ten times that expected for dipolar excitation, the amplitude (at $z \simeq 50$) is one-thousand times that expected for monopolar excitation! The introduction of collisions (which are not negligible for such a plasma) by Virmont (1972) did not significantly reduce this factor. Such a large discrepancy casts some doubt on the interpretation of the observed signal as a Landau-damped ion-acoustic wave, to say the least.

An alternative interpretation of this experiment has been given by Jahns and Van Hoven (1972) who did similar experiments. They interpreted the observed phenomenon as electron plasma waves that are able to propagate at these low frequencies because of finite-geometry effects. In support of this interpretation, the phase velocities measured by Jahns and Van Hoven were found to be in good agreement with the dispersion relation of Malmberg and

Wharton (1966) for the propagation of electron plasma waves in a finite-size, isothermal, warm-plasma column.

10.3.4. Collective Effects in an Isothermal Plasma

The collective effect is characterized by the ion-acoustic wave contribution. As shown in Fig. 10.8 for an isothermal plasma, the collective effect dominates at low frequencies, since, to first order, the other contributions are negligible in this range. However, the other contributions can still be measured as is shown in the following example: Suppose we choose $f = \omega_0/\omega_{pi} = 3$ and $x = 10\lambda_d$ (corresponding to $z \simeq 50$). We are then in the ballistic regime, and from Fig. 10.6

$$n_{iI}(x = 10\lambda_D) \sim \frac{eV_0}{KT_i} 10^{-6}. \tag{10.94}$$

For a plasma having a density $n_{e0} \simeq 10^{11}$ cm^{-3}, this density fluctuation corresponds to an ion-current-density perturbation of the order of

$$j_i(A/cm^2) \sim n_i q_i a_i \sim 10^{-9}\left(\frac{eV_0}{KT_i}\right), \tag{10.95}$$

since $a_i \sim 500$ m/s in a Q-machine. Even though this is a weak perturbation, Doucet and Gresillon (1970) have shown that it is measurable. This fact, plus the fact that an examination of Figs. 10.5 and 10.6 shows that the difference between the amplitudes of the collective and ballistic contributions is not large, justifies the remark of Hirshfield and Jacob (1968) concerning the possibility of there being more than one contribution to the signal observed by Wong *et al.* in their experiment.

In Figs. 10.9 and 10.10 we give the results of some numerical computations that allow us to establish some criteria by which it is possible to distinguish between collective and ballistic effects in a forced-oscillations experiment is an isothermal plasma for the case of dipolar excitation. Using Eqs. (8.85) we have calculated the local phase velocity $\omega/k_r a_i$ and damping rate k_i/k_r for both the potential and density perturbations. Figure 10.9 gives the low-frequency results, while Fig. 10.10 gives the high-frequency results.

In Fig. 10.9, we see two different regimes. In one, the phase velocity and the damping rate are given approximatively by the first root of the dispersion relation. In the second, the phase velocity reaches very high values due to the electron-wave contribution. We note that the ion-wave contribution is dominant over a range which includes four wavelengths for the density perturbation, but only two wavelengths for the potential perturbation.

In Fig. 10.10, we note that the domain in which the ion wave was dominant at low frequencies is now almost completely ballistic at high frequencies,

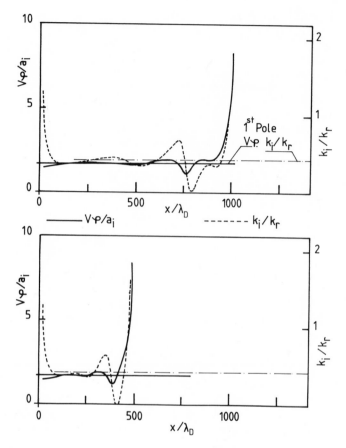

Fig. 10.9. Local values of the phase velocity v_φ/a_i and damping k_i/k_r as a function of distance for the perturbations produced in an isothermal cesium plasma by low-frequency ($\omega_0/\omega_p = 0.1$) dipole excitation. Upper figure: as computed from the ion-density perturbation. Lower figure: as computed from the potential perturbation.

as can be seen from the fact that the phase velocity $v_\varphi \simeq [2/(3^{1/2})](z/2)^{1/3}$ is approximately equal to the free-streaming value given by Eq. (8.89).

We conclude that if we desire to study the collective effect in an isothermal plasma, it is better to measure the density perturbations than to measure the potential perturbations. In fact, when the signal can be detected over a distance sufficiently large that more than one wavelength can be observed, it is possible to distinguish between collective and ballistic effects from the fact that, in the collective case, the wavelength is independant of the distance, while in the ballistic case the wavelength depends on the distance.

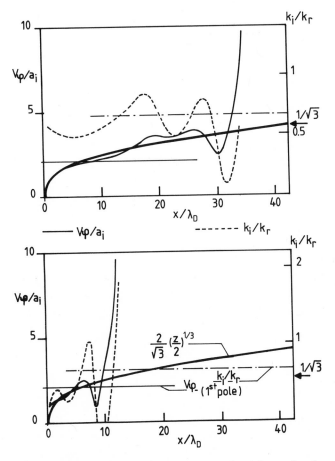

Fig. 10.10. Local values of the phase velocity v_φ/a_i and damping k_i/k_r as a function of distance for the perturbations produced in an isothermal cesium plasma by high-frequency $(\omega_0/\omega_p = 3.0)$ dipole excitation. Upper figure: as computed from the ion-density perturbation. Lower figure: as computed from the potential perturbation.

We see therefore that the isothermal-plasma case corresponds to a marginal situation where the distinction between collective and ballistic effects is possible but, as can be appreciated from the discussion presented here, is quite a difficult problem. On the other hand, for $T_e \geqslant 2T_i$ the problem is much less difficult. This can be seen from Fig. 10.11, which shows the variation of the ion-density perturbation at a low frequency as a function of distance for several electron–ion temperature ratios. Now, the ion-density perturbation in an isothermal plasma, at low frequencies and for the distances shown, has

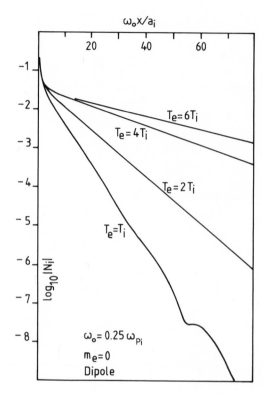

Fig. 10.11. Amplitude variation of the normalized density perturbation $N_i(z)$ produced in an isothermal cesium plasma by dipole excitation for several values of T_e/T_i as a function of distance from the antenna.

been seen to result predominantly from the ion wave and ballistic contributions (see Fig. 10.5). Moreover, it can be seen from Eq. (10.75) that the ballistic contribution is almost constant with respect to variations of the temperature ratio. Thus, we can conclude from Fig. 10.11 that for $T_e \geqslant 2T_i$ the ion-density perturbation is strongly dominated by the ion-acoustic wave contribution for low frequencies. Therefore, making the distinction between collective and ballistic effects in a *non*isothermal plasma is, in principle, not a difficult problem in low-frequency experiments.

10.4. SELECTED EXPERIMENT: LANDAU DAMPING OF ION-ACOUSTIC WAVES IN A NONISOTHERMAL PLASMA

Alexeff *et al.* (1967) were able to produce at will Landau damping of ion-acoustic waves in a nonisothermal plasma without changing any of the gross parameters—electron temperature, ion temperature, wave velocity, or wave frequency—of the wave–plasma system. Figure 10.12 shows a schematic of the apparatus used in their experiment. As the reader will recognize,

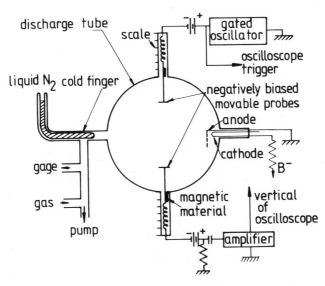

Fig. 10.12. Schematic of the apparatus used (see also Fig. 3.3). (After Alexeff *et al.*, 1967.)

this system is the same as the one shown in Fig. 3.3 and is repeated here only for convenience. The basic operation of the system and the techniques of wave propagation and measurement are described in detail in Section 3.4.1 and will not be repeated here. The procedure used in the experiment was as follows: A plasma of heavy ions (xenon; mass = 131 amu) was produced, and ion-acoustic wave trains were generated by the applications of sine wave bursts to a negatively biased planar probe (see Section 3.4.1). Figure 10.13 shows some typical data. Under these conditions, the wave speed [see Eq. (3.28)] is much higher than the mean thermal speed of the xenon ions, and Landau damping was neither predicted nor observed. However, a trace (less than 1%) of light ions (helium; mass = 4 amu) was then added to the plasma

Fig. 10.13. Typical transmitted (bottom trace) and received (upper trace) signals. The vertical axis corresponds to amplitude (in relative units), and the horizontal axis corresponds to time (10 μs/cm). (See also Fig. 3.4.) (After Alexeff *et al.*, 1967.)

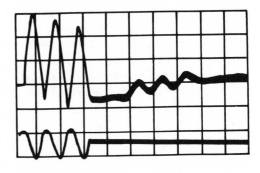

by adding helium gas to the system. The mean thermal speed of the light ions was comparable to the ion-acoustic wave speed, and Landau damping was both predicted and observed. Coherent detection was used to improve the signal-to-noise ratio. The experimental measurements were made in the center of the spherical tube in order to minimize radial plasma drifts and plasma-boundary effects. Typical plasma–wave parameters were: $T_e = 1.2$ eV, $T_i = 0.052$ eV, $n_e = 5 \times 10^8 \, \text{cm}^{-3}$, discharged current = 150 mA, applied voltage = 22 V, and measured wave speed = 7.6×10^4 cm/s.

The Landau damping measurements were made in the following manner. With no helium present, the amplitude of the signal was measured as a function of frequency and propagation distance. The signal strength far from the source was observed to decrease as $1/r^2$, showing that the voltage output of the receiving probe was proportional to energy. The rate of decrease of amplitude with distance was observed to be frequency independent. Next, helium was added. The amplitude of the received signal now was observed to decrease more rapidly with distance than before. Also, the rate of decrease was observed to be more rapid with increasing frequency. If the plot of received signal versus amplitude was extrapolated back to zero transmission distance, the source strength of the signal was observed not to change appreciably when helium was added. The wave speed and electron temperate changed only about 10%. Thus, it was assumed that the addition of helium did not appreciably perturb the system, other than to provide Landau damping.

The data were analyzed as follows. First, the observed amplitude of the received signal for each electrode spacing and frequency, with helium present, was divided by the corresponding signal without helium. This division, for a given frequency yielded the relative amplitude decrease as a function of distance due to helium alone. (Note that this technique eliminated any geometry factors from the result.) The relative decrease of amplitude with distance was then fitted by least squares with an exponential function of the form $\exp(-X/X_0)$, where X was the electrode spacing and X_0 was the observed damping distance for that frequency. Finally, the observed damping distance X_0 was plotted as a function of frequency, as shown in Fig. 10.14.

In Fig. 10.14, we see that the observed damping distance X_0 varies inversely with frequency. Thus, it was concluded that the damping could not have been due to collisions, because collisional damping of weakly damped waves in free flight is frequency independent. To compare the observed damping with theory, a simple linear approximation, valid for small contamination and based on the Landau-damping calculations of Fried and Gould (1961) and Gould (1964), was used. The theoretical Landau-damping factor $X_0^{-1/2}$ was found to be given by

$$X_0^{-1/2} = \frac{\varepsilon}{\lambda} \frac{\pi^{3/2}}{2^{1/2}} \left(\frac{T_e}{T_i}\right)^{3/2} \left(\frac{m_2}{m_1}\right)^{1/2} \exp\left[-\frac{m_2 T_e}{2 m_1 T_i}\right], \qquad (10.96)$$

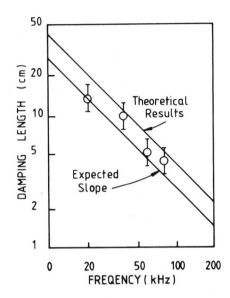

Fig. 10.14. Experimentally observed and theoretically computed damping distances as a function of frequency. No adjustable parameters are involved in the theoretical calculation. The curve marked "expected slope" corresponds to assuming that $T_i = 1/30$ eV. The curve marked "theoretical results" corresponds to the measured value of $T_i = 0.052$ eV. Data were taken at 20, 40, 60, and 80 kHz. (After Alexeff *et al.*, 1967.)

where ε is the ratio of the number density of light to heavy ions, λ is the wavelength of the ion wave, T_e is the electron temperature, T_i is the ion temperature, m_1 is the mass of the heavy ion, and m_2 is the mass of the light ion. As was noted from this equation, the damping distance X_0 varies directly with wavelength or, for their system, which had negligible dispersion, inversely with frequency. Thus—see Fig. 10.14—the predicted dependence of damping on frequency was observed. Not only is the observed damping observed to be qualitatively correct, but, within a factor of two, is also seen to be quantitatively correct. Considering the uncertainty in the calculation of ε, which was calculated typically to be 3×10^{-3}, the agreement between the observed damping and the theoretically predicted Landau damping is good.

Further support for assuming that the observed damping was Landau damping came from earlier, preliminary experiments done by these same researchers (1967) using pulses rather than sinusoidal voltages on the transmitter probe. For this type of excitation, the excited wave would have consisted of a mixture of frequencies rather than a single frequency. As in the later experiments, a small amount of helium was observed to produce strong damping. However, for this case, the damped wave packet was observed to spread in time as it propagated. But, as these researchers noted, this is just what would be expected if the damping were Landau damping since, as already noted, the high-frequency components are predicted to be more strongly damped than the low-frequency components.

NUMERICAL METHODS

11.1. INTRODUCTION

The broad use of microcomputers now makes it possible for students to solve physical problems that just a few years ago could be considered only from a conceptual point of view. In this chapter we present some numerical methods that the reader may use on a microcomputer with two benefits: (1) By looking at the numerical solution of a particular problem, the reader may reach a better understanding of some of the theoretical subtleties discussed in this book; and (2) the reader will become more familiar with some of the numerical techniques currently used in plasma physics research.

We consider two topics in this chapter: (1) The calculation of Hilbert transforms; and (2) the numerical solution of dispersion relations. These problems have been selected because of their importance, both in wave propagation phenomena as well as in many other problems in science and engineering.

11.2. NUMERICAL EVALUATION OF HILBERT TRANSFORMS

We consider the Hilbert transform $Q^\pm$ of an entire function $\varphi_0(u)$ defined by

$$Q^\pm(z) = \int_{L^\pm} \frac{\varphi_0(u)}{u - z} du, \tag{11.1}$$

where $L^\pm$ are the Landau paths defined in Fig. 1.8. We calculate $Q^+(z)$ directly, but since Q^+ and $Q^-(z)$ are related by the equation

$$Q^+(z) = Q^-(z) + 2\pi i \varphi_0(z), \tag{11.2}$$

once we calculate Q^+, Q^- can then be easily found. We look at the case where $\varphi_0(u)$ is the Maxwellian distribution function

$$\varphi_0(u) = \pi^{-1/2} \exp(-u^2), \tag{11.3}$$

i.e., we compute the plasma dispersion function $Z^+(z)$. As shown in Chapter 1, Z^+ is related to the error function and in large computer facilities is probably implemented as a FORTRAN subroutine in the library of the computer. However, for microcomputers, this function is not available and will have to be programmed. However, if $\varphi_0(u)$ is not a Maxwellian, then it will likely be necessary to compute $Q^+(z)$ since the non-Maxwellian case will not likely be in the library of even a large computing facility. Therefore, we present a method for computing Z^+ that may be used both for small and large computers, even if $\varphi_0(u)$ is not a Maxwellian.

11.2.1. Calculation of $Q^+(z)$ by Integration

Equation (11.1) is a simple integral, and numerical integration is very simple to implement on any computer. However, the numerical evaluation of Eq. (11.1) encounters two numerical problems:

1. The integration interval is infinite. This is not adapted to the computer, which can manipulate only finite numbers.
2. The integrand in Eq. (11.1) is singular for $u = z$. Again, the computer is not adapted to handle this kind of situation.

In order to avoid these difficulties, we have to do some algebra in order to have a finite integration path and to avoid the singular behavior of $\varphi_0(u)/(u - z)$.

The first procedure is to make a change of variable. For example, we can define

$$u = \tan(x). \tag{11.4}$$

Then, when u varies from $-\infty$ to $+\infty$, x varies from $-\pi/2$ to $+\pi/2$, and Eq. (11.1) becomes

$$Q^+(z) = \int_{-\pi/2}^{+\pi/2} \frac{\varphi_0[u(x)]}{u(x) - z} [1 + u^2(x)] \, dx, \qquad \text{for } \text{Im}(z) > 0. \tag{11.5}$$

Now the integration path is finite, but the integrand can still be irregular when $\text{Im}(z) \to 0$. To solve this problem we will use the technique given in Section 1.5.2.1:

$$Q^+(z) = \int_{-\infty}^{+\infty} \frac{\varphi_0(u) - \varphi_0(z)}{u - z} \, du + i\pi\varphi_0(z), \qquad \text{for } \text{Im}(z) \geqslant 0. \tag{11.6}$$

Now the integrand is no longer singular and, using the change of variable defined by Eq. (11.4), Eq. (11.6) becomes

$$Q^+(z) = \int_{-\pi/2}^{+\pi/2} \frac{\varphi_0[u(x)] - \varphi_0(z)}{u(x) - z} [1 + u^2(x)] \, dx + i\pi\varphi_0(z),$$

$$\text{for } \text{Im}(z) \geqslant 0. \tag{11.7}$$

The integrand in Eq. (11.7) can still present a numerical difficulty when $u(x) \simeq z$. In this case we may make a Taylor expansion of the integrand, given by

$$\frac{\varphi_0(u) - \varphi_0(z)}{u - z} \simeq \varphi_0'(z) + \varphi_0''(z) \frac{u - z}{2} + \cdots \tag{11.8}$$

At this point, we have only to integrate numerically a regular integrand over a finite integration path. This is easily done with a good accuracy using an integration equation of the Gaussian type

$$\int_a^b f(u) \, du = \frac{b - a}{2} \sum_{i=1}^n W_i f(u_i),$$

with $\tag{11.9}$

$$u_i = \frac{b - a}{2} X_i + \frac{b + a}{2},$$

where the X_i are the zeros of the Legendre polynomials $P_n(X)$, and the weight factors W_i are given by

$$W_i = \frac{2}{(1 - X_i)^2} [P_n'(X_i)]^2. \tag{11.10}$$

The abscissas X_i and the weights W_i are tabulated in standard mathematical handbooks such as the one by Abramowitz and Stegun (1972).

11.2.2. Series Expansion

The *straight-forward* numerical integration of Eq. (11.1) is convenient because it can be applied to any entire function $\varphi_0(u)$ which can represent a distribution function, because it requires very little algebra. However, if a higher computing speed is needed, one can use more clever methods. The first method used, generally, is the series expansion.

If $\varphi_0(u)$ is specified, then a Taylor expansion of $Q^+(z)$ is possible, since Q^+ is an entire function. Such an expansion around $z = 0$ is given by

$$Q^+(z) = Q^+(0) + Q^{+\prime}(0)z + Q^{+\prime\prime}(0) \frac{z^2}{2} + \cdots, \tag{11.11}$$

where the $Q^{+(n)}(z)$ may be computed analytically. The use of a series expansion such as given by Eq. (11.11) greatly increases the speed of the calculation for $|z| < a$.

11.2.3. Asymptotic Expansion

For *large* values of $|z|$ the asymptotic expansion defined by Eq. (1.162) may be used when the moments of the distribution function can be calculated analytically. We use this technique for speeding up the computation for large values of $|z|$ for the case where φ_0 is Maxwellian.

11.2.4. Other Methods

Depending on the particular function $\varphi_0(u)$, other numerical methods may be used. In the case of a Maxwellian distribution the continued fraction (see, for instance, Abramowitz and Stegun, 1972) can be interesting for $\text{Im}(z) > a$, or the Salzer formula (Abramowitz and Stegun, 1972) for moderate values of $|z|$.

If $\varphi_0(u)$ is not Maxwellian, similar analytical formulas may be found that will speed up the calculations. In any case, the method presented here, using straight-forward integration, can provide a first approach, allowing, at least, the production of numerical tables that can be used for checking the other methods.

11.2.5. The Subroutine ZNDEZ

We now describe a subprogram that is very useful for computing the zeros of dispersion relations in warm plasmas. This subroutine computes the function $Z^{+\prime}(z)$ and its derivatives $Z^{+\prime\prime}$ and $Z^{+\prime\prime\prime}$. These derivatives are needed for the determination of the roots of the dispersion relation, as will be shown in the next section.

The subroutine called ZNDEZ is presented in this chapter using an English-like language in order to give a clear understanding of its structure. A BASIC version is given in the appendix of this chapter. ZNDEZ is easy to translate from BASIC into FORTRAN because the two languages are very similar for scientific applications.

We define $z = X + i * Y$, where $i^2 = -1$. We wish to compute

$$Z^{+\prime}(z) = \int_{L^+} \frac{\varphi_0'(u)}{u - z}\, du,$$

where

$$\varphi_0(u) = (\pi)^{-1/2} \exp(-u^2),$$

(11.12)

as well as its first and second derivatives. For $\text{Im}(z) > 0$ we obtain

$$Z^{+\prime}(z) = \int_{-\infty}^{+\infty} \frac{\varphi_0'(u)}{u - z}\, du,$$

(11.13)

and for $\text{Im}(z) < 0$

$$\bar{Z}^{+\prime}(z) = [\bar{Z}^{+\prime}(z^*) + 2i\pi\varphi_0'(z^*)]^*. \tag{11.14}$$

Therefore, our subroutine ZNDEZ can be symbolically written as follows:

- Subroutine ZNDEZ(X,Y,Z1,Z2,Z3,F$,KX)

```
*Comments: Calculation of Z⁺'(z), Z⁺''(z), and Z⁺'''(z)
*z=X+I*Y
*w=X+I*|Y|
*Z1(1)=Re(Z'(z))
*Z1(2)=Im(Z'(z))
*Z2(1)=Re(Z''(z))
*Z2(2)=Im(Z''(z))
*Z3(1)=Re(Z'''(z))
*Z3(2)=Im(Z'''(z))
*F$ initialization flag:
*F$=''D'' at the first CALL of ZNDEZ.
*KX flag=0/1. The value is 1 if F1, F2, and F3 are
             calculated.
*F1(1)=Re(φ₀'(w))
*F1(2)=Im(φ₀'(w))
*F2(1)=Re(φ₀''(w))
*F2(2)=Im(φ₀''(w))
*F3(1)=Re(φ₀'''(w))
*F3(2)=Im(φ₀'''(w))
*End of comments.

KX=0 (F1, F2, and F3 not yet calculated)
XX=X
YY=Y
If Y<0 THEN YY=|Y|
   CALL ZNXY(XX,YY,Z1,Z2,Z3)

IF Y=YY THEN RETURN
IF KX=0 THEN CALL FN(XX,YY,F1,F2,F3,KX)
   Z1=(Z1+2πiF1)*
   Z2=(Z2+2πiF2)*
   Z3=(Z3+2πiF3)*
RETURN
```

The subroutine FN is defined as follows:

- Subroutine FN(XX,YY,F1,F2,F3,KX)

```
*Comments:  w=XX+iYY
```

$$F1(1)=\text{Re}(\varphi_0{}'(w))$$
$$F1(2)=\text{Im}(\varphi_0{}'(w))$$
$$F2(1)=\text{Re}(\varphi_0{}''(w))$$
$$F2(2)=\text{Im}(\varphi_0{}''(w))$$
$$F3(1)=\text{Re}(\varphi_0{}'''(w))$$
$$F3(2)=\text{Im}(\varphi_0{}'''(w))$$

```
KX=1(F1, F2, and F3 calculated)
RETURN
```

Let us examine now the subprogram ZNXY:

- Subroutine ZNXY(XX,YY,Z1,Z2,Z3,F$)

```
*Comments:  w=XX+i*YY
*Calculate Z⁺'(w), Z⁺''(w), and Z⁺'''(w) for YY>0.
*Initialize the constants for Gaussian integration.
*End of comments.
      IF F$=''F'' THEN SKIP
      CALL INITIALIZE
      F$=''F''
SKIP: CALL ZN(XX,YY,Z1,Z2,Z3,KX)
      RETURN
```

The purpose of the subroutine INITIALIZE is to assign a value to all constants used by ZNDEZ. These constants are: π, $\pi^{1/2}$, the Gaussian abscissae $XI(K)$, and weights $WI(K)$, as well as $\varphi^{(n)}\{u[XI(K)]\}$. The degree of the Legendre polynomials has been chosen to be equal to 12, and we use the fact that

```
XI(2*L+1)= -XI(L)
WI(2*L+1)=WI(L)
```

Moreover, in order to increase the precision, the interval of integration $(-\pi/2, +\pi/2)$ is split into $LL = 4$ subintervals.

- Subroutine INITIALIZE

```
Read Values of XI and WI
Initialize π,  π^(1/2), and LL=4
```

```
Q=π/LL (sub-interval length)
QS=Q/2 (normalization factor)
K=0
A=-π/2-Q
REPEAT FOR I=1 TO LL
     A=A+Q (sub-interval start)
     B=A+Q (sub-interval end)
     QP=(A+B)/2

     REPEAT FOR J=1 TO 6
          K=K+1
          U(K)=QS*XI(J)+QP
          CALL GN(U,G1,G2,G3)
          K=K+1
          U(K)=-QS*XI(J)+ QP
          CALL GN(U,G1,G2,G3)
     END OF LOOP FOR J

END OF LOOP FOR I
RETURN
```

The subroutine GN takes into account the change of variable $u = \tan(x)$:

- Subroutine GN(U,G1,G2,G3)

```
U=TAN(U)
G1=φ₀'(U)
G2=φ₀''(U)
G3=φ₀'''(U)
RETURN
```

The subprogram ZN has the following structure:

- Subroutine ZN(XX,YY,Z1,Z2,Z3,KX)

```
*Comment: w=XX+i*YY
*Calculate Z⁺'(w), Z⁺''(w), and Z⁺'''(w) for YY<0.
*If needed, compute F1, F2, and F3 and sets KX=1 in
 that case.
*End of comments.
          D=|XX+i*YY|²
          IF D>25 THEN ASYMPTOTIC
          CALL FN(XX,YY,F1,F2,F3,KX)
          IF D<4 THEN SERIES
```

```
GAUSS:  .........
            RETURN
ASYMPTOTIC:..........
            RETURN
SERIES:..........
            RETURN
```

Now we examine the block called **SERIES**, where $w = XX + i*YY$:

```
SERIES:  Z1=-2+4w²
         Z2=8*w
         Z3=8
         R=8
         I=1
LOOP:  I=I+1
         IF I>9 THEN OUT

            J=2*I
            AO=W²J/(J-3)/(I-1)
            R=AO*R*w²
            T=R*w/(J-1)
            V=w*T/J
            Z1=Z1+V
            Z2=Z2+T
            Z3=Z3+R
            DD=|Z3|²
            IF DD=0 THEN LOOP
         IF |R|²/DD>10⁻¹² THEN LOOP
OUT:  Z1=Z1+iπF1
      Z2=Z2+iπF2
      Z3=Z3+iπF3
      RETURN
```

The recurrence relations used for Z1, Z2, and Z3 are obviously valid for a Maxwellian distribution. As regards the block called **ASYMPTOTIC** we have for a Maxwellian distribution:

```
ASYMPTOTIC:  R=1/w²
             Z1=1/w²
             Z2=-2/w³
             Z3=6/w²
             REPEAT FOR I=2 TO 10
                 J=2*I
                 R=(J-1)*R*/2/w²
                 T=-J*R/w
```

```
                  V=-(J+1)*T/w
                  Z1=Z1+R
                  Z2=Z2+T
                  Z3=Z3+V
            END OF LOOP FOR I
            RETURN
```

Finally, the last block to define is GAUSS:

```
GAUSS:  Z1=Z2=Z3=0 (Initialize)
        KJ=0
        K=0
        IF YY>1 THEN KJ=1
        REPEAT FOR I=1 TO LL
              REPEAT FOR J=1 TO 6
                      KI=0
              AGAIN:  K=K+1
                      CALL INTEGRAND
                      IF KI=2 THEN SUM
                      R=GZ
                      T=HZ
                      V=FZ
                      KI=2
                      GOTO AGAIN
                SUM:  Z1=Z1+WI(K)*(R+GZ)
                      Z2=Z2+WI(K)*(T+HZ)
                      Z3=Z3+WI(K)*(V+FZ)
              END OF LOOP J
        END OF LOOP I
        Z1=Z1*QS
        Z2=Z2*QS
        Z3=Z3*QS
        IF KJ=1 THEN RETURN
        Z1=Z1+iπF1
        Z2=Z2+iπF2
        Z3=Z3+iπF3
        RETURN
```

For the calculation of the integrand we have:

- Subroutine INTEGRAND(XX,YY,K,GZ,HZ,FZ)

```
*Comment: w=XX+i*YY

         DD=|U(K)-w|²
         IF DD<10⁻⁴ THEN TAYLOR
```

```
        IF KJ=O THEN BRANCH2

BRANCH1:  H1=(φ'₀(U(K))-φ'₀(w))/(U(K)-w)(YY=<1)
          H2=(φ'₀'(U(K))-φ'₀'(w))/(U(K)-w)
          H3=(φ'₀''(U(K))-φ'₀''(w))/(U(K)-w)
          GOTO MULT

BRANCH2:  H1=φ'₀(U(K))/(U(K)-w)(YY>1)
          H2=φ'₀'(U(K))/(U(K)-w)
          H3=φ'₀''(U(K))/(U(K)-w)
          GOTO MULT

TAYLOR:   H1=φ'₀'(w)+φ'₀''(w)*(U(K)-w)+···+(|U(K)-w|<ε)
          H2=···
          H3=···

MULT:     DD=1+U(K)*U(K) (Change of variable u=tan(x))
          GZ=H1*DD
          HZ=H2*DD
          FZ=H3*DD
          RETURN
```

11.2.6. Checking ZNDEZ

When you have written the subroutine ZNDEZ you have to check the program for correctness and precision. The best way is to compute a table of $Z^{+\prime}(z)$ and to check the results obtained against the table of the same function given by Fried and Conte (1961). It is also recommended to check for $Z^{+(2)}$ and $Z^{+(3)}$, but instead of printing these values it is sufficient to verify that these functions satisfy the recurrence relation

$$Z^{(n+1)} = -2[\xi Z^{(n)} + nZ^{(n-1)}] \tag{11.15}$$

Here is an example of a BASIC program printing the table of Z^+ and the relative error E on the recurrence relation (11.15)

$$E = \frac{|Z^{+\prime\prime\prime} + 2(\xi Z^{+\prime\prime} + 2Z^{+\prime})|}{|Z^{+\prime\prime\prime}|} \tag{11.16}$$

Listing of the program CHECKZN:

```
100 CLS: CLEAR 1000
110 DEFINT I-O
120 DIM XI(6),WI(6),Z1(2),Z2(2),Z3(2)
130 DIM X1(2),X2(2),X3(2),X4(2),X5(2)
```

```
140 DIM F0(2),F1(2),F2(2),F3(2),F4(2),F5(2)
150 DIM W1(2),W2(2),W3(2),W4(2),W5(2),H1(2),H2(2),H3(2)
160 DIM D1(2),D2(2),GZ(2),HZ(2),FZ(2),R(2),T(2),V(2)
170 DIM U(48),G1(48),G2(48),G3(48)
180 '
190 'TABLE OF Z'
200 '
210 F$=''D''                              'INITIALIZATION FLAG
220 '
230 FOR Y=-5 TO 5 STEP 1                  'LOOP FOR Y
240 '
250     LPRINT                            'PRINT THE HEADER
260     LPRINT''                           Y='';Y
270     LPRINT
280     LPRINT '' X'','' Re(Z')'' ,'' IM(Z')'','' E''
290 '
300     FOR X=0 TO 9 STEP 1               'LOOP FOR X
310         GOSUB 10070                   'CALL ZNDEZ========>
320         E1=-2*(X*Z2(1)-Y*Z2(2))-4*Z1(1)-Z3(1)
330         E2=-2*(X*Z2(2)+Y*Z2(1))-4*Z1(2)-Z3(2)
340         E=SQR(E1*E1+E2*E2)
350         E1=SQR(Z3(1)*Z3(1)+Z3(2)*Z3(2))
360         IF E1<1.E-6 THEN E1=1.E-1   'AVOID DIVISION BY 0
370         LPRINTX,Z1(1),Z1(2),E/E1
380     NEXT X
390 '
400 NEXT Y
410 END
```

If we compare the table calculated by CHECKZN to the table of Fried and Conte (1961), then we should verify that the relative error of $Z^{+'}$ does not exceed 5×10^{-5} in the upper complex-z half-plane. It should be noted also that the subroutine used for the calculation of $\exp(-z^2)$ by Fried and Conte was not very precise, particularly in the lower complex half-plane.

11.3. HUNTING THE ROOTS OF A DISPERSION RELATION

In this section we examine the problem of the numerical determination of the roots of a dispersion relation. To be more specific, we consider a particular dispersion relation, the dispersion of ion waves in the approximation of Boltzmann electrons. As shown in Chapter 10, this dispersion relation may be written as

$$D(f,\xi) = f^2 - \xi^2 \left[Z^{+\prime}(\xi) - 2\frac{T_e}{T_i} \right], \tag{11.17}$$

where $f = \omega/\omega_{pi}$, $\xi = \omega/ka_i$, and $a_i^2 = (2KT_i/m_e)$. The problem we consider is to find the roots of Eq. (11.17) in the complex-ξ plane for real frequencies ω. If we define the reduced wave number η by

$$\eta = \frac{1}{\xi} = \frac{ka_i}{\omega}, \tag{11.18}$$

we see that if ξ_n is a root of D in the complex ξ-plane, $\eta_n = 1/\xi_n$ is also a root in the complex η-plane, and that we are looking for the roots of the ion wave dispersion relation for real frequencies in the complex-k plane. We characterize a root ξ_n by a phase velocity $V_{\varphi n}$ and a relative damping $(k_i/k_r)_n$:

$$
\begin{aligned}
\frac{V_{\varphi n}}{a_i} &= \frac{\omega}{k_r a_i} = \frac{1}{\text{Re}(\xi_n)}, \\
\left(\frac{k_i}{k_r}\right)_n &= \frac{\text{Im}(k)}{\text{Re}(k)} = \frac{-\text{Im}(\xi_n)}{\text{Re}(\xi_n)}.
\end{aligned}
\tag{11.19}
$$

11.3.1. Newton's Method

The technique we use for the determination of the roots is that of Newton. This method is of the iterative type: starting from an approximate value for the root ξ_κ we have

$$D(f,\xi_{\kappa+1}) \simeq d + ad\xi_\kappa + bd\xi_\kappa^2, \tag{11.20}$$

where

$$d = D(f,\xi_\kappa), \qquad a = \left[\frac{\partial D(f,\xi)}{\partial \xi}\right]_{\xi=\xi_\kappa}, \qquad b = \frac{1}{2}\left[\frac{\partial^2 D(f,\xi)}{\partial \xi^2}\right]_{\xi=\xi_\kappa}, \tag{11.21}$$

and $d\xi_\kappa = (\xi_{\kappa+1} - \xi_\kappa)$.

Assuming that $D(f,\xi_{\kappa+1}) = 0$, and using the formula for reversion of series [Abramowitz and Stegun, p. 16 (1964)], we obtain for $d\xi_{\kappa+1}$

$$d\xi_{\kappa+1} = -\frac{1}{a}d - \frac{b}{a^3}d^2. \tag{11.22}$$

If $|d\xi_{\kappa+1}| > \varepsilon$, then one may iterate again until the required precision is obtained.

Obviously, the reversion formula fails if $a = 0$, i.e., if the root is a double root. In this case, Eq. (11.22) may be replaced by one of the exact solutions of the quadratic equation, Eq. (11.20).

11.3.2. Initialization of the Roots

Newton's method is easy to use for a large class of dispersion relations. However, there is still a difficulty: one has to furnish an estimate of the root. This estimation can be done analytically in some cases, but in general this estimation has to be done numerically.

As an example of a direct numerical determination of a root, we use the analytical properties of $D(f,\xi)$. Assuming that D is analytic and has a simple root at $\xi = \xi_0$, then from the residues equation

$$\frac{1}{2\pi i} \int_C \frac{D'(f,\xi)}{D(f,\xi)} \, \xi \, d\xi = \xi_0, \tag{11.23}$$

where C is a closed path around ξ_0.

Therefore, one could design a program computing the integral of Eq. (11.23) and by a proper scanning of the complex-ξ plane, it is possible, in principle, to obtain an automatic determination of the roots of D. However, such a procedure is complex and time consuming. A more practical approach is to divide the problem in two parts: (1) estimation of the roots for a particular set of parameters $(f, T_e/T_i)$; and (2) use of Newton's method for a precise determination and slow changes of the parameters.

As a crude estimation of the roots, we draw an altitude chart of $|D|$ in the complex-ξ plane. However, on a microcomputer, an automatic altitude-chart drawing program is not very easy to implement. We prefer another method founded on the "principle of the argument."

We consider the function $D(\xi)$, analytic in a domain D and for simplicity we assume that there are only simple roots in D. Let C be a closed path in D. Then, from the residues theorem,

$$\frac{1}{2\pi i} \int_C \frac{dD(\xi)/d\xi}{D(\xi)} \, d\xi = \mathcal{N} \tag{11.24}$$

where $\mathcal{N}$ is the number of simple roots inside C. Moreover,

$$\int_C \frac{dD(\xi)/d\xi}{D(\xi)} \, d\xi = \int_C \frac{d\log[D(\xi)]}{d\xi} \, d\xi = i(\vartheta_2 - \vartheta_1), \tag{11.25}$$

The angles ϑ_1 and ϑ_2 can be different because the log function is a multi-valued function. Comparison of Eqs. (11.24) and (11.25) shows that

$$\vartheta_2 - \vartheta_1 = 2\pi \mathcal{N}. \tag{11.26}$$

Hence, if one represents in the ξ plane, for a point ξ, an arrow of unit length, starting at ξ and pointing in the direction ϑ_1 defined by

$$D(\xi) = |D(\xi)| e^{i\vartheta_1}, \tag{11.27}$$

then if ξ follows a closed path around a single root, the arrow makes a full turn when one comes back to the original position.

This provides a very simple way for finding the roots of a complex function in the complex plane: if on each point of a grid in the complex plane one draws an arrow of unit length and with a direction given by the argument of the function, then one can easily visualize the root by noting the vortex around

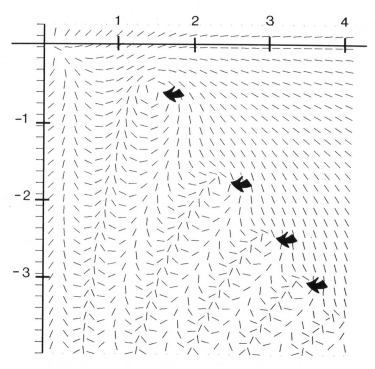

Figure. 11.1. Scanning the complex-ξ plane using the "principle of the argument" to hunt the roots. The vortex formed by the arrows indicates the approximate position of a root of Eq. (11.17).

each root. Fig. 11.1 gives an example of a scanning of the complex plane by this method applied to Eq. (11.7), and we can estimate without difficulty the approximate position of the roots of the ion wave dispersion relation.

11.3.3. Numerical Solution

The reader will find in the appendix to this chapter the listing of the program HUNT. This program computes the roots of the ion-wave dispersion relation of Eq. (10.17). The root is initialized to the first root (the least damped) of the ion wave for $T_i/T_e = 1$,) i.e., $XO = 1.5$ and $YO = -0.6$. For $T_e/T_i = 1$ and $f = 0$, the program reaches a more precise value after three iterations. For reference, this value is for the reduced phase velocity and damping, 1.697 and 0.4161, respectively. Using a similar program, the reader may draw with a computer having graphic capabilities the full dispersions curves V_φ and k_i/k_r as a function of ω/ω_{pi}.

11.4. APPENDIX

11.4.1. Program ZNDEZ

Listing of the program ZNDEZ written in BASIC.

```
10000 '*********************************************
10010 ' SUBROUTINE ZNDEZ(X,Y,Z1,Z2,Z3,KX)          *
10020 ' Z1(1)=RE(Z'(Z)),Z1(2)=IM(Z'(Z))            *
10030 ' Z2(1)=RE(Z''(Z)),Z2(2)=IM(Z''(Z))          *
10040 ' Z3(1)=RE(Z'''(Z)),Z3(2)=IM(Z'''(Z))        *
10050 ' Z=X+I*Y                                     *
10060 '*********************************************
10070 KX=0
10080 XX=X:YY=Y: IF Y<0 THEN YY=ABS(YY)
10090     GOSUB 10220                'CALL ZNXY ==========>
10100 IF Y=YY THEN RETURN
10110 IF KX=0 THEN GOSUB 11550       'CALL FN ============>
10120     Z1(1)=Z1(1)+2*PI*F1(2)     'ANALYTIC CONTINUATION
10130     Z1(2)=-Z1(2)+2*PI*F1(1)
10140     Z2(1)=Z2(1)+2*PI*F2(2)
10150     Z2(2)=-Z2(2)+2*PI*F2(1)
10160     Z3(1)=Z3(1)+2*PI*F3(2)
10170     Z3(2)=-Z3(2)+2*PI*F3(1)
10180 RETURN
10190 '*********************************************
10200 ' SUBROUTINE ZNXY(XX,YY,Z1,Z2,Z3)            *
```

```
10210  '*****************************************************
10220     IF F$=''F'' THEN 10260        ' JUMP TO SKIP =====>
10230     GOSUB 11920                   ' CALL INITIALIZE ==>
10240     F$=''F''
10250  'SKIP
10260     GOSUB 10310                   ' CALL ZN ==========>
10270     RETURN
10280  '*****************************************************
10290  ' SUBROUTINE ZN(XX,YY,Z1,Z2,Z3)                      *
10300  '*****************************************************
10310     D=XX*XX+YY*YY
10320     IF D>25 THEN 10860            ' JUMP ASYMPTOTIC ==>
10330  '
10340  ' NON ASYMPTOTIC CASE
10350  '
10360     GOSUB 11550                   ' CALL FN ==========>
10370     IF D<4 THEN 11100             ' JUMP SERIES ======>
10380  ' GAUSS INTEGRATION
10390     FOR N=1 TO 2
10400        Z1(N)=0
10410        Z2(N)=0
10420        Z3(N)=0
10430     NEXT N
10440     KJ=0
10450     A=-PI/2-Q
10460     K=0
10470     IF YY>1 THEN KJ=1
10480     FOR I=1 TO LL
10490        A=A+Q
10500        B=A+Q
10510        QP=(A+B)/2
10520        FOR J=1 TO 6
10530           KI=1
10540  ' AGAIN
10550           K=K+1
10560     GOSUB 12300                   ' CALL INTEGRAND ===>
10570     IF KI=2 THEN 10640            ' JUMP SUM =========>
10580           R(1)=GZ(1):R(2)=GZ(2)
10590           T(1)=HZ(1):T(2)=HZ(2)
10600           V(1)=FZ(1):V(2)=FZ(2)
10610           KI=2
10620           GOTO 10550              ' JUMP AGAIN =======>
10630  ' SUM
```

```
10640              FOR N=1 TO 2
10650                  Z1(N)=Z1(N)+WI(J)*(R(N)+GZ(N))
10660                  Z2(N)=Z2(N)+WI(J)*(T(N)+HZ(N))
10670                  Z3(N)=Z3(N)+WI(J)*(V(N)+FZ(N))
10680              NEXT N
10690          NEXT J
10700      NEXT I
10710      Z1(1)=Z1(1)*QS
10720      Z1(2)=Z1(2)*QS
10730      Z2(1)=Z2(1)*QS
10740      Z2(2)=Z2(2)*QS
10750      Z3(1)=Z3(1)*QS
10760      Z3(2)=Z3(2)*QS
10770      IF KJ=1 THEN RETURN
10780      Z1(1)=Z1(1)-PI*F1(2)
10790      Z1(2)=Z1(2)+PI*F1(1)
10800      Z2(1)=Z2(1)-PI*F2(2)
10810      Z2(2)=Z2(2)+PI*F2(1)
10820      Z3(1)=Z3(1)-PI*F3(2)
10830      Z3(2)=Z3(2)+PI*F3(1)
10840 RETURN
10850 ' ASYMPTOTIC EXPANSION
10860      X2(1)=XX*XX-YY*YY: X2(2)=2*XX*YY
10870      DD=X2(1)*X2(1)+X2(2)*X2(2)
10880      R(1)=X2(1)/DD: R(2)=-X2(2)/DD
10890      X2(1)=R(1): X2(2)=R(2)
10900      DD=XX*XX+YY*YY
10910      H1(1)=XX/DD: H1(2)=-YY/DD
10920      Z1(1)=R(1): Z1(2)=R(2)
10930      Z2(1)=-2*(R(1)*H1(1)-R(2)*H1(2))
10940      Z2(2)=-2*(R(1)*H1(2)+R(2)*H1(1))
10950      Z3(1)=6*(R(1)*R(1)-R(2)*R(2)): Z3(2)=12*R(1)*R(2)
10960      FOR I=2 TO 10
10970          J=I+I
10980          S=(J-1)/2*(R(1)*X2(1)-R(2)*X2(2))
10990          R(2)=(J-1)/2*(R(1)*X2(2)+R(2)*X2(1)): R(1)=S
11000          S=-J*(R(1)*H1(1)-R(2)*H1(2))
11010          T(2)=-J*(R(1)*H1(2)+R(2)*H1(1)): T(1)=S
11020          S=-(J+1)*(T(1)*H1(1)-T(2)*H1(2))
11030          V(2)=-(J+1)*(T(1)*H1(2)+T(2)*H1(1)): V(1)=S
11040          Z3(1)=Z3(1)+V(1): Z3(2)=Z3(2)+V(2)
11050          Z2(1)=Z2(1)+T(1): Z2(2)=Z2(2)+T(2)
11060          Z1(1)=Z1(1)+R(1): Z1(2)=Z1(2)+R(2)
```

```
11070     NEXT I
11080   RETURN
11090 ' SERIES EXPANSION
11100     Z1(1)=-2+4*X2(1):Z1(2)=4*X2(2)
11110     Z2(1)=8*XX:Z2(2)=8*YY
11120     Z3(1)=8:Z3(2)=0
11130     R(1)=8:R(2)=0
11140     I=1
11150 ' LOOP
11160     I=I+1
11170     IF I>10 THEN 11360 'JUMP OUT ===============>
11180         J=I+I
11190         A0=J/((J-3)*(I-1))
11200         S=-A0*(R(1)*X2(1)-R(2)*X2(2))
11210         R(2)=-A0*(R(1)*X2(2)+R(2)*X2(1)):R(1)=S
11220         S=(R(1)*XX-R(2)*YY)/(J-1)
11230         T(2)=(R(1)*YY+R(2)*XX)/(J-1):T(1)=S
11240         S=(T(1)*XX-T(2)*YY)/J
11250         V(2)=(T(1)*YY+T(2)*XX)/J:V(1)=S
11260         Z1(1)=Z1(1)+V(1)
11270         Z1(2)=Z1(2)+V(2)
11280         Z2(1)=Z2(1)+T(1)
11290         Z2(2)=Z2(2)+T(2)
11300         Z3(1)=Z3(1)+R(1)
11310         Z3(2)=Z3(2)+R(2)
11320         DD=Z3(1)*Z3(1)+Z3(2)*Z3(2)
11330         IF DD=0 THEN 11160 'JUMP LOOP ===============>
11340         IF (R(1)*R(1)+R(2)*R(2))/DD>1.E-12 THEN 11160
11350 'OUT
11360     Z1(1)=Z1(1)-PI*F1(2)
11370     Z1(2)=Z1(2)+PI*F1(1)
11380     Z2(1)=Z2(1)-PI*F2(2)
11390     Z2(2)=Z2(2)+PI*F2(1)
11400     Z3(1)=Z3(1)-PI*F3(2)
11410     Z3(2)=Z3(2)+PI*F3(1)
11420 RETURN
11430 '*****************************************************
11440 ' SUBROUTINE GN(U,K,G1,G2,G3)                        *
11450 '*****************************************************
11460 U(K)=TAN(U(K))
11470 VV=EXP(-U(K)*U(K))/RP
11480 G1(K)=-2*U(K)*VV
11490 G2(K)=(4*U(K)*U(K)-2)*VV
```

```
11500 G3(K)=(-8*U(K)*U(K)*U(K)+12*U(K))*VV
11510 RETURN
11520 '***********************************************
11530 ' SUBROUTINE FN(XX,YY,F1,F2,F3,KX)              *
11540 '***********************************************
11550 X1(1)=XX
11560 X1(2)=YY
11570 X2(1)=XX*XX-YY*YY
11580 X2(2)=2*XX*YY
11590 X3(1)=X2(1)*XX-X2(2)*YY
11600 X3(2)=X2(1)*YY+X2(2)*XX
11610 X4(1)=X2(1)*X2(1)-X2(2)*X2(2)
11620 X4(2)=2*X2(1)*X2(2)
11630 X5(1)=X2(1)*X3(1)-X2(2)*X3(2)
11640 X5(2)=X2(1)*X3(2)+X2(2)*X3(1)
11650 F0(1)=EXP(-X2(1))/RP
11660 F0(2)=-F0(1)*SIN(X2(2))
11670 F0(1)=F0(1)*COS(X2(2))
11680 W1(1)=-2*X1(1)
11690 W1(2)=-2*X1(2)
11700 W2(1)=4*X2(1)-2
11710 W2(2)=4*X2(2)
11720 W3(1)=-8*X3(1)+12*X1(1)
11730 W3(2)=-8*X3(2)+12*X1(2)
11740 W4(1)=16*X4(1)-48*X2(1)+12
11750 W4(2)=16*X4(2)-48*X2(2)
11760 W5(1)=-32*X5(1)+160*X3(1)-120*X1(1)
11770 W5(2)=-32*X5(2)+160*X3(2)-120*X1(2)
11780 F1(1)=W1(1)*F0(1)-W1(2)*F0(2)
11790 F1(2)=W1(1)*F0(2)+W1(2)*F0(1)
11800 F2(1)=W2(1)*F0(1)-W2(2)*F0(2)
11810 F2(2)=W2(1)*F0(2)+W2(2)*F0(1)
11820 F3(1)=W3(1)*F0(1)-W3(2)*F0(2)
11830 F3(2)=W3(1)*F0(2)+W3(2)*F0(1)
11840 F4(1)=W4(1)*F0(1)-W4(2)*F0(2)
11850 F4(2)=W4(1)*F0(2)+W4(2)*F0(1)
11860 F5(1)=W5(1)*F0(1)-W5(2)*F0(2)
11870 F5(2)=W5(1)*F0(2)+W5(2)*F0(1)
11880 KX=1:RETURN
11890 '***********************************************
11900 ' SUBROUTINE INITIALIZE                         *
11910 '***********************************************
11920 DATA 0.1252334085,0.3678314989,0.5873179542
```

```
11930 DATA 0.7699026741,0.9041172563,0.9815606342
11940 DATA 0.2491470458,0.2334925365,0.2031674267
11950 DATA 0.1600783285,0.1069393259,0.0471753363
11960 '
11970 ' READ THE GAUSSIAN ABSCISSAE
11980 FOR I=1 TO 6
11990    READ XI(I)
12000 NEXT I
12010 ' READ THE GAUSSIAN WEIGHT
12020 FOR I=1 TO 6
12030    READ WI(I)
12040 NEXT I
12050 PI=3.14159265
12060 RP=1.77245385
12070 ' NUMBER OF INTEGRATION INTERVALS: LL
12080 LL=4
12090 Q=PI/LL
12100 QS=Q/2
12110 'COMPUTE U(K) AND GN(K)
12120 K=0
12130 FOR I=1 TO LL
12140    A=A+Q
12150    B=A+Q
12160    QP=(A+B)/2
12170    FOR J=1 TO 6
12180        K=K+1
12190        U(K)=QS*XI(J)+QP
12200        GOSUB 11460           ' CALL GN =========>
12210        K=K+1
12220        U(K)=-QS*XI(J)+QP
12230        GOSUB 11460           ' CALL GN =========>
12240    NEXT J
12250 NEXT I
12260 RETURN
12270 '*************************************************
12280 ' SUBROUTINE INTEGRAND                          *
12290 '*************************************************
12300    DX=(U(K)-XX)
12310    DD=DX*DX+YY*YY
12320    IF DD <1.E-4 THEN 12560   ' JUMP TAYLOR =====>
12330    D1(1)=DX/DD
12340    D1(2)=YY/DD
12350    IF KJ=1 THEN 12490        ' JUMP BRANCH2 ====>
```

```
12360 '
12370 ' CASE YY=<1
12380 ' BRANCH1
12390    H1(1)=(G1(K)-F1(1))*D1(1)+F1(2)*D1(2)
12400    H1(2)=(G1(K)-F1(1))*D1(2)-F1(2)*D1(1)
12410    H2(1)=(G2(K)-F2(1))*D1(1)+F2(2)*D1(2)
12420    H2(2)=(G2(K)-F2(1))*D1(2)-F2(2)*D1(1)
12430    H3(1)=(G3(K)-F3(1))*D1(1)+F3(2)*D1(2)
12440    H3(2)=(G3(K)-F3(1))*D1(2)-F3(2)*D1(1)
12450    GOTO 12750               ' JUMP MULT ========>
12460 '
12470 ' CASE YY>1
12480 ' BRANCH2
12490    S=G1(K):H1(1)=S*D1(1):H1(2)=S*D1(2)
12500    S=G2(K):H2(1)=S*D1(1):H2(2)=S*D1(2)
12510    S=G3(K):H3(1)=S*D1(1):H3(2)=S*D1(2)
12520    GOTO 12750               ' JUMP MULT ========>
12530 '
12540 ' TAYLOR EXPANSION OF THE INTEGRAND
12550 ' TAYLOR
12560    X1(1)=DX/2
12570    X1(2)=-YY/2
12580    X2(1)=(X1(1)*X1(1)-X1(2)*X1(2))/6
12590    X2(2)=X1(1)*X1(2)/3
12600    X3(1)=(X2(1)*X1(1)-X2(2)*X1(2))/24
12610    X3(2)=(X2(1)*X1(2)+X2(2)*X1(1))/24
12620    H1(1)=F2(1)+F3(1)*X1(1)-F3(2)*X1(2)+F4(1)*X2(1)
12630    H1(1)=H1(1)-F4(2)*X2(2)+F5(1)*X3(1)-F5(2)*X3(2)
12640    H1(2)=F2(2)+F3(1)*X1(2)+F3(2)*X1(1)+F4(1)*X2(2)
12650    H1(2)=H1(2)+F4(2)*X2(1)+F5(1)*X3(2)+F5(2)*X3(1)
12660    H2(1)=F3(1)+F4(1)*X1(1)-F4(2)*X1(2)+F5(1)*X2(1)
12670    H2(1)=H2(1)-F5(2)*X2(2)
12680    H2(2)=F3(2)+F4(1)*X1(2)+F4(2)*X1(1)+F5(1)*X2(2)
12690    H2(2)=H2(2)+F5(2)*X2(1)
12700    H3(1)=F4(1)+F5(1)*X1(1)-F5(2)*X1(2)
12710    H3(2)=F4(2)+F5(1)*X1(2)+F5(2)*X1(1)
12720 '
12730 ' MULTIPLY BY (1+U*U)
12740 ' MULT
12750    DD=1+U(K)*U(K)
12760    GZ(1)=DD*H1(1)
12770    GZ(2)=DD*H1(2)
12780    HZ(1)=DD*H2(1)
```

```
12790     HZ(2)=DD*H2(2)
12800     FZ(1)=DD*H3(1)
12810     FZ(2)=DD*H3(2)
12820 RETURN
```

11.4.2. Listing of the Program HUNT in BASIC.

```
10 CLS: CLEAR 1000
20 DEFINT I-O
30 DIM XI(6),WI(6),Z1(2),Z2(2),Z3(2)
40 DIM X1(2),X2(2),X3(2),X4(2),X5(2)
50 DIM F0(2),F1(2),F2(2),F3(2),F4(2),F5(2)
60 DIM W1(2),W2(2),W3(2),W4(2),W5(2),H1(2),H2(2),H3(2)
70 DIM D1(2),D2(2),GZ(2),HZ(2),FZ(2),R(2),T(2),V(2)
80 DIM U(48),G1(48),G2(48),G3(48)
90 '
100 ' FIRST ESTIMATION OF THE ROOT
110 '
120 XO=1.5:YO=-.6
130 F$=''D''
140 TE=1:F=0
150 '
160 ' INPUT OF FREQUENCY AND TE/TI
170 '
180 INPUT''TE/TI:''; TE
190 INPUT''F/FPI:''; F:FF=F*F
200 GOSUB 370              ' =====>  FIND THE ROOT
210 IF NE=1 THEN PRINT ''ROOT NOT FOUND'':END
220 '
230 ' COMPUTE THE REDUCED PHASE VELOCITY AND DAMPING
240 '
250     VP=(XO*XO+YO*YO)/XO: AM=-YO/XO
260     PRINT''F/PI= '';F;'' VPHI/A= '';VP;''
        KI/KR= '';AM;'' NQ= '';NQ
270 GOTO 190
280 ' END OF PROGRAM
290 '************************************************
300 '
310 ' SEARCH FOR A ROOT XO,YO FOR FF,TE/TI AND ME/MI
      CONSTANT
320 ' INITIAL VALUE XO,YO
330 ' METHOD : NEWTON, SECOND ORDER
340 ' IF NE=1 ROOT NOT FOUND
```

```
350 ' IF NE=O ROOT FOUND XO,YO
360 '*********************************************************
370 NE=1
380 NQ=O
390 '
400 ' ITERATION LOOP
410 '
420 NQ=NQ+1:IF NQ>100 THEN RETURN
430     GOSUB 630                    ' =====> COMPUTE D,D' AND D''
440     DD=AA(1)*AA(1)+AA(2)*AA(2)
450     S=AA(1)/DD:AA(2)=-AA(2)/DD:AA(1)=S
460     X2(1)=AA(1)*AA(1)-AA(2)*AA(2):X2(2)=2*AA(1)*AA(2)
470     S=-(BB(1)*AA(1)-BB(2)*AA(2)):BB(2)=-(BB(1)*AA(2)
        +BB(2)*AA(1))
480     BB(1)=S
490     S=BB(1)*X2(1)-BB(2)*X2(2):BB(2)=BB(1)*X2(2)+BB(2)*X2(1)
500     BB(1)=S
510     S=D1(1)*D1(1)-D1(2)*D1(2):D2(2)=2*D1(1)*D1(2):D2(1)=S
520     DX=-(AA(1)*D1(1)-AA(2)*D1(2))+(BB(1)*D2(1)-BB(2)*D2(2))
530     DY=-(AA(1)*D1(2)+AA(2)*D1(1))+(BB(1)*D2(2)+BB(2)*D2(1))
540     XO=XO+DX:YO=YO+DY
550     DD=DX*DX+DY*DY
560 IF DD<1.E-10 THEN NE=O:RETURN
570 GOTO 420
580 '*********************************************************
590 ' CALCULATION OF D, D' AND D''
600 ' INPUT : FF, XO,YO
610 ' OUTPUT: AA=D',BB=.5*D'' AND D1=D
620 '*********************************************************
630 X=XO:Y=YO:GOSUB 10070
640 E1=Z1(1)-2*TE:E2=Z1(2)
650 X2(1)=X*X-Y*Y:X2(2)=2*X*Y
660 AA(1)=-2*(X*E1-Y*E2)-(X2(1)*Z2(1)-X2(2)*Z2(2))
670 AA(2)=-2*(X*E2+Y*E1)-(X2(1)*Z2(2)+X2(2)*Z2(1))
680 BB(1)=-E1-2*(X*Z2(1)-Y*Z2(2))-(X2(1)*Z3(1)-X2(2)*Z3(2))/2
690 BB(2)=-E2-2*(X*Z2(2)+Y*Z2(1))-(X2(1)*Z3(2)+X2(2)*Z3(1))/2
700 D1(1)=FF-(X2(1)*E1-X2(2)*E2)
710 D1(2)=-(X2(1)*E2+X2(2)*E1)
720 RETURN
```

REFERENCES

CHAPTER 1

B. D. FRIED and S. D. CONTE, *The Plasma Dispersion Function*, Academic, New York (1961).

G. E. ROBERTS and H. KAUFMAN, *Table of Laplace Transforms*, W. B. Saunders, Philadelphia, PA (1966).

B. W. ROOS, *Analytic Functions and Distributions in Physics and Engineering*, Wiley, New York (1969).

W. RUDIN, *Real and Complex Analysis*, McGraw-Hill, New York (1966); Second Edition (1974).

E. T. WHITTAKER and G. N. WATSON, *A Course of Modern Analysis*, University Press, Cambridge (1902); Fourth Edition (1965).

CHAPTER 2

W. P. ALLIS, S. J. BUCHSBAUM and A. BERS, *Waves in Anisotropic Plasmas*, M.I.T. Press (1963).

F. F. CHEN, *Introduction to Plasma Physics*, Plenum, New York (1974).

J. F. DENISSE and J. L. DELCROIX, *Théorie des Ondes dans les Plasmas*, Dunod, Paris (1961).

M. A. HEALD and C. B. WHARTON, *Plasma Diagnostics with Microwaves*, Wiley, New York (1965).

P. MILLS and J. P. M. SCHMITT, Density measurements by microwaves in a Q-machine, Laboratory P.M.I. Report. 431, Ecole Polytechnique, Palaiseau, France (1969).

D. R. NICHOLSON, *Introduction to Plasma Theory*, Wiley, New York (1983).

D. QUEMADA, *Ondes dans les Plasmas*, Hermann, Paris (1968).

T. H. STIX, *Theory of Plasma Waves*, McGraw-Hill, New York (1962)

CHAPTER 3

I. ALEXEFF, W. D. JONES, and D. MONTGOMERY, Effects of electron temperature variation on ion-acoustic waves, *Phys. Fluids.* **11**, 167–173 (1968).

N. D'ANGELO, S. VON GOELER, and T. OHE, Propagation and damping of ion waves in a plasma with negative ions, *Phys. Fluids* **9**, 1605 (1966).

D. BOHM and E. P. GROSS, Theory of plasma oscillations: Part-A, Origin of mediumlike behavior, *Phys. Rev.* **75**, 1851 (1949); Part-B, Excitation and damping of oscillations, *Phys. Rev.* **75**, 1864 (1949).

H. DERFLER and T. C. SIMONEN, Landau waves: an experimental fact, *Phys. Rev. Lett.* **17**, 172–175 (1966).

H. J. DOUCET, Production of a quasi-electron-free plasma using electronic attachment, *Phys. Lett.* **33A**, 5, 283 (1970).

H. J. DOUCET, I. ALEXEFF, and W. D. JONES, Simultaneous measurement of ion-acoustic wave potential and plasma density perturbation to yield γ_e, *Phys. Fluids* **11**, 2451–2453 (1968).

W. D. JONES and I. ALEXEFF, A study of the properties of ionic sound waves, *Proceedings of the Seventh International Conference on Phenomena in Ionized Gases*, (B. Perovic and D. Tosic, eds.), Vol. II, pp. 330–335, Gradevinska Knjiga, Beograd (1966).

G. Joyce, K. Lonngren, I. Alexeff, and W. D. Jones, Dispersion of ion-acoustic waves, *Phys. Fluids* **12**, 2592–2599 (1969).

A. Y. Wong, R. W. Motley, and N. D'Angelo, Landau damping of ion-acoustic waves in highly ionized plasmas, *Phys. Rev. A* **133**, 436–442 (1964).

CHAPTER 4

Y. Hatta and N. Sato, The experimental study of the ionic sound wave in the dark plasma, *Proceedings of the Fifth International Conference on Ionization Phenomena on Gases*, Munich, 1961, Vol. I, pp. 478–484, North-Holland, Amsterdam (1962).

CHAPTER 5

M. Abramowitz and I A. Stegun, *Handbook of Mathematical Functions*, Dover, New York (1965); Tenth Edition (1972).

P. F. Little, Acoustic waves in a plasma, *Proceedings of the Fifth International Conference on Ionization Phenomena in Gases*, Munich, 1961, Vol. II, pp. 1440–1455, North-Holland, Amsterdam (1962).

P. F. Little, Ion waves in a bounded plasma, *Nature* **194**, 1137–1139 (1962).

P. Mills, and H. J. Doucet, Excitation d'onde ionique pseudosonore dans un plasma cylindrique par un faisceau électronique, *C. R. Acad. Sci.* **266**(*B*), 149–152 (1968).

P. Mills, Propagation d'une onde ionique dans une colonne de plasma, Thèse de Doctorat de 3ème Cycle, 101 pp., Université de Paris, Orsay (1968).

R. W. Motley, *Q Machines*, Academic, New York (1975).

A. Y. Wong, Propagation of ion-acoustic waves along cylindrical plasma columns, *Phys. Fluids* **9**, 1261–1262 (1966).

A. Y. Wong, R. W. Motley, and N. D'Angelo, Landau damping of ion-acoustic waves in highly ionized plasmas, *Phys. Rev. A* **133**, 436–442 (1964).

CHAPTER 6

L. Brillouin, La mécanique ondulatoire de Schrödinger; une méthode générale de résolution par approximations successives. *C. R. Acad. Sci.* **183**, 24–26 (1926).

H. J. Doucet, W. D. Jones, and I. Alexeff, Linear ion-acoustic waves in a density gradient, *Phys. Fluids* **17**, 1738–1743 (1974).

H. J. Doucet and M. Feix, De l'effet des gradients et des courbures sur la reflexion d'une onde electrostatique en plasma hétérogène. *J. Phys. (Paris)* **36**, 37 (1975).

O. Ishihara, I. Alexeff, H. J. Doucet, and W. D. Jones, Reflection and absorption of ion-acoustic waves in a density gradient, *Phys. Fluids* **21**, 2211–2217 (1978).

W. D. Jones, C. B. Mattson, and A. Lee, Reflection of ion-acoustic waves by a density gradient, *Bull. Am. Phys. Soc.* **21**(9), 1075 (1976).

H. A. Kramers, *Z. Phys.* **39**, 828 (1926).

C. Mattson, Observation of ion-acoustic-wave reflection from the presheath density gradient near a negatively biased plasma boundary, Master's Thesis, University of South Florida, Tampa (1976).

G. Wentzel, *Z. Phys.* **38**, 518 (1926).
K. C. Yeh and C. H. Liu, *Theory of Ionospheric Waves*, Academic, New York (1972).

CHAPTER 7

M. Abramowitz and I. A. Stegun, *Handbook of Mathematical Functions*, Dover, New York (1965); Tenth Edition (1972).
B. D. Fried and S. D. Conte, *The Plasma Dispersion Function*, Academic, New York (1961).
B. D. Fried and R. W. Gould, Longitudinal Ion Oscillations in a Hot Plasma, *Phys. Fluids* **4**, 139 (1961).
R. W. Motley, *Q Machines*, Academic, New York (1975).
D. R. Nicholson, *Introduction to Plasma Theory*, Wiley, New York (1983).
B. W. Roos, *Analytic Functions and Distributions in Physics and Engineering*, Wiley, New York (1969).
G. M. Sessler and G. Pearson, Propagation of ion waves in weakly ionized gases, *Phys. Rev.* **162**, 108 (1967).
E. T. Whittaker and G. N. Watson, *A Course of Modern Analysis*, Cambridge University Press, Cambridge (1902); Fourth Edition (1965).

CHAPTER 8

I. Alexeff, W. D. Jones and D. Montgomery, Effects of electron-temperature variation on ion-acoustic waves, *Phys. Fluids.* **11**, 167–173 (1968).
M. Abramowitz and I. A. Stegun, *Handbook of Mathematical Functions*, p. 1001, Dover, New York (1965); Tenth Edition (1972).
K. G. Budden, *Radio Waves in the Ionosphere*, p. 297, Cambridge University Press (1966).
J. M. Buzzi, Etude théorique et expérimentale des perturbations électrostatiques en plasma chaud unidimensionnel, Thèse de Doctorat d'Etat, Université Paris-Sud, Orsay (1974).
H. J. Doucet and D. Gresillon, Grid excitation of ion waves at frequencies above the ion plasma frequency, *Phys. Fluids* **13**, 773 (1970).
M. R. Feix, Impedance of RF grids and plasma condensers, *Phys. Lett.* **9**, 123 (1964).
R. W. Gould, Excitation of ion-acoustic waves, *Phys. Rev. A* **136**, 991–997 (1964).
J. L. Hirshfield and J. H. Jacob, Free-streaming and spatial Landau damping, *Phys. Fluids* **11**, 411(1968).
U. E. Kruse and N, F. Ramsey, The integral $\int_0^\infty y^3 \exp(-y^2 + ix/y)\,dy$, *J. Math. and Phys.* **30**, 40 (1951)
L. Landau, On the vibrations of the electronic plasma, J. Phys. USSR **10**, 25–34 (1946).
O. Laporte, Absorption coefficients for thermal neutrons, *Phys. Rev.* **52**, 72 (1937).
P. M. Morse and H. Feshbach, *Methods of Theoretical Physics*, McGraw-Hill, New York (1953).
G. M. Sessler and G. Pearson, Propagation of ion waves in weakly ionized gases, *Phys. Rev.* **162**, 108, (1967).
C. H. Torrey, Notes on intensities of radiofrequency spectra, *Phys. Rev.* **59**, 293 (1941).
C. T. Zahn, Absorption coefficients for thermal neutrons, *Phys. Rev.* **52**, 67 (1937).

CHAPTER 9

M. Abramowitz and I. A. Stegun, *Handbook of Mathematical Functions*, Dover, New York, (1965); Tenth Edition (1972).
I. Alexeff, Private Communication. (1963).

J. M. Buzzi, Etude théorique et expérimentale des perturbations électrostatiques en plasma chaud unidimensionnel, Thèse de Doctorat d'Etat, Université Paris-Sud, Orsay (1974).

H. Derfler, in *Proceedings of the Seventh International Conference on Phenomena in Ionized Gases*, Vol. 2, p. 282, (D. Perovich and D. Toshich, eds.), Gradevinska Knjiga, Beograd, Yugoslavia (1966).

H. Derfler and T. C. Simonen, Landau waves: an experimental fact, *Phys. Rev. Lett.* **17**, 172–175 (1966).

B. D. Fried and S. D. Conte, *The Plasma Dispersion Function*, Academic, New York (1961).

R. W. Gould, Excitation of Ion-Acoustic Waves, *Phys. Rev. A* **136**, 991–997 (1964).

J. L. Hirshfield, J. H. Jacob, and D. E. Baldwin, Interpretation of spatially decaying ion-acoustic waves, *Phys. Fluids* **14**, 615 (1971).

G. L. Johnston, Ph.D. Thesis, University of California, Los Angeles (1967).

G. A. Massel, Ph.D. Thesis, Raleigh University, Raleigh, NC (1967).

D. Middleton, *An Introduction to Statistical Communication Theory*, McGraw-Hill, New York (1960).

P. M. Morse and H. Feshbach, *Methods of Theoretical Physics*, McGraw-Hill, New York (1953).

B. W. Roos, *Analytic Functions and Distributions in Physics and Engineering*, Wiley, New York (1969).

T. C. Simonen, SUIPR Report No. 100, Stanford University, Stanford, CA (1966).

CHAPTER 10

I. Alexeff, W. D. Jones and D. Montgomery, Controlled Landau damping of ion-acoustic waves., *Phys. Rev. Lett.* **19**, 422–425 (1967).

L. Brillouin, *Wave Propagation and Group Velocity*, Academic, New York (1960).

J. M. Buzzi, Etude théorique et expérimentale des perturbations électrostatiques en plasma chaud unidimensionnel, Thèse de Doctorat d'Etat, Université Paris-Sud, Orsay (1974).

J. M. Buzzi and H. J. Doucet, On the existence or nonexistence of ion-acoustic waves in a single-ended Q machine, *Bull. Am. Phys. Soc.* **17**, 1050 (1972).

J. M. Buzzi and J. Henry, *Ondes Electrostatiques Dans un Plasma d'Iodure de Thallium*, Internal Report PMI No. 576, Laboratoire de Physique des Milieux Ionisés, Ecole Polytechnique, France (1972).

P. Davidovits and J. L. Hirshfield, Dielectric constant of an electron-free $Tl^+ I^-$ plasma, *Appl. Phys. Lett.* **15**, 290–292 (1969).

H. J. Doucet, I. Alexeff, and W. D. Jones, Simultaneous measurement of ion-acoustic wave potential and plasma density perturbation to yield γ_e, *Phys. Fluids* **11**, 2451–2453 (1968).

H. J. Doucet and D. Gresillon, Grid excitation of ion waves at frequencies above the ion plasma frequency, *Phys. Fluids* **13**, 773 (1970).

K. Estabrook and I. Alexeff, Nonexistence of ion-acoustic waves and Landau damping driven electrostatically in an ideal Q machine, *Phys. Rev. Lett.* **29**, 573 (1972).

M. R. Feix, Impedance of RF grids and plasma condensers, *Phys. Lett.* **12**, 316 (1964).

B. D. Fried and R. W. Gould, Longitudinal ion oscillations in a hot plasma, *Phys. Fluids* **4**, 139–147 (1961).

R. W. Gould, Excitation of ion-acoustic waves, *Phys. Rev. A* **136**, 991–997 (1964).

D. Gresillon, Etude des ondes ioniques excitées par une grille dans un plasma sans collision, *J. Phys. (Paris)* **32**, 269 (1971).

J. L. Hirshfield and J. H. Jacob, Free-streaming and spatial Landau damping, *Phys. Fluids* **11**, 411 (1968).

G. Jahns and Van Hoven, Low-frequency grid excitation in a magnetized plasma column, *Phys. Rev.* **5**, 2622 (1972).

J. H. Malmberg and C. B. Wharton, Dispersion of electron plasma waves, *Phys. Rev. Lett.* **17**, 175 (1966).

R. W. Motley, *Q-Machines*, Academic, New York (1975).

G. M. Sessler and G. Pearson, Propagation of ion waves in weakly ionized gases, *Phys. Rev.* **162**, 108 (1967).

J. Virmont, Ondes ioniques en conditions aux limites avec des collisons ion-neutre et électron-neutre, Internal Report PMI No. 545, Laboratoire de Physique des Milieux Ionisés, Ecole Polytechnique, France (1972).

A. Y. Wong, Observation of the electron contribution to ion-acoustic waves *Phys. Rev. Lett.* **14**, 252 (1965).

A. Y. Wong, R. W. Motley, and N. D'Angelo, Landau damping of ion-acoustic waves in highly ionized plasmas, *Phys. Rev. A* **133**, 436–442 (1964).

CHAPTER 11

M. Abramowitz and I. A. Stegun, *Handbook of Mathematical Functions*, Dover, New York (1965); Tenth Edition (1972).

B. D. Fried and S. D. Conte, *The Plasma Dispersion Function*, Academic, New York (1961).

INDEX